T0213594

Undergraduate Texts in Mathematics

Undergraduate Texts in Mathematics

Undergraduate Texts in Mathematics are generally aimed at third- and fourth-year undergraduate mathematics students at North American universities. These texts strive to provide students and teachers with new perspectives and novel approaches. The books include motivation that guides the reader to an appreciation of interrelations among different aspects of the subject. They feature examples that illustrate key concepts as well as exercises that strengthen understanding.

Cam McLeman • Erin McNicholas •
Colin Starr

Explorations in Number Theory

Commuting through the Numberverse

 Springer

Cam McLeman
Department of Mathematics
University of Michigan–Flint
Flint, MI, USA

Erin McNicholas
Department of Mathematics
Willamette University
Salem, OR, USA

Colin Starr
Department of Mathematics
Willamette University
Salem, OR, USA

ISSN 0172-6056   ISSN 2197-5604   (electronic)
Undergraduate Texts in Mathematics
ISBN 978-3-030-98933-0   ISBN 978-3-030-98931-6   (eBook)
https://doi.org/10.1007/978-3-030-98931-6

Mathematics Subject Classification: 11-01

This Springer imprint is published by the registered company Springer Nature Switzerland AG
The registered company address is: Gewerbestrasse 11, 6330 Cham, Switzerland

Preface

...wherein we welcome readers of all sorts to our journey.

Who is This Text's Audience?

We have written this book for you, our inquisitive, delightful reader! Your discerning eye for mathematics, interest in numbers, willingness to explore, and desire to understand underlying structures deeply are commendable. While this text examines number systems from an algebraic lens, no previous exposure to abstract algebra is assumed. We do, however, assume some familiarity with vector spaces and basic proof techniques. *Calculus: Multivariable* by McCallum et al., and *Proofs and Fundamentals: A First Course in Abstract Mathematics* by Ethan Bloch are excellent references for the assumed background material.

To the Student:

Dear Student,

Welcome to the start of your journey! Consider this book as your official companion guide—it is in some ways a traditional mathematics textbook (it has, for example, lots of math in it), but deviates from standard practice in several notable ways as well. First, it has multiple authors coming from a variety of backgrounds and experiences within the general realm of algebra and number theory. You might see sparks of more excited writing as one of us encountered a favorite piece of number theory, for example, though we generally attempted to conform our writing to a single voice[1].

Second, and more importantly, we call particular attention to the existence of "Explorations" sprinkled through the chapters. These worksheets are the manifestations of the authors' zeal for inquiry-based learning—the process of figuring out mathematics for yourself like our mathematical forebears out in the unexplored

---

[1] Well, *some* of us did.

wilderness. There are no solutions provided to the worksheets, as their point is not typically to "get the answer" but rather to bring up questions that we'll spend time discussing in the following section. So don't skip them! In fact, you might want to photocopy those pages for repeated use. Or better yet, buy several copies of the book, just to be safe.

To the Instructor:

Dear Instructor,

Phew! Now that the students have stopped reading, we can be frank. Two adjectives that stand out as highly desirable when choosing a textbook are *comprehensive* and *self-contained*. This book does not attempt to achieve either of those ambitions. In exchange, we stake a claim to the partially opposite adjectives of being *modern* and *flexible*. As to comprehensiveness, the over-arching goal of the text has been to hook the reader's interests in the mystery of numbers, and it is the opinion of the authors that no book this size can do justice to the vast wealth of classical number-theoretic pearls while simultaneously working toward contemporary number theory. Consequently, our "to include or not to include" philosophy has been to focus on developing a few core themes that persist through the study of number theory at all levels:

- The study of Diophantine equations, partly for their own sake, but more typically as catalysts for introducing and developing bigger structural ideas.
- The notion of what a "number" is, and the premise that it takes familiarity in quite a large variety of number systems to fully explore number theory.
- The use of abstract algebra in number theory, and in particular the extent to which it provides the "Fundamental Theorem of Arithmetic" for various new number systems.

In addition to the core themes, other aspects of modern number theory are present in smaller but persistent threads woven through chapters and exercise sets, e.g., the study of elliptic curves, the analogies between integer and polynomial arithmetic, p-adic valuations, relationships between the spectrum of primes in various rings, etc.

The proper sequencing of number theory and abstract algebra is a conundrum we have chosen to embrace. While an introduction to number theory provides a good preparation for the study of abstract algebra, students who have had abstract algebra are better able to fully grasp the concepts in an introductory number theory course. Our solution is to develop and define algebraic concepts as needed to summarize and formalize the patterns observed in various number systems. In this text, students encounter groups, rings, fields, ideals, and more. While we have taken uncharacteristic care to make the coverage of these concepts self-contained, it is certainly not comprehensive. In our experience, the result is a text that simultaneously serves both students with and without any previous exposure to abstract algebra. Students with little to no abstract algebra experience are able to grasp the salient details by applying the intuition and experience gained from the adjacent number theory. These students gain a solid foundation on which to build in a subsequent abstract algebra course. For students who have had a course in abstract algebra, the pithy and targeted coverage of algebraic concepts is a good refresher that helps them apply their algebraic knowledge to gain a deeper understanding of the number theory content. For these students, the text provides not only an introduction to number theory but also a strengthened appreciation for the power of an algebraic perspective.

As to self-containedness more broadly, we have been content to assume some reasonable prerequisites and avoid compulsively writing appendices to provide basic familiarity with complex numbers, matrix algebra, and vector spaces. In addition to directing students to good text sources for this background material, we take advantage of a thoroughly modern approach, namely, the ability of readers to go online and look things up. The skill of finding and processing mathematics online is an increasingly crucial one and should be fostered rather than discouraged. This philosophy makes itself explicit in the form of a section of each chapter's homework problems ("General Number Theory Awareness" problems) whose goal is to get students to conduct research online on biographical information about mathematicians and mathematical content that space prevented us from including, or pieces of mathematical culture that don't fit strictly in the traditional academic view of what a mathematics textbook should cover. Exercises in these sections also lead students to practice one of the most fundamental mathematical research skills: seeing, upon making a mathematical discovery, if someone else has already done it.

The broader claim to modernity is an over-arching belief in inquiry-based learning. In addition to these research-based homework questions, a second sub-section of exercises has students numerically investigate conjectures and make their own. Sometimes, the tools to prove these conjectures will be within a student's grasp, and sometimes they will be cutting-edge research questions to which humanity does not know the answer. While we do not prescribe a programming language for these numerical investigations, the book's companion website has Python worksheets designed to walk students with little to no programming experience through those exercises. Finally, and perhaps most importantly, inquiry-based learning is built into the text itself: exploration worksheets lovingly placed throughout the chapters let students lead the way by coming up with the pivotal ideas for the upcoming sections while simultaneously improving their fluency in previously covered topics. The final chapter closely mirrors that of a purely inquiry-based textbook, allowing students to work individually or in groups—perhaps as a final project—through problem-driven sections covering material begun in the main text (Fermat's Last Theorem, quaternions, real quadratic units, elliptic curves, cryptography, ideal theory). Exercises designed to pique students' interest in these topics are interspersed throughout earlier chapters.

The pacing of the course is also rather flexible. The default approach, we believe, should begin with the standard theoretical development of $\mathbb{Z}$ and $\mathbb{Z}/(n)$ but using the explicit language of groups and rings, rather than shunting this perspective off as advanced material near the end of the course. Incorporating abstract algebra into this process necessarily proceeds more slowly, but has the benefit of being quickly generalizable: once you prove that $\mathbb{Z}[i]$ has a Euclidean algorithm, for example, proving that it enjoys unique factorization is essentially a matter of copy–pasting the argument from $\mathbb{Z}$. Going quickly through Chapters 5 and 6 leaves enough time to do both culminating chapters, on quadratic reciprocity and p-adic numbers, in a single semester. On the other hand, an instructor who wanted to use the book to fill two semesters covering both abstract algebra and number theory could easily do so by using all of the in-class explorations, covering the details of abstract Euclidean domains, and providing in-class time for student presentations from Chapter 9.

Finally, some closing remarks on the use of the book. As mentioned in the student preface, we highly value the inclusion of the explorations appearing throughout the book. They provide an inquiry-based break from the more traditional class structure during which an instructor can engage in informal formative assessment of how the class is doing and allow students to engage in independent or group discovery. Questions and answers to these problems are typically developed more carefully in the sections that follow. Relatedly, the homework addresses a variety of ways in which one becomes a master of number theory. There are computational exercises to promote fluency in arithmetic calculations, proof-based problems to develop theoretical understanding, the aforementioned online research problems, and problems that lend themselves to computer-aided exploration.

Suggested Pacing and Content Coverage

A note on the use of explorations: we find that different classes take significantly different amounts of time to work through the items on each exploration. Accordingly, we do not necessarily recommend trying to get through a full exploration in class. Parts could be omitted or assigned as additional homework as needed.

For a 14-week semester, 3 hours per week, we suggest the following pacing:

- Week 1: Sections 1.1–2.3 (leaving most details for students to read), Exploration A.
- Week 2: Section 3.1, Exploration B, and Section 3.2.
- Week 3: Exploration C, Sections 3.3 and 3.4, Exploration D, and Section 3.5.
- Week 4: Sections 4.1–4.3, Exploration E, and Section 4.4.
- Week 5: Exploration F, Section 4.5, Exploration G, and Section 4.6.
- Week 6: Exploration H, Sections 4.7 and 4.8.
- Week 7: Sections 5.1 and 5.2, Exploration I, Sections 5.3 and 5.4, and Exploration J.
- Week 8: Section 5.5, Exam 1, and Section 5.6.
- Week 9: Sections 5.6–6.2, Exploration K, and Section 6.3.
- Week 10: Sections 6.3 and 6.4, Exploration L, and Sections 6.5 and 6.6.

- Week 11: Sections 7.1 and 7.2, Exploration M, Sections 7.3 and 7.5.
- Week 12: Sections 7.5–7.7.
- Week 13: Chapter 9 topics.
- Week 14: Chapter 9 topics.

Flint, USA   Cam McLeman
Salem, USA   Erin McNicholas
Salem, USA   Colin Starr

Acknowledgments We appreciate the efforts of our editors to keep us on track and the helpful and insightful comments of the reviewers. Additionally, each of the authors would like to express their heartfelt appreciation to the other two authors for each doing approximately 25% of the work.

Contents

What is a Number?

...wherein the Reader becomes acquainted with a familiar cast of characters, we ask a silly question, and we get a silly answer.

1.1 Human conception of numbers

Roughly 37,000 years ago, for reasons not completely understood, one of the first humans scratched 55 tally marks into a wolf bone (now called the *Lebombo bone* after the mountain range in which it was found). Perhaps they were tallying kills, marking time as a primitive calendar, or merely expressing an early interest in number theory as a hobby. Regardless of the intent, other bone artifacts from the Paleolithic era show a clear progression of this system, moving from individual marks to groups of marks organized in a way similar to the modern system of tallying in groups of five.

This progression from tally marks to groups is the first of countless small steps in the evolution of human understanding of numbers. Further milestones tie inexorably with the needs of culture, and interlace with other developmental breakthroughs in society and technology. For example, agricultural advancements allowed the cultivation of much larger quantities of crops and the congregation of much larger populations, which necessitated the ability to describe and record significantly larger numbers than is feasible using tally marks. Imagine the thrill of the first intrepid (perhaps literal) bean counter to recognize that instead of maintaining a pile of tokens to keep track of a client's inventory they could employ *cipherization*, the use of written symbols to represent specific numerical values. This breakthrough of the fourth millennium BCE predates—and indeed, may have instigated [4]—the advent of written language.

The Egyptian system for doing this, *hieroglyphs*, is an *additive* system, meaning that each symbol represents a set value and a number is represented by a collection of symbols whose sum has the desired value. By introducing new symbols for powers of 10 as the need arose, they could abstractly express any natural number.

Supplementary Information The online version contains supplementary material available at https://doi.org/10.1007/978-3-030-98931-6_1.

C. McLeman et al., *Explorations in Number Theory*, Undergraduate Texts in Mathematics, https://doi.org/10.1007/978-3-030-98931-6_1

▶ **Example 1.1.1** Some numeral hieroglyphs and their modern equivalents:

I	∩	ℓ	𝕀	𝕁	👑
1	10	100	1, 000	10, 000	1, 000, 000

Hieroglyphic representation of $78, 557 = 7 \cdot 10, 000 + 8 \cdot 1, 000 + 5 \cdot 100 + 5 \cdot 10 + 7$

$$𝕁𝕁𝕁𝕁𝕁𝕁𝕀𝕀𝕀𝕀𝕀𝕀𝕀𝕀ℓℓℓℓℓ∩∩∩∩∩IIIIIII$$

Later number systems, like the Hindu-Arabic system we use today, often express numbers using a finite number of symbols whose associated value is dependent on their *placement* in the representation. These *positional* number systems were first developed by the Ancient Mesopotamians around 3000 BCE. They employed a sexagesimal system—not *quite* as exciting as it sounds—meaning that instead of a shorthand based on sums of powers of 10 they used sums of powers of 60. That is, whereas we would write 7389 as shorthand for the base-10 expansion $7 \cdot 10^3 + 3 \cdot 10^2 + 8 \cdot 10^1 + 9 \cdot 10^0$, they would write "239" for the same number, representing the base-60 expansion $2 \cdot 60^2 + 3 \cdot 60^1 + 9 \cdot 60^0$. Except, of course, that they wouldn't, because they didn't use the numerals 1, 2, 3, etc. Instead:

▶ **Example 1.1.2** Cuneiform, the script of Ancient Mesopotamia, used the symbol 𝕀 to represent 1 and ⟨ to represent 10. To write down a number with multiple "digits," we leave a space between them (as opposed to writing them next to each other — 7 3 8 9 vs. 7389). For example, the number $1513 = 25 \cdot 60 + 13$ would have been written as follows:

Difficulties in using the Cuneiform system, still an early manifestation of humanity's experimentation with place values, quickly become apparent. In particular, the actual places (the specific powers of 60) were typically deduced implicitly from the context and left ambiguous in the notation—the numbers $25 \cdot 60^3 + 13 \cdot 60$ and $25 \cdot 60 + 13 \cdot 60^0$ and $25 \cdot 60^{-1} + 13 \cdot 60^{-4}$ would have all had the same Cuneiform representation as the one in the example. One explanation for this difficulty is that it stems from an evolving notion of the number zero. With the inclusion of a numeral 0 and a convention that we write the base-60 expansion of a natural number in descending powers of 60 with some notation to mark the 60^0 term, a sexagesimal system is just as expressive as a decimal one.

It is significant that these systems have hardwired into them the ambient world-view that the only numbers that need to be described are all positive and rational. The discovery, therefore, of the existence of *irrational numbers*, commonly attributed to the fifth-century BCE Greek mathematician Hippasus of Metapontum, represented a watershed moment in our historical understanding of numbers. Their mere existence presented a notational and philosophical challenge to the Ancient Greeks, who had no way of symbolically representing these "incommensurable" ratios. In response, the Greeks considered numbers as being of one of two types: a number either represents a discrete quantity or a continuous quantity. Discrete entries, like tally marks, were used to count things, whereas continuous quantities were endowed with geometric meaning and thought to represent either lengths, areas, or volumes. In this way, the ancient Greeks were able to geometrically represent and perform operations on positive numbers both rational and irrational. This geometric perspective on numberhood permeates ancient Greek mathematics and largely explains why they did not investigate polynomial relationships of degree four or higher (as such relationships do not have a clear geometric representation in three-dimensional space).

And yet the journey through numberhood presses on. While negative numbers were accepted and used in China as early as the third century CE, and in India in the seventh century CE, they were still met with suspicion by European mathematicians as late as the eighteenth century. Their acceptance in Europe coincided with a growing acceptance of *complex numbers*: numbers of the form $a + b\sqrt{-1}$, where a and b are real numbers. In 1545, Italian mathematician Gerolamo Cardano published his treatise *Ars Magna* outlining solutions to cubic and quartic polynomial equations. These techniques naturally led to the consideration of square roots of negative quantities. While Cardano laid out the rules for working with negative numbers, he called their square roots "truly imaginary since operations may not be performed with [them] as with a pure negative number, nor as in other numbers." It was almost 30 years later that Rafael Bombelli showed how arithmetic with imaginary numbers could provide real roots of polynomials with real coefficients. It would be another 200 years before negative and complex numbers were fully accepted by the European mathematical community.

This journey, from the humble beginnings of enumerating tally marks to the solution of polynomial equations using complex roots, represents a radical expansion of humanity's working notion of a number. A modern encoding of this transformation, and indeed the principal method of investigation used in this book, proceeds not by the study of the individual numbers themselves, but rather by packaging them into sets of numbers and then studying those sets. This algebraic approach to the study of number, taken up in the next section, very cleanly encodes both the formal mathematical objects and, progressing from one set to another, the evolution of humanity's conception of number described above.

1.2 Algebraic Number Systems

The various types of numbers that humanity has considered can be categorized into systems, many of which you'll have become familiar with over your mathematical career thus far. The algebraic approach to studying numbers is to observe that these systems share many of the same abstract properties. These commonalities give us great power: whatever we can prove for a general number system will immediately apply to all similar number systems. We shall adopt the standard notations for the most commonly encountered sets of numbers. Pay particular attention to how each set compares to the previous ones:

Definition 1.2.1 (Commonly encountered sets of numbers)

$\mathbb{N} = \{1, 2, 3, \ldots\}$, the set of **natural numbers**. A case *could* be made for including 0 as a natural number, but we'll maintain the current definition for this book. We'll let you decide which convention is more morally defensible[1].

$\mathbb{Z} = \{\ldots, -3, -2, -1, 0, 1, 2, 3, \ldots\}$, the set of **integers**, consisting of the natural numbers, their negatives, and zero. Note that the integers are closed under subtraction (meaning that the difference of two integers is always an integer), whereas the natural numbers are not. Visit the appendix for a brief look at an axiomatic development of $\mathbb{N}$ and $\mathbb{Z}$.

$\mathbb{Q} = \{\frac{a}{b} : a, b \in \mathbb{Z}, b \neq 0\}$, the set of **rational numbers**, or "the rationals," as their friends call them. Note that the rationals are closed under (non-zero) division, whereas the integers are not.

$\mathbb{R} = \{x : x \text{ is a real number}\}$, the set of **real numbers**. Too circular? The definition is improved by defining them as limits of convergent sequences of rationals, but that is not particularly crucial to hash out now. As a consequence, the real numbers are closed under taking appropriate limits, whereas the rationals are not.

$\mathbb{C} = \{a + bi : a, b \in \mathbb{R}\}$, where $i = \sqrt{-1}$, the set of **complex numbers**. We will talk further about this mysterious entity i shortly. Note that, among other things, the complex numbers are closed under taking square roots, whereas the real numbers are not. ◄

All of these sets share some fundamental algebraic properties. For example, they all come equipped with a notion of addition and multiplication of their elements, and these operations are structurally quite similar: addition and multiplication are **commutative** ($a+b = b+a$ and $ab = ba$) and **associative** ($(a+b)+c = a+(b+c)$ and $(ab)c = a(bc)$), and multiplication **distributes** over addition ($a(b + c) = ab + ac$). All these sets include the number 1 (a **multiplicative identity**: $1 \cdot a = a \cdot 1 = a$), and, except for the natural numbers, they all include 0 (an **additive identity**: $a + 0 = 0 + a = a$) and an **additive inverse** for each element (for each element a, there is an element b such that $a + b = 0 = b + a$). The existence of

---

[1] See Appendix I for a formal construction of this set and a more nuanced discussion on the inclusion of 0.

additive inverses further endows the set with a notion of subtraction (by defining $a - b = a + (-b)$).

At the same time, the descriptions of the sets highlight some substantial structural differences. For example, in $\mathbb{R}$, every element except 0 has a **multiplicative inverse** (an element b is the multiplicative inverse of a if $ab = ba = 1$). In $\mathbb{Z}$, only the numbers ± 1 have such an inverse. Distinguishing which algebraic properties are satisfied by a number system (and the consequences of possessing those properties) turns out to be a rather significant component of our story—a component that will rise periodically to the forefront of our attention through the introduction of words like *rings* and *fields*. These words typically play the lead roles in a course on abstract algebra but will earn a "Best Supporting Idea" award for their role in this book in helping us come to terms with new and increasingly exotic types of numbers.

1.3   New Numbers, New Worlds

The synthesis of the previous two sections is that we have, on the one hand, an intuitive but informal understanding of what a number is, and on the other, a formal but thus far undeveloped notion of an algebraic structure. As with any instance of trying to find the right definition of a term, it is the borderline cases that really test our understanding. The role of a definition is to unilaterally mandate on which side of the border various exceptional cases fall. Is zero a natural number? Is 1 prime? Is the empty set connected? Do pants come in pairs? Is a hot dog a sandwich?[2]

In this vein, we close this preliminary chapter by introducing two potentially borderline candidates for inclusion under the umbrella of numberhood, hopefully highlighting the difficulty in delineating precisely the difference between numbers and non-numbers. While you have likely met these candidates before, this quick introduction might cast them in a new, unfamiliar, light. Don't be alarmed if you end the chapter feeling less confident in your knowledge of these familiar friends than you were at the start. We will encounter these sets throughout the text, gaining a deeper understanding of them as we go.

Candidate 1: An Imaginary Number?

We have already come into contact with the historically radical step of introducing a new "number" called i whose square is -1. Of course, no such number exists in $\mathbb{R}$, so one might reasonably question whence this i comes, but we can nevertheless treat it as a formal addition to our collection of numbers and extend algebraic rules to do arithmetic with it. Insisting that we have the number i and also reasonable rules of arithmetic in place forces the inclusion of yet more numbers. If we want

---

[2] Discuss in small groups. *Calmly.*

our set of numbers to be closed under multiplication, we will need to include ± 1 (as $i \cdot i = -1$ and $(-1) \cdot (-1) = 1$). The existence of new numbers quickly follows. For example, it seems reasonable to insist that, if we add i to itself, we should get the new number, $2i$, and likewise for adding i and 7 to get, well, $i + 7$. We might further insist that the commutative, associative, and distributive laws continue to hold, so that, for example,

$$(2i) \cdot (i + 7) = (2i) \cdot (i) + (2i) \cdot (7) = 2(i^2) + (2 \cdot 7)i = -2 + 14i.$$

Various sets of numbers of the form $a + bi$ provide interesting new worlds to play in. Particularly notable for this book will be the case where a and b are required to be integers.

Definition 1.3.1 (Various types of numbers of the form $a + bi$)

The set of **Gaussian integers**, denoted $\mathbb{Z}[i]$, consists of the numbers of the form $a + bi$ where a and b are integers:

$$\mathbb{Z}[i] = \{a + bi : a, b \in \mathbb{Z}, i^2 = -1\}.$$

Analogously, the **Gaussian rationals**, are the set

$$\mathbb{Q}[i] = \{a + bi : a, b \in \mathbb{Q}, i^2 = -1\}.$$

Finally, taking a and b from $\mathbb{R}$, we get what in principle could be called the "Gaussian reals" but is instead called the set of **complex numbers**, denoted $\mathbb{C}$:

$$\mathbb{C} = \mathbb{R}[i] = \{a + bi : a, b \in \mathbb{R}, i^2 = -1\}. \qquad \blacktriangleleft$$

It is notable that in contemporary mathematical circles, it is unquestioned that the moniker "number" applies to elements of any of these sets—the complex numbers simply build upon the real numbers by the inclusion of a handy missing element. From an algebraic perspective, the introduction of i provides a root of the polynomial $x^2 + 1$, which previously had no roots, thereby plugging a hole in the world of real numbers. While one could imagine that we might need to throw in a new number for each polynomial without a real root (e.g., $x^2 + 2$ or $x^4 - 2x^3 + 4x + 12$), the fact is—the Fundamental Theorem of Algebra—that the buck stops at $\mathbb{C}$: all polynomials with real (or complex) coefficients have all of their roots in $\mathbb{C}$, thereby solidifying the complex numbers as a world of tremendous import.

Further, $\mathbb{Z}[i]$, $\mathbb{Q}[i]$, and $\mathbb{C}$ also enjoy all (or nearly all) of the formal algebraic properties that we described in the last section. To review for later use, extending our first example of multiplication above via the distributive law[3] gives the standard arithmetic definitions:

---

[3] *FOIL*-ing, if you will.

Definition 1.3.2 (Addition and multiplication in $\mathbb{C}$)

$$(a + bi) + (c + di) = (a + c) + (b + d)i$$

and

$$(a + bi)(c + di) = ac + adi + bic + bdi^2 = (ac - bd) + (ad + bc)i. \quad \blacktriangleleft$$

These two operations admit a particularly compelling geometric interpretation. In Figure 1.1, we have the **complex plane**, where the real and imaginary axes have replaced the standard xy-axes. We envision a complex number $z = a + bi$ as located where we usually put the point (a, b) in the Cartesian plane—the figure shows the complex numbers $z_1 = 2 + i$ and $z_2 = 1 - i$ depicted as the points $(2, 1)$ and $(1, -1)$ in the plane. Note that $\mathbb{Z}[i]$, the collection of Gaussian integers, corresponds to the square lattice of points with integer coordinates, marked as white dots in the figure.

The first of our operations, addition of complex numbers, is visualized fairly trivially by adding component-wise, the same as in vector addition: the sum $z_1 + z_2 = (2 + i) + (1 - i) = 3 + 0i$ corresponds to the sum $(2, 1) + (1, -1) = (3, 0)$ in the complex plane. Multiplication is only slightly more involved, and is best visualized

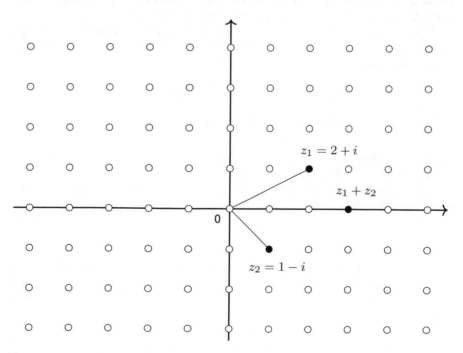

Fig. 1.1 The Gaussian Integers inside $\mathbb{C}$

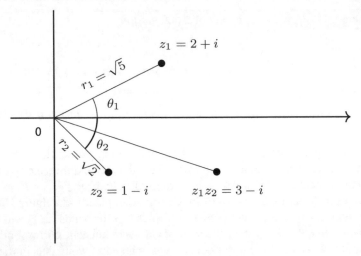

Fig. 1.2 Polar coordinates for complex numbers

through our understanding of polar coordinates, which we briefly recall here: Given $z \in \mathbb{C}$, let r be defined as the length of the line segment from z to the origin in the plane, and $\theta \in [0, 2\pi)$ be defined as the (counterclockwise) angle from the positive $\mathbb{R}$-axis to this segment. While we will not use this notation extensively, it is handy to use the notation $r \angle \theta$ to refer to the complex number with that value of r and θ (including extending the notation by periodicity to all $\theta \in \mathbb{R}$).

Trigonometry provides the familiar dictionary between representations of a complex number. Given the two representations $z = a + bi$ and $z = r \angle \theta$ of a complex number, we have:

$$r^2 = a^2 + b^2 \qquad a = r \cos(\theta)$$

$$\theta = \tan^{-1}\left(\frac{b}{a}\right) \qquad b = r \sin(\theta),$$

when $a > 0$. When $a \leq 0$, the relationship between a, b and θ is instead given by $\theta = \tan^{-1}(b/a) \pm \pi$ depending on the quadrant.

As a consequence, we can link together the polar and Cartesian descriptions of a given complex number:

$$r \angle \theta = (r \cos \theta) + (r \sin \theta)i$$

Continuing with our example z_1 and z_2 from above, we can compute for z_1 its radius $r_1 = \sqrt{2^2 + 1^2} = \sqrt{5}$ and angle $\theta_1 = \tan^{-1}(1/2)$, and similarly for z_2, we have $r_2 = \sqrt{2}$ and, since z_2 is in the 4th quadrant, we have $\theta_2 = \tan^{-1}(-1)$, as in Figure 1.2.

The language of polar coordinates allows us to describe multiplication of complex numbers with remarkable ease: If $z_1 = r_1 \angle \theta_1$ and $z_2 = r_2 \angle \theta_2$, then we can compute their product using the angle sum formulas for sine and cosine:

$$z_1 z_2 = r_1(\cos\theta_1 + i\sin\theta_1) \cdot r_2(\cos\theta_2 + i\sin\theta_2)$$
$$= r_1 r_2[(\cos\theta_1\cos\theta_2 - \sin\theta_1\sin\theta_2) + i(\sin\theta_1\cos\theta_2 + \cos\theta_1\sin\theta_2)]$$
$$= r_1 r_2(\cos(\theta_1 + \theta_2) + i\sin(\theta_1 + \theta_2)).$$

That is,

$$(r_1 \angle \theta_1)(r_2 \angle \theta_2) = (r_1 r_2) \angle (\theta_1 + \theta_2).$$

In words, *radii multiply, but angles add*. A special case of particular relevance is the squaring formula $z^2 = r^2 \angle (2\theta)$, and the natural generalization

$$z^n = r^n \angle (n\theta).$$

One of the most famous equations in all of mathematics, easily verified by comparing the Taylor expansions of both sides, is *Euler's formula*

$$e^{i\theta} = \cos(\theta) + i\sin(\theta).$$

This equality provides a commonly used alternative to the $z = r\angle\theta$ or $z = r\cos(\theta) + ir\sin(\theta)$ notations for complex numbers, namely $z = re^{i\theta}$. Multiplication of complex numbers has a particularly appealing form under this notation.

> **Theorem 1.3.3**
> Given complex numbers $z = re^{i\theta}$, $z_1 = r_1 e^{i\theta_1}$, and $z_2 = r_2 e^{i\theta_2}$, we have
>
> $$z_1 z_2 = r_1 r_2 e^{i(\theta_1 + \theta_2)} \qquad \text{and} \qquad z^n = r^n e^{in\theta} \quad (n \in \mathbb{Z}).$$

We will make important use of such identities once we reach our more substantial investigation of complex numbers. For example, in Chapter 5, we use these identities to verify that multiplying a complex number by i ($= e^{i\frac{\pi}{2}}$) is equivalent to rotating the number through a right angle in the counterclockwise direction; and in Chapter 7, we note that the complex number $z = 1e^{i\frac{2\pi}{n}}$ has the curious property that

$$z^n = 1^n e^{\left(i\frac{2\pi}{n}\right)n} = 1e^{i2\pi} = 1,$$

providing for each $n \in \mathbb{Z}$ a complex number whose nth-power is precisely 1.

Before moving on, we close with some standard jargon for complex numbers:

> **Definition 1.3.4**

Given a complex number $z = a + bi$, a is the **real part** of z and b is the **imaginary part** of z. Note that the imaginary part of z does not include the "i," so the imaginary part of z is real. Two complex numbers $z_1 = a + bi$ and $z_2 = c + di$

are **equal** if they have the same real and imaginary parts (i.e., $a = c$ and $b = d$). The **complex conjugate** (or just **conjugate**) of z is denoted $\bar{z}$ and defined by $\overline{a + bi} = a - bi$. ◄

Geometrically, $\bar{z}$ is the point you get when you reflect z about the $\mathbb{R}$-axis. We will see that the complex conjugate often simplifies algebraic calculations in $\mathbb{C}$. As a first instance, consider division, which makes use of the observation that if $z = a + bi \in \mathbb{C}$, then $z\bar{z} = (a + bi)(a - bi) = a^2 + b^2$ is a non-negative real number, called the **norm** $N(z)$ of z. To divide complex numbers, multiply the numerator and denominator by the conjugate of the denominator to get an answer in standard form:

$$\frac{a + bi}{c + di} = \frac{a + bi}{c + di} \cdot \frac{c - di}{c - di} = \frac{(ac + bd) + (bc - ad)i}{c^2 + d^2} = \frac{ac + bd}{c^2 + d^2} + \frac{bc - ad}{c^2 + d^2}i.$$

From this expression, we can see that the quotient of two complex numbers is always a complex number (assuming the denominator isn't zero), but the quotient of two Gaussian integers may not itself be a Gaussian integer.

Finally, we must woefully observe that the terminology conflicts with the closely related notion of the norm from complex analysis, defined as its radial coordinate

$$|z| = r = \sqrt{a^2 + b^2} = \sqrt{N(z)}.$$

We will stick to using the word *norm* to reference $N(z)$.

All told, the inclusion of i results in a reasonably robust number system. Once we overcome the psychological hesitancy toward allowing square roots of negative numbers, we find ourselves in a new world where meaningful new arithmetic flows like water out of a pure mountain spring. Enlarging our notion of number to accommodate this world seems a desirable step, and so, to take the metaphor clearly too far, we slap a hydroelectric power plant onto our mountain spring to harness its numerical potential. But why stop there?

Candidate 2: An Indeterminate Number?

A second candidate for a new "number," again testing the line defining the property of numberhood, is the indeterminate "x" of polynomial fame—the same x that graces us with its presence in polynomial expressions like $x^4 - 2x^3 + 4x + 12$. As with i, the inclusion of x and some algebraic governance force the inclusion of many new things—we can add x to itself to get $2x$, or x to 7 to get, well, $x + 7$. Like with complex numbers, we have a notion of polynomial addition and multiplication that exhibits all of our standard algebraic operations, allowing arithmetic calculations like

$$(2x) \cdot (x + 7) = (2x) \cdot (x) + (2x) \cdot (7) = 2x^2 + 14x.$$

The one and only difference between this computation and the analogous computation with complex numbers is the absence of a rule concerning the square of x. Instead, as we add and multiply expressions of this form, we simply get higher and higher powers of x, leading to the set $\mathbb{R}[x]$ of all polynomials with real coefficients:

$$\mathbb{R}[x] = \{a_0 + a_1 x + a_2 x^2 + \cdots + a_n x^n : n \in \mathbb{Z}_{\geq 0} \text{ and } a_0, a_1, \ldots, a_n \in \mathbb{R}\}.$$

To introduce some standard terminology, the a_i are the **coefficients** of the polynomial, and the largest exponent n for which x^n has a nonzero coefficient is its **degree**, denoted for a polynomial f by $\deg(f)$. In the case that f is the zero polynomial, we set $\deg(f) = -\infty$. Two polynomials are **equal** if and only if they have the same degree and sequence of coefficients. As with the distinction between $\mathbb{Z}[i]$, $\mathbb{Q}[i]$, and $\mathbb{C}$, we can also vary the coefficients in this construction to produce $\mathbb{Z}[x]$, the set of polynomials with coefficients in $\mathbb{Z}$, and likewise for many other sets of coefficients (e.g., $\mathbb{Q}[x]$, $\mathbb{R}[x]$, $\mathbb{C}[x]$, etc.).

You are likely familiar with the standard operations on polynomials, so let's phrase them in a fashion slightly more sophisticated than necessary.

Definition 1.3.5 (Polynomial addition and multiplication)

Given two polynomials f and g defined by the expressions

$$f = a_0 + a_1 x + a_2 x^2 + \ldots + a_m x^m = \sum_{i=0}^{m} a_i x^i$$

$$g = b_0 + b_1 x + b_2 x^2 + \ldots + b_n x^n = \sum_{j=0}^{n} b_j x^j,$$

we define their sum term by term:

$$f + g = (a_0 + b_0) + (a_1 + b_1)x + \cdots + (a_M + b_M)x^M = \sum_{i=0}^{M} (a_i + b_i)x^i,$$

where $M = \max\{m, n\}$ and we take any missing coefficients to be zero as necessary. Likewise, their product is defined via the distributive law[4] ; i.e., fg is defined as the polynomial

$$fg = \left(\sum_{i=0}^{m} a_i x^i\right)\left(\sum_{j=0}^{n} b_j x^j\right) = \sum_{k=0}^{m+n} \left(\sum_{\ell=0}^{k} a_\ell b_{k-\ell}\right) x^k.$$

◀

---

[4] If you "FOIL" to multiply two complex numbers, don't forget to "FMOMMMIML" two quadratic polynomials!

A consequence of this definition is that the degree of the product is $m + n$, which shows that $\deg fg = \deg f + \deg g$ for non-zero polynomials $f, g \in \mathbb{Z}[x]$ (or $\mathbb{Q}[x]$, $\mathbb{R}[x]$, $\mathbb{C}[x]$). (What happens if either or both of f or g is the zero polynomial?) Notice that *unlike* the real or complex numbers, but *like* the integers, most elements of $\mathbb{R}[x]$ do not have multiplicative inverses. For example, x has no multiplicative inverse in $\mathbb{R}[x]$ since the condition $xg = 1$ immediately leads to a contradiction (the left-hand side has degree at least 1, while the right-hand side has degree 0).

The Conclusion, or Lack Thereof

All of this brings us to the question asked in the chapter title. You are likely far enough along in your mathematical career to understand the pervasive role that precise definitions play in the formulation of mathematics, so it may come as a bit of a shock to realize that you likely lack a working definition for the fundamental term "number." This section has observed that the notion has not been historically constant, with new and interesting additions appearing over time. But if you were to propose an all-encompassing definition of your current understanding of the term, how would you do it? What makes a mathematical object a *number*? Despite the close parallels of their construction, why is it that $3 + 5i$ is commonly considered a number but $3 + 5x$ is not? Are $\mathbb{R}[i]$ and $\mathbb{R}[x]$ really so different? Are imaginary numbers so superior to indeterminate ones?

Though it sadly renders the section title rather clickbaity, we are neither willing nor prepared to answer this question here[5] . Instead, we hope the question serves as an invisible ether through which the rest of the content of this book floats. As you progress through this text, continuing to encounter newer and more enigmatic number systems that may stretch your internal understanding of the term *number*, take heart that as was the case for mathematicians before you, fluency in arithmetic in these exotic realms develops with practice. Throughout the text, we will introduce you to new number systems by showing ways in which they are natural extensions of existing systems, the ways in which their common properties allow us to piggyback on intuition in one world to develop it in another, and the ways in which their differences spur mathematical growth and excitement. But, lest we get too far ahead of ourselves, there is no better place to begin than the beginning. Thus, it is fitting that the next chapter launches our journey by exploring the natural numbers $\mathbb{N} = \{1, 2, 3, \ldots\}$, those quantities that our paleolithic forebears so fastidiously and revolutionarily began to record.

---

[5] You may have assumed that three people writing a text on the theory of "numbers" would know what one is. Hopefully, by the end of the book, you will agree the answer is not as straightforward as it first appears.

1.4 Exercises

Calculation & Short Answer

Exercise 1.1 Perform the following operations. Give your answers in hieroglyphic form.

a. 𐦜 𐋇 ∩ ∩ ∩ ∩ ∩ ∩ ∩ |||| plus ∩ ∩ |||||||||
b. 𐋇 𐋇 𐋇 ||| minus 𐋇 ∩ ||||||

Exercise 1.2 The ambiguity in place value in cuneiform makes it difficult to unambiguously perform symbolic arithmetic. Perform the following operations, noting instances where there could be ambiguity. Give your answers in cuneiform script[6].

a. ⟪𒐖𒐖 ⟪⟪⟪𒐖𒐖 plus ⟪𒐖𒐖
b. ⟪𒐖𒐖 times 𝖨 𝖨
c. ⟪𒐖𒐖 times ⟨𒐖𒐖

Exercise 1.3 For $z = 2 + 3i$ and $w = 4 - 2i$, calculate $z + w$, $z - w$, zw, $\frac{z}{w}$, $\bar{z}$, and $|z|$.

Exercise 1.4 What is i^{17}? i^{35}? i^{428347}?

Exercise 1.5 Let $\omega = \frac{-1+\sqrt{3}i}{2}$. Find $\omega^2, \bar{\omega}, \frac{1}{\omega}, \omega^3, \bar{\omega}^3, \omega^{428347}$. Comment on anything noteworthy.

Exercise 1.6 Describe the solution sets to each of the following algebraic equations ($z \in \mathbb{C}$):

a. $\bar{z} = z$ c. $\bar{z} = \frac{1}{z}$

b. $\bar{z} = -z$ d. $\bar{z} = -\frac{1}{z}$

Exercise 1.7 In *Ars Magna*, Cardano gave the following equation for the root r of a polynomial of the form $x^3 + mx = n$:

$$r = \sqrt[3]{\frac{n}{2} + \sqrt{\frac{n^2}{4} + \frac{m^3}{27}}} - \sqrt[3]{-\frac{n}{2} + \sqrt{\frac{n^2}{4} + \frac{m^3}{27}}}.$$

a. Verify using graphing software (or by deducing via calculus) that the polynomial $x^3 - 15x - 4$ has three real roots.
b. Use the formula above to find a root of the polynomial $x^3 - 15x - 4$ whose expression involves complex number arithmetic.

---

[6] Preferably imprinted on clay tablets, and submitted to your professor once dried.

14   1 What is a Number?

c. Show as Bombelli did, that if you entertain the idea of complex numbers, you
   can recognize this root as a real number. It may help to first calculate $(2 + i)^3$
   and $(-2 + i)^3$.

Exercise 1.8 Let $f = 16t^2 - 12$ and $g = 64t^3 - 72t \in \mathbb{Z}[t]$. Compute the degrees
$\deg(f + g)$, $\deg(fg)$, $\deg(f^3)$, $\deg(g^2)$, and $\deg(g^2 - f^3)$.

Exercise 1.9 Determine which polynomials in $\mathbb{R}[x]$ have multiplicative inverses.
What about in $\mathbb{Q}[x]$ and $\mathbb{Z}[x]$?

Exercise 1.10 Express an opinion as to whether 1×1 real matrices, like [7], are
numbers. Repeat for constant polynomials in $\mathbb{R}[x]$. Then move on to 2×2 matrices
and linear polynomials.

Exercise 1.11 One number system that extends the complex numbers is the set
$\mathbb{R}[\mathbf{i}, \mathbf{j}, \mathbf{k}]$ of *quaternions*. These are numbers of the form $a + b\mathbf{i} + c\mathbf{j} + d\mathbf{k}$, where
$a, b, c, d \in \mathbb{R}$, and we define operations on quaternions by component-wise addition,
setting the basic products

$$\mathbf{i}^2 = \mathbf{j}^2 = \mathbf{k}^2 = \mathbf{ijk} = -1,$$

and calculating arbitrary products using the distributive law.

a. Assuming only associativity, use the relations above to fill in the following
   multiplication table. For example,

$$\mathbf{ij} = \mathbf{ij(kk)}(-1) = (\mathbf{ijk})(-\mathbf{k}) = (-1)(-\mathbf{k}) = \mathbf{k}.$$

·	1	i	j	k
1	1	i	j	k
i	i	−1	k	
j	j		−1	
k	k			−1

b. Is multiplication of quaternions commutative? Explain.
c. Perform the following quaternionic operations:

   i. $(5 + 3\mathbf{j} + 3\mathbf{k}) - (3 + 2\mathbf{i} - 4\mathbf{j} + \mathbf{k})$
   ii. $(2\mathbf{j} + 3\mathbf{k}) \cdot (5 + \mathbf{i} - 2\mathbf{k})$

Exercise 1.12 The *symmetric difference* is an operation defined on the set of all
subsets of a given set. Given two sets A and B, their symmetric difference, denoted
$A \triangle B$, is defined as

$$A \triangle B = (A - B) \cup (B - A) = \{x : x \in A \text{ and } x \notin B, \text{ or } x \in B \text{ and } x \notin A\}.$$

a. What set E acts as the $\triangle$-identity? That is, what E has the property that $A \triangle E = E \triangle A = A$ for all sets A?

b. Given a set A, must there be a $\triangle$-inverse of A? That is, does there necessarily exist a set B such that $A \triangle B = B \triangle A = E$, where E is the $\triangle$-identity from the previous part?

Formal Proofs

Exercise 1.13 Prove that if x and y are rational numbers, then so are xy and $x + y$. Show that this statement is false if you replace "rational" with "irrational."

The ancient Egyptians represented any *unit fractions*—that is, any fraction of the form $\frac{1}{n}$ for some natural number n—by placing the symbol � before the number. Thus �||| represented the quantity $\frac{1}{3}$, and ∩∩|�||⌀||| represented $21 + \frac{1}{2} + \frac{1}{3}$ or $21\frac{5}{6}$.

Exercise 1.14 An *Egyptian fraction representation* of a rational number r is an expression for r as a sum of distinct unit fractions, e.g., $\frac{5}{6} = \frac{1}{2} + \frac{1}{3}$. Find Egyptian fraction representations for each of

$$\frac{7}{12} \qquad \frac{4}{5} \qquad \frac{110}{2701}.$$

Exercise 1.15 Let $n \in \mathbb{N}$. Verify the identity

$$\frac{1}{n} = \frac{1}{n+1} + \frac{1}{n(n+1)}$$

and use it to prove that every positive rational number has a Egyptian fraction representation.

Exercise 1.16 The decimal representations of irrational numbers (and some rational ones) require an infinite number of terms.

a. Find a rational number that has an infinite decimal expansion in our base-ten system, but that can be represented by a finite number of symbols in cuneiform script. Are there any numbers that have a finite decimal expansion in base 10, but require infinitely many terms to express in cuneiform?
b. Prove that an irrational number can not be represented by a finite number of terms in cuneiform script.

Exercise 1.17 Let $w, z \in \mathbb{C}$. Prove that $\overline{z + w} = \overline{z} + \overline{w}$ and $\overline{zw} = \overline{z} \cdot \overline{w}$.

Exercise 1.18 Let $w, z \in \mathbb{C}$. Prove or find a counterexample for each equation:

$$N(wz) = N(w)N(z) \qquad \text{and} \qquad N(w + z) = N(w) + N(z)$$

Exercise 1.19 Use induction and the previous exercise to prove the identity $\overline{z^n} = \overline{z}^n$ for $n \in \mathbb{N}$.

Exercise 1.20 Prove that if z is a complex root of a polynomial $f \in \mathbb{R}[x]$, then so is $\bar{z}$. What happens if instead $f \in \mathbb{C}[x]$?

Exercise 1.21 Prove that addition of polynomials in $\mathbb{R}[x]$ is commutative and associative.

Exercise 1.22 Prove that multiplication of polynomials in $\mathbb{R}[x]$ is commutative and associative.

Exercise 1.23 The text suggests that i is quite special for generating the complex numbers, but it is not unique in doing so. Prove that if we introduce to $\mathbb{R}$ the number $j = \sqrt{-13} = \sqrt{13}i$ and form the set

$$\mathbb{C}' = \{a + bj : a, b \in \mathbb{R}\},$$

then $\mathbb{C} = \mathbb{C}'$. Then check that with the analogous construction, $\mathbb{Z}[i] \neq \mathbb{Z}[j]$.

Exercise 1.24 Let $\triangle$ denote the symmetric difference (see Exercise 1.12).

a. Prove that the symmetric difference is associative on the set of all subsets of a set S; i.e., for all sets $A, B, C \subseteq S$,

$$A \triangle (B \triangle C) = (A \triangle B) \triangle C.$$

b. Prove set intersection distributes over the symmetric difference, i.e., for all sets A, B, and C,
$$A \cap (B \triangle C) = (A \cap B) \triangle (A \cap C).$$

c. Does set union distribute over set difference? Prove it does or give a counterexample.

Computation and Experimentation

Exercise 1.25 Get familiar with a programming language of your choice so that you will be ready to tackle the exercises in later sections. The text website has an introductory worksheet for Python ("Getting to know Python") to help you get started with that language if you so choose (or if your instructor so chooses).

Exercise 1.26 Write a short program to multiply quaternions.

Exercise 1.27 Write a short program to implement the cubic formula.

General Number Theory Awareness

Exercise 1.28 As alluded to in the opening paragraph of this chapter, human understanding of what it means to be "a number" has undergone a lengthy evolution. A famous example is the reverberations in Pythagoras' teaching academy when it was first proven to them that $\sqrt{2}$ is irrational. Find the famous story involving a debateaufication[7] and then do a fact-check: how much of it is currently believed to be apocryphal?

Exercise 1.29 Repeat the previous problem for the historical significance and adoption of the number $i = \sqrt{-1}$. Were cults involved? Witchcraft? Beans? Nicholas Cage?

Exercise 1.30 Look up the history of Geralomo Cardano and his relationship with Niccolò Fontana and Lodovico Ferrari. Who says mathematicians are boring!?

Exercise 1.31 As with most ancient history, much of our understanding of ancient mathematics comes from a mix of historical documents and informed speculation. Do some research on "Plimpton 322"—what do we definitively learn about ancient mathematics from it? What are some of the more controversial claims it has prompted?

Exercise 1.32 Research William Rowan Hamilton's discovery of the quaternions. Where did he scratch their defining relations? What are some applications of the quaternions?

Exercise 1.33 What is our current state of understanding of Egyptian fraction representations? What can you say about the length and sizes in the "best" representation? Are there surprising representations of, say, 2?

Exercise 1.34 One of the most valuable tools for the experimenting number-theorist is the *On-Line Encyclopedia of Integer Sequences*, a tool that allows you to input a sequence of integers that you've come across, and the pattern matches it to previously researched sequences. Try it out on the sequence

$$1, 1, 2, 3, 5$$

What famous sequence does the OEIS imagine you're probably encountering? Report back some interesting results from the page. Finally, repeat this process if in your sequence the 5 was followed by a 7.

Exercise 1.35 Gaussian integers are named after a mathematician named Gauss, which makes sense. Read up on Gauss' exploits. Be sure to track down the correct pronunciation so you don't accidentally send your professor into conniptions[8] .

---

[7] Like a defenestration but for boats.

[8] This is not meant to be a guide on *intentionally* sending your professor into conniptions.

References

1. Boyer, C.B.: Fundamental steps in the development of numeration. Isis **35**(2), 153–168 (1944)
2. Burton, D.M.: The history of mathematics: An introduction. In: International Series in Pure and Applied Mathematics, McGraw-Hill (1997)
3. Dunham, W.: Journey through genius: the great theorems of mathematics. Penguin Books, Wiley science editions (1991)
4. Ivars, P.: From counting to writing. *Science News*, March 8 (2006)
5. Smith, D.E.: A Source Book in Mathematics, vol. 3. Dover Publications (1959)

A Quick Survey of the Last Two Millennia

2

...wherein the reader meets the problems of Diophantus of Alexandria and their many offspring.

2.1 Fermat, Wiles, and The Father of Algebra

In 1994, mathematician Andrew Wiles, aided by his colleague Richard Taylor, forged the last link in an extensive chain of mathematical reasoning needed to resolve one of the most infamously difficult problems in all of mathematics. The history of this result, commonly referred to as *Fermat's Last Theorem* after the seventeenth-century French mathematician Pierre de Fermat, has all the hallmarks of an epic tale: an arduous quest spanning centuries and continents taken on by a legion of mathematicians; a resilient foe that bested many of the greatest minds for 300 years; and the classic template of human anguish, redemption, and then ultimate triumph. Wiles' breakthrough garnered him fame and fortune: his proof was heralded across the globe in newspaper headlines and TV reports, and he was even listed as one of People Magazine's "25 Most Intriguing People of the Year."

It is tempting to imagine that a proof that merits such praise, and that confounded mathematicians for centuries, must stem from an impenetrably intricate theorem. On the contrary, it is a hallmark of number theory that some of the deepest and hardest-to-prove results need little more than an elementary school education to understand, and Fermat's Last Theorem fits this mold perfectly. It says quite simply that if a, b, and c are natural numbers, and n is a natural number greater than 2, then $a^n + b^n \neq c^n$. In Fermat's own words (translated to English from its original French [2]):

> It is impossible to separate a cube into two cubes, or a fourth power into two fourth powers, or in general, any power higher than the second, into two like powers. I have discovered a truly marvelous proof of this, which this margin is too narrow to contain.

Supplementary Information The online version contains supplementary material available at https://doi.org/10.1007/978-3-030-98931-6_2.

Fermat scrawled this statement in the margin of one of his books, but this purported proof was never found, and the prevailing modern consensus is that either he was unknowingly in possession of a flawed proof or that he was engaging in a bit of cavalier mathematical bravado[1]. Nevertheless, the tantalizing claim of "a truly marvelous proof" succeeded in piquing the curiosity of mathematicians for over three centuries. Mathematical legends like Leonhard Euler, Sophie Germaine, Adrien-Marie Legendre, Johann Lejeune Dirichlet, Gabriel Lamé, Srinivasa Ramanujan, and Ernst Kummer all took aim at the result, often proving special cases along the way, but ultimately falling short of a complete proof.

It seems safe to hypothesize that the failure of these eminent historical figures to solve this problem had an understandably chilling effect on its study by later mathematicians. David Hilbert, a renowned mathematician of the early twentieth century, once responded to why he had not tried his hand at the theorem by saying [1], "Before beginning I should put in three years of intensive study, and I haven't that much time to squander on a probable failure." By 1963, when 10-year-old Andrew Wiles stumbled across a statement of the theorem in a library book and found himself intrigued, most mathematicians had given up hope of finding a proof. Indeed, up until 1995, Fermat's Last Theorem held the Guinness World Record for the longest standing open math problem. But filled with the curiosity and audacity of youth, young Andrew dreamed of one day finding the proof and, 31 years later, rose to vanquish his childhood foe.

Somewhat ironically, the *result* of Fermat's Last Theorem is not a particularly useful one, at least in terms of applicability to other problems or disciplines. So why the big fuss? A second hallmark of number-theoretic investigations stems from a dogged tenacity in the belief that true facts deserve proof, no matter the difficulty or applicability. There is immense merit in this perspective—the aforementioned failed attempts by mathematicians through the ages led to new insights into the realms of algebra and number theory that transcended the single application for which they were built. Indeed, the mathematical breakthroughs and extensions developed in pursuit of a proof of Fermat's Last Theorem have since become staples in vast swaths of modern mathematics, including ties to the study of groups, cryptology, elliptic curves, modular forms, and lots more fancy words besides.

Thus, the culminating argument put forward by Wiles and Taylor is the denouement of a centuries-long story of mathematical progress and understanding, one that began long before even Fermat. We can trace the history of the question at least as far back as the book in which Fermat wrote his famous margin note, the text *Arithmetica* by the third-century Greek mathematician Diophantus of Alexandria. Diophantus is often referred to as "the father of algebra" in recognition of his use of symbols to represent variables, an easily overlooked revolution in the history of human thought. Before Diophantus, equations were expressed entirely in words, as in the quote by Fermat above—to see the revolutionary nature of such a step, imagine trying to set up and solve even a simple calculus problem without using symbolic notation!

---

[1] Classic Pierre.

Diophantus's text was concerned specifically with finding solutions to polynomial equations, of which Fermat's Last Theorem was just one of many examples. As we begin our explorations in number theory, there is likely no better place to begin than with those same questions that delighted Diophantus, fascinated Fermat, and wowed Wiles. Along the way, we will see how the study of something as simple as quadratic equations reveals hidden structures in numbers and stimulates the development of a fantastic wealth of mathematics.

2.2 Quadratic Equations

A quick warm-up to get us started on solution sets of quadratic equations:

Question 2.2.1 What are the solutions to the equation $x^2 + y^2 = 3$? What about $x^2 + y^2 = 5$?

Perhaps your first thought was to envision the circles in the plane of respective radii $\sqrt{3}$ and $\sqrt{5}$. If this is the case, then you, the tragic hero of this story, have fallen into the trap set by us, the duplicitous authors. Indeed, you've been tricked into answering an ambiguous question, in that we have left unspecified *where* we want our solutions to live.

> A recurring theme of the course – and perhaps *the* theme of the course – is that *the context of which types of solutions you're interested in will greatly affect the quantity and structure of said solutions.*

A relatively modern perspective is that you should think of an equation as an abstract entity, which provides different solution sets upon being handed different solution domains. That is, if we have an equation (e.g., $x^2 + y^2 = 3$), we can as usual think of its real solutions as defining a curve in the plane (in this case, a circle), but could also think of its solutions in a variety of number sets. Notationally, if we denote such a "curve" by C, then its set of solutions with coordinates in $\mathbb{Z}$ is $C(\mathbb{Z})$, its set of rational solutions is $C(\mathbb{Q})$, etc.

Table 2.1 shows some solution sets of the two curves from Question 2.2.1 over different domains. The first row consists of the real solutions, visualized as circles in the plane. The next two rows are easily justifiable by brute-force exhaustion (e.g.,

Table 2.1 Solutions to Two Curves over Varying Sets

Set \ Curve C	$x^2 + y^2 = 3$	$x^2 + y^2 = 5$
$C(\mathbb{R})$	Circle of radius $\sqrt{3}$	Circle of radius $\sqrt{5}$
$C(\mathbb{N})$	None!	$(1, 2)$ and $(2, 1)$.
$C(\mathbb{Z})$	None!	$(\pm 1, \pm 2)$ and $(\pm 2, \pm 1)$.
$C(\mathbb{Q})$	None!	Infinitely many!!!

if either x or y were an integer with an absolute value bigger than 2, then $x^2 + y^2 > 5$). But the last row, asking for rational solutions, exhibits a fairly remarkable phenomenon. It claims (without justification—for now) that there are no rational solutions at all to the equation $x^2 + y^2 = 3$, but that in contrast, there are infinitely many such solutions for $x^2 + y^2 = 5$. A few rational solutions suggest themselves immediately (e.g., $(\pm 1, \pm 2)$), but $C(\mathbb{Q})$ also includes less obvious elements like

$$\left(\frac{11}{5}, \frac{2}{5}\right) \quad \text{and} \quad \left(\frac{316378}{172393}, \frac{220231}{172393}\right).$$

Given the similarity (both algebraic and geometric) of the two curves, the difference between the two columns is notable. Also relevant is the relationship between the rows in a given column. For example, if there are no rational points on a curve, then there aren't any integer points either, for the simple reason that every integer solution would also be a rational solution. Similarly, if there aren't any real solutions, there can't be any rational solutions: the general observation is that if $A \subseteq B$, then $C(A) \subseteq C(B)$. An integral part of number theory is to pass between different domains in search of answers to questions—maybe you would like to know $C(\mathbb{Z})$, but it's easier to compute $C(A)$ for some other set A, and knowledge of $C(A)$ somehow helps you understand $C(\mathbb{Z})$ better, etc. This perspective requires determining which sets lead to the best results for which problems and expanding our options for the sets of numbers at our disposal. As a preview of things to come, we might also consider (pending careful definitions of these terms) algebraic solutions, transcendental solutions, complex solutions, irrational solutions, solutions modulo primes, or even solutions in more exotic places (ooooh...).

To illustrate the power of being able to move between these worlds, let us tackle a famous problem of antiquity, introducing a novel[2] technique, due to Diophantus himself, exploiting the presence of geometric techniques for real solutions for one equation to find integral solutions to another.

The Diophantus Chord Method

A first approach to systematically studying solution sets to equations like $x^2 + y^2 = 3$ might be to do so geometrically, considering the geometric properties of the corresponding circle in the plane. But there are also hints that this approach might run into trouble—after all, from the perspective of the Euclidean plane, our two circles in Table 2.1 are geometrically "similar," and consequently hard to distinguish using geometric ideas. In particular, there is an easy bijection between their sets of real solutions (the top row in Table 2.1), but certainly *not*, given the table, their rational ones. Let's use the unit circle to illustrate the process of translating between domains.

▶ **Example 2.2.2**   Find *all* rational solutions to $x^2 + y^2 = 1$.

---

[2] As least, as of a couple of millennia ago

The slick classical idea, the *Diophantus chord method*, that we employ to resolve this stalemate between algebra and geometry is a beautiful blend of the two approaches.

Solution We have one obvious solution of $(1, 0)$. Here's the approach we pursue to find more:

> If we take a line through $(1, 0)$ with a rational slope m, it intersects the circle in one other point (a, b) – that point is another rational solution. Further, every rational solution arises from such a line. That is, in Figure 2.1, the slope m is rational if and only if a and b are.

Let's check the details: suppose m is a rational number. Then the equation for the line through $(1, 0)$ with slope m is given by $y = m(x - 1)$. If (a, b) is the other intersection point, then $b = m(a - 1)$, and substituting $a = \frac{b+m}{m}$ for y in the equation for the unit circle gives $\left(\frac{b+m}{m}\right)^2 + b^2 = 1$. Simple algebra converts this to

$$b((m^2 + 1)b + 2m) = 0.$$

The trivial solution to this equation, $b = 0$, corresponds to the point $(1, 0)$ we started with. The other, $b = -\frac{2m}{m^2+1}$, must thus represent the y-coordinate of our second intersection point, and is rational whenever m is (Figure 2.1).

Likewise,

$$a = \frac{b + m}{m} = \frac{\frac{-2m}{m^2+1} + m}{m} = \frac{-2}{m^2 + 1} + 1 = \frac{m^2 - 1}{m^2 + 1}$$

Fig. 2.1 A Hunt for Points

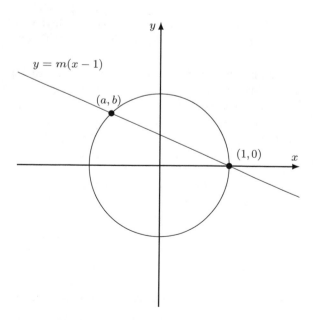

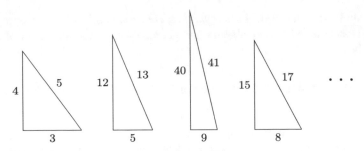

Fig. 2.2 ♫ These are a few of my favorite things... ♫

is also rational. Finally, the rather trivial converse: if (a, b) is the second intersection point of a line through $(1, 0)$ and the circle $x^2 + y^2 = 1$ and a and b are both rational, then so too is the slope $m = \frac{b-0}{a-1}$ of the line containing them.   □

   The conclusion of this discussion is that we now have **all** of the rational solutions—there is $(1, 0)$, and then there is one additional point for every rational number m. (Note the identical claim holds for real solutions and real slopes m). For example, if $m = -\frac{3}{7}$, the example chosen to generate Figure 2.1, then we get $(a, b) = \left(-\frac{20}{29}, \frac{21}{29}\right)$.

   As it turns out, the solution to this problem holds the key to one of the most significant classically solved problems in number theory: the identification of Pythagorean triples.

Definition 2.2.3

A **Pythagorean triple** (a, b, c) consists of three natural numbers that can form the side lengths of a right triangle (the last of which we assume to be the hypotenuse); i.e., that are constrained by the Pythagorean Theorem to satisfy $a^2 + b^2 = c^2$. ◄

   It is amusing to contrast this equation with that of Fermat's Last Theorem: while it may be impossible to separate a cube into two cubes, and likewise, for higher powers, it is very much possible to separate a square into two squares. In fact, you likely already know some examples of Pythagorean triples: $(3, 4, 5)$ is a big hit at parties, and only slightly less known are triples like $(5, 12, 13)$, $(9, 40, 41)$, $(8, 15, 17)$, and so on. But this "and so on" is slightly unsatisfying—does the list continue forever? Can we find as many as we want?

   Right off the bat, there is a trivial way of constructing an infinite supply of Pythagorean triples. Namely, if you take the $(3, 4, 5)$ right triangle and scale each side by a factor of 2, you get a similar right triangle with side lengths 6, 8, and 10, giving rise to the new Pythagorean triple $(6, 8, 10)$. Likewise, the triple $(3k, 4k, 5k)$ is a Pythagorean triple for every natural number k, and the same goes for the families of triples $(5k, 12k, 13k)$, $(9k, 40k, 41k)$, $(8k, 15k, 17k)$, etc. Nevertheless, we push forward and ask not only for a mere infinitude of Pythagorean triples but also *all* of them. Remarkably, we did most of the hard work already when we found all of the rational points on the unit circle. The key observation is the following: If (a, b, c) is a Pythagorean triple, then after dividing by c^2, the Pythagorean Theorem tells us that

$$\left(\frac{a}{c}\right)^2 + \left(\frac{b}{c}\right)^2 = 1,$$

which says that $\left(\frac{a}{c}, \frac{b}{c}\right)$ is a rational point on the unit circle. Conversely, if we have a rational point on the unit circle, then by clearing denominators we get a Pythagorean triple, e.g.,

$$\left(\frac{20}{29}\right)^2 + \left(\frac{21}{29}\right)^2 = 1 \quad \longrightarrow \quad 20^2 + 21^2 = 29^2.$$

The conclusion is that since Example 2.2.2 successfully parameterizes all of the rational points on the unit circle, we have a corresponding parameterization of every single Pythagorean triple. Note that this correspondence is not bijective: the pairs $(3/5, 4/5)$ and $(6/10, 8/10)$ are the same point on the unit circle, but $(3, 4, 5)$ and $(6, 8, 10)$ are different Pythagorean triples. So, any point on the unit circle gives, by scaling, infinitely many different Pythagorean triangles. Let's put this idea to use:

> **Theorem 2.2.4**
> For all $u, v \in \mathbb{N}$ with $u > v$, the triple $(u^2 - v^2, 2uv, u^2 + v^2)$ is a Pythagorean triple. Further, every Pythagorean triple $(a, b, c) \in \mathbb{N}^3$ is a multiple of one of this form.

Proof Start by verifying the identity

$$(u^2 - v^2)^2 + (2uv)^2 = (u^2 + v^2)^2,$$

for any integers[3] u, v. Now suppose (a, b, c) is a Pythagorean triple. Then, as above, (a, b, c) corresponds to a rational solution $(x, y) = (\frac{a}{c}, \frac{b}{c})$ of $x^2 + y^2 = 1$. By the solution to Example 2.2.2, we can write

$$\left(\frac{a}{c}, \frac{b}{c}\right) = \left(\frac{m^2 - 1}{m^2 + 1}, \frac{-2m}{m^2 + 1}\right),$$

for some rational number m. Further, since this point is in the first quadrant, we can take m to be negative (see Figure 2.1), and so write $m = -\frac{u}{v}$ for positive integers u and v. Substituting gives

$$\left(\frac{a}{c}, \frac{b}{c}\right) = \left(\frac{\left(\frac{u}{v}\right)^2 - 1}{\left(\frac{u}{v}\right)^2 + 1}, \frac{2\left(\frac{u}{v}\right)}{\left(\frac{u}{v}\right)^2 + 1}\right) = \left(\frac{u^2 - v^2}{u^2 + v^2}, \frac{2uv}{u^2 + v^2}\right).$$

---

[3] Or indeed, as will be important in Exploration A, for any reasonable objects u, v at all!

We conclude that for some integer k,

$$a = k(u^2 - v^2), \quad b = k(2uv), \quad \text{and } c = k(u^2 + v^2),$$

establishing the result.   □

▶ **Remark 2.2.5** In Chapter 3, once we have established some basic properties of divisibility, primes, and common factors, we will be able to refine the parameterization: every Pythagorean triple is a multiple of one coming from a pair (u, v) with one of u, v even, one odd, and u and v having no factors in common.

Return to Quadratic Equations

Let us now reverse our chronology and imagine that we found ourselves in the situation, as Pythagoras did in 500 BCE, of having deduced the remarkable relationship $a^2 + b^2 = c^2$ among the sides of a right triangle, and wanting to exhaust the integer solutions to this equation. By the previous section, all we need is a couple of centuries of patience waiting for the mathematics of Euclid and Diophantus, and the conceptualization of a rational number. Then we need only combine two observations:

- The *integer* solutions to $a^2 + b^2 = c^2$ are closely related to the *rational* solutions to $x^2 + y^2 = 1$.
- The *rational* solutions of $x^2 + y^2 = 1$ are a subset of the *real* solutions of $x^2 + y^2 = 1$, which by virtue of being graphically represented as the unit circle, are subject to study using the tools of geometry (the chord method).

As a result we arrive at Theorem 2.2.4, whose evident power raises our hopes that we can systematically generate solutions to equations as we run across them. For example, we can now generate new Pythagorean triples with ease: taking $(u, v) = (2, 1)$ gives the Pythagorean triple $(a, b, c) = (3, 4, 5)$, the pair $(3, 2)$ gives the triple $(5, 12, 13)$, the pair $(4, 1)$ gives the triple $(8, 15, 17)$, and the pair $(u, v) = (4, 2)$ gives the multiple $(12, 16, 20)$ of $(3, 4, 5)$. In Exercise 2.2, you will find the pair (u, v) that rediscovers the Pythagorean identity

$$13500^2 + 12709^2 = 18541^2,$$

known to the Babylonians as early as 1800 BCE [3].

And indeed the technique *does* generalize nicely to an extent: there is nothing particularly special about the specific curve $x^2 + y^2 = 1$, and we could repeat this for any "conic" curve, as you'll do for $x^2 + y^2 = 5$ in Exercise 2.3. Except...there must be *something* special about these curves, as we've already promised that the very similar curve $x^2 + y^2 = 3$ not only doesn't have infinitely many rational solutions but also has precisely zero. As it turns out, the difference between zero and infinitely many solutions for such curves is precisely the difference between zero and one solution. That is, once we have found one rational solution point on the curve, we can conclude there must be infinitely many. For $x^2 + y^2 = 1$, we had the point $(1, 0)$

serve in this role of an initial rational solution; but for $x^2 + y^2 = 3$, there is no such point (though again, we stress that this claim is not *yet* justified). The mechanism for finding all rational solutions on conics with such a point is precisely the Diophantus chord method that worked for the unit circle: given one solution, a line with rational slope will hit precisely one more rational point on the conic, and as we vary over all rational slopes, we get all of the rational points. Likewise, clearing denominators allows us to find all integral solutions to the corresponding Diophantine equation. In the case of the curve $x^2 + y^2 = 1$, the corresponding Diophantine equation is the Pythagorean equation $a^2 + b^2 = c^2$; for $x^2 + y^2 = 5$, it is the equation $a^2 + b^2 = 5c^2$.

Theorem 2.2.6

Suppose a, b, c are non-zero rational numbers. If the equation $ax^2 + by^2 = c$ has one rational solution, it has infinitely many, and each such solution provides an infinite family of integer solutions to the equation $ax^2 + by^2 = cz^2$.

Proof The technique is entirely analogous to that applied to the pair $x^2 + y^2 = 1$ and $x^2 + y^2 = z^2$. The details are left to Exercise 2.8. □

We see from this that the last remaining question for equations of this type is the existence of even just one rational solution. We close this discussion by noting that this is a pretty tough question. Indeed, the lack of integer solutions doesn't preclude the existence of a rational solution. For example, you can solve $43 = x^2 + 26y^2$ for $x, y \in \mathbb{Q}$—e.g., by taking $x = 1/7$ and $y = 9/7$—but certainly not for $x, y \in \mathbb{Z}$ (why? Hint: y?). We will return to this question in Chapter 8, only after we have developed significantly more mathematical machinery.

Moving on, while it is the case that quadratic equations are well handled by the theorem above, this represents only a small sliver of the possible types of equations we might run across. Let us give a name to the generalization.

2.3 Diophantine Equations

Definition 2.3.1

A **Diophantine equation** is a polynomial equation with integer coefficients in which the variables represent integers; i.e., where integer solutions are sought. ◄

Though the study of Diophantine equations is done largely with rather abstract goals in mind (we will talk about applications to other branches of mathematics as we go), there is a certain comforting concreteness stemming from considering only integer or natural number solutions. A somewhat famous example of a problem in which only natural number solutions are sought is the so-called *cannonball problem*, which can be modernized into a problem somewhat like this:

▶ **Example 2.3.2** Vineas and Herb have a truly marvelous collection of rubber balls. When Vineas takes possession of the balls, he arranges them in a perfect square pyramid. When Herb takes possession of the balls, he arranges them in a perfect flat square. How many balls are there?[4]

The problem seems at first glance to be woefully underspecified, but in fact, these geometric constraints on the configurations of the balls turn out to be very restrictive. For example, there could not be 14 balls, because 14 is not the square of a natural number (Figure 2.3). Similarly, while 25 is indeed a perfect square, it is not possible to arrange 25 balls into a perfect square pyramid, since such a pyramid with three layers uses too few balls ($1 + 4 + 9 = 14$), and one with four layers uses too many ($1+4+9+16 = 30$). So the number of balls, n, must be both a perfect square ($n = b^2$ for some $b \in \mathbb{N}$) and a "perfect square pyramidal number" ($n = 1+4+9+\cdots+a^2$ for some $a \in \mathbb{N}$). We will leave to Exercise 2.6 the induction proof of the identity $1 + 4 + 9 + \cdots + a^2 = \frac{a(a+1)(2a+1)}{6}$, but we will use it here in the meantime to succinctly encode both of these conditions in one fell swoop: for the number n to be expressible in both of these forms, we need there to exist natural numbers a and b such that

$$6b^2 = a(a + 1)(2a + 1), \tag{2.1}$$

a particularly clean-looking Diophantine equation. Now, it is a remarkable fact that there is, in the entirety of natural numbers, only one pair[5] (a, b) that form a solution

Fig. 2.3 14 balls in a square pyramid, 25 balls in a square

---

[4] I know what we're gonna do today!

[5] Ignoring the trivial solution $a = 1, b = 1$, which we feel is justifiably ruled out by the plural "rubber balls" in the problem statement. "Look at this cool pyramid I made out of my one rubber ball!" Come on. You're better than this.

to this Diophantine equation, and hence only one possible number n of rubber balls that satisfy both Vineas's and Herb's needs. (Note that this differs from the quadratic case: one solution does not imply infinitely many.) While the proof of the uniqueness of this solution is somewhat challenging, we won't spoil the surprise as to the exact value—Exercise 2.12 has you cut your computational teeth on finding it.

As it turns out, equations, like (2.1), encompass an enormous swath of modern mathematics, and we will be seeing them several times over the course of the book. To name such curves early on:

Definition 2.3.3

An **elliptic curve** E over $\mathbb{Q}$ is an equation of the form

$$y^2 = x^3 + sx + t \qquad (s, t \in \mathbb{Q}),$$

where the cubic polynomial $x^3 + sx + t$ has no repeated roots, or equivalently, the discriminant $4s^3 + 27t^2 \neq 0$ (see Exercise 2.10). ◄

The change of variables $x = 12a + 6$, $y = 72b$ transforms the equation $6b^2 = a(a+1)(2a+1)$ into the form $y^2 = x^3 - 36x$, the elliptic curve E with $s = -36$ and $t = 0$. For example, corresponding to our trivial square pyramid solution $(a, b) = (1, 1)$ is the point $(x, y) = (18, 72)$ on E. Note that the reverse change of variables involves division and so can introduce denominators (though, notably, only divisors of 72), so it is plausible that the elliptic curve has more integer points than the square pyramid equation.

A less brute force solution to the square pyramid problem is then to find more points on this elliptic curve, and specifically integer points (x, y) that correspond to integer pairs (a, b). As it turns out, there is exactly one more ± pair of integer points on E—how do we find it? The Diophantus Chord method just keeps on giving! While it is not true that a line with a rational slope through a rational point on an elliptic curve intersects only other rational points, the following variant is true: given a line that intersects an elliptic curve in three points, two of which are rational, the third must also be rational. For example, given any two of the points $(-3, 9)$, $(-2, 8)$, or $(6, 0)$ on $y^2 = x^3 - 36x$, we could find the third by simple algebra and could do the same with combinations of points including $(0, 0)$, $(-6, 0)$, and any points we demonstrate along the way. Our missing integer point happens to be the third point of intersection of the elliptic curve with the line through $(-2, 8)$ and $(-3, -9)$ (Exercise 2.5). The significance of elliptic curves in modern mathematics is largely thanks to this process, taking two points on a curve and producing a third. The consequences of this observation surpass even the classification of Pythagorean triples, with a notable highlight being an entire cryptographic protocol based on the premise. We discuss this remarkable application in Section 9.3.2.

While there is enough mathematics in studying our conics and cubic equations to fill dozens of textbooks, we note that there is a rich and expansive world of even more involved Diophantine equations that we will not have space to get to. For example, one could also consider *systems* of Diophantine equations. One such system comes

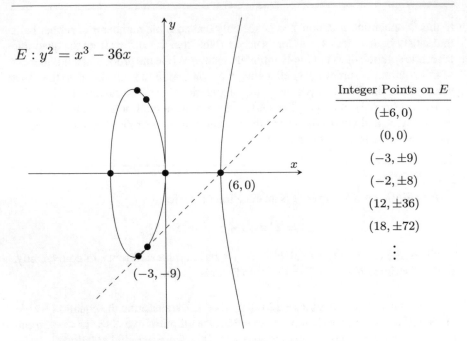

$$E : y^2 = x^3 - 36x$$

(6, 0)

(−3, −9)

Integer Points on E
$(\pm 6, 0)$
$(0, 0)$
$(-3, \pm 9)$
$(-2, \pm 8)$
$(12, \pm 36)$
$(18, \pm 72)$
$\vdots$

Fig. 2.4 The elliptic curve E defined by $y^2 = x^3 - 36x$

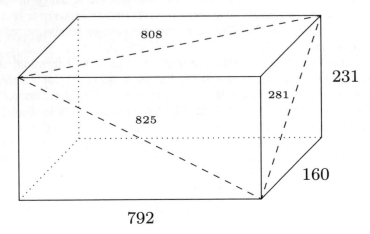

Fig. 2.5 An Euler Brick

from a natural generalization of Pythagorean triples in the study of *Euler Bricks*, rectangular prisms in which the sides *and* the diagonals of each face are all integers. Figure 2.5 shows an example, and in the exercises, you will construct infinitely many more. Note that each side length of the prism is involved in two different Pythagorean triples, leading to a Diophantine system.

Much like the raw equation $6b^2 = a(a + 1)(2a + 1)$ inspires little insight as to its potential uses for entertaining kids on summer vacation, so do most Diophantine equations begin life looking rather innocuous—indeed, even the central equation of

Fermat's Last Theorem seems more a curiosity than an instigator for centuries of mathematical progress. We offer one more tantalizing taste: the equation $x^2+y^2 = n$, or equivalently the question of which natural numbers can be written as the sum of two squares—that 5 *can* but 3 can *not*—was of significance at the start of this chapter. There is an obvious generalization to this question in the equation

$$x^3 + y^3 + z^3 = n,$$

i.e., that of which natural numbers can be written as the sum of three cubes. When $n = 6$, for example, we have the easy solution $(2, -1, -1)$ since $2^3+(-1)^3+(-1)^3 = 6$. Some other values of n are equally easily dispatched, but then others prove more challenging, even for nearby values for n. For example, $n = 43$ admits the easy solution $(3, 2, 2)$, and for $n = 44$, there is the much less obvious solution $(8, -7, -5)$, whereas for $n = 42$, no solution was known until 2019 when, during the writing of this very Guide to the Numberverse, mathematicians Andrew Booker and Andrew Sutherland found [4] the solution

$$80435758145817515^3 + 12602123297335631^3 + (-80538738812075974)^3 = 42.$$

This solution came from millions of hours of deep thought and computation on a distributed network of idle computers across the globe, in what surely must be the first instance in fact or fiction of the ultimate question of 42 being resolved by a world-scale supercomputer.

This solution, and more generally the history of these types of problems, highlights well the nature of working with Diophantine equations. Given any specific equation, a host of natural questions presents itself: are there any solutions? Can we find them by brute-force search? By a more clever search? Are there finitely or infinitely many? Can we count them? Describe them geometrically? Parameterize them? Should we move to a different world of numbers and ask the question there? In the notation introduced earlier, if C denotes the equation or system of equations, then we can recode this as asking if $C(\mathbb{Z})$ is empty, finite, infinite, etc., and then ask the same questions for $C(\mathbb{N})$, $C(\mathbb{Q})$, $C(\mathbb{R})$, etc., and for other worlds of numbers beyond even these. If all this fails, can we deduce the abstract existence of solutions? Or is it possible that the Diophantine equation is *undecidable*, i.e., that it could be provably impossible to resolve the question of whether or not the equation has solutions (see Exercise 2.19)? We stress once more that it is not always even the case that the equation itself is of interest, but rather the mathematics to which its study gives rise. Questions like the above have driven us to delve into the mysteries of the integers, discover new and exotic number systems, and develop along the way the mathematical structures that govern our modern electronic communications and commerce. These questions have tantalized some of humanity's greatest minds[6] and through their history form a link from the ancient civilizations to today.

---

[6] And ours, too!

One Idea, Many Worlds ◀

The goal of this Exploration is to investigate the analogs of Pythagorean Triples in other worlds. Just as important as seeing how different these various worlds are is seeing how much they have in common. For example, the identity

$$(u^2 - v^2)^2 + (2uv)^2 = (u^2 + v^2)^2$$

would seem to hold not only for *integers u* and *v* but also a whole host of mathematical constructions.

Gaussian Integers

A.1 Find a "Gausso-Pythagorean triple" in $\mathbb{Z}[i]$ that is not in $\mathbb{Z}$. That is, find $\alpha, \beta, \gamma \in \mathbb{Z}[i]$, not all in $\mathbb{Z}$, such that $\alpha^2 + \beta^2 = \gamma^2$.

A.2 Explore other Diophantine equations in the context of the Gaussian integers ("Gaussophantine"?). For example, revisit $x^2 + y^2 = 3$ and $x^2 + y^2 = 5$: are there new solutions to either?

A.3 Factor 2 in $\mathbb{Z}[i]$. That is, find two Gaussian integers $a + bi$ and $c + di$ not in $\mathbb{Z}$ such that $(a + bi)(c + di) = 2$. How many distinct factorizations can you find? Repeat for factoring 3 and 5.

Producing Pythagorean Polynomials

A.4 Use the identity above to find some "Pythagonomial triples" of polynomials (with at least two terms). That is, find interesting $f, g, h \in \mathbb{R}[x]$ such that $f^2 + g^2 = h^2$.

A.5 Recall from calculus that if a curve C is parameterized by $r(t) = \langle x(t), y(t) \rangle$ for $t \in [a, b]$, where f and g have continuous derivatives on $[a, b]$ and C is traversed only once, then its length L can be computed by

$$L = \int_a^b \sqrt{(x'(t))^2 + (y'(t))^2}\, dt.$$

Such integrals are typically very hard to compute, but a clever choice of $x(t)$ and $y(t)$ will leave a perfect square under the radical.

1. Make use of the polynomials f, g from the previous problem to write and solve an arclength problem in which the integrand simplifies so that there is no radical.
2. Write and solve a non-trivial arc length problem in which f, g are *not* polynomials and the integrand simplifies so that there is no radical.

In technical mathematical parlance, such problems are called "rigged."

2.4 Exercises

Calculation & Short Answer

Exercise 2.1 Find a Pythagorean triple that can not be written in the form $(u^2 - v^2, 2uv, u^2 + v^2)$ for any integers u, v.

Exercise 2.2 From Theorem 2.2.4, we learn that the Pythagorean triple (12709, 13500, 18541) known to the Babylonians must arise as a multiple of one having the form given in that result. In fact, it equals $(u^2 - v^2, 2uv, u^2 + v^2)$ for some $u, v \in \mathbb{N}$. Find the u and v that generate this triple, and more generally describe a process for finding the u and v corresponding to a triple of that form.

Exercise 2.3 Returning to our opening Diophantine equation, use the chord method to find all rational solutions to $x^2 + y^2 = 5$. You should have one solution for each rational line m. Use this to find one non-obvious integer solution to the Diophantine equation $a^2 + b^2 = 5c^2$.

Exercise 2.4 Find an infinite family of integer solutions to the Diophantine equation $x^2 + y^2 = 2z^2$.

Exercise 2.5 Use algebra to find the third point on the dashed line in Figure 2.4 and back-substitute to solve Example 2.3.2.

Formal Proofs

Exercise 2.6 Prove by induction that for every natural number a, we have

$$1^2 + 2^2 + \cdots + a^2 = \frac{a(a + 1)(2a + 1)}{6}.$$

Exercise 2.7 Prove that for all natural numbers n, there is a Pythagorean triple with leg lengths $2n + 1$ and $2n^2 + 2n$. How can we see this as a special case of Theorem 2.2.4? Does this family account for all Pythagorean triples?

Exercise 2.8 Prove Theorem 2.2.6. Suppose a, b, c are non-zero rational numbers and that (x_0, y_0) is one rational solution to the conic $ax^2 + by^2 = c$. Prove that this equation has infinitely many rational solutions.

Exercise 2.9 Determine (with proof) whether the diagonals of an Euler brick (as shown in Figure 2.5) can be a Pythagorean triple.

Exercise 2.10 Check that the polynomial $x^3 + sx + t \in \mathbb{R}[x]$ has a repeated root if and only if the discriminant $4s^3 + 27t^2 = 0$. (Hint: Write $x^3 + sx + t = (x - r_1)(x - r_2)(x - r_3)$ and relate s, t to r_1, r_2, r_3).

Computation and Experimentation

Exercise 2.11 Build a Pythagorean triple generator based on Theorem 2.2.4. The Python worksheet Pythagorean Triples is available on the text website.

Exercise 2.12 Write a program to deduce how many rubber balls Vineas and Herb must have had in Example 2.3.2; i.e., find a natural number solution to the equation $6b^2 = a(a + 1)(2a + 1)$. Sample code might begin

```
for a in range(1,1000):
    for b in range(1,1000):
        if ...
```

The Python worksheet Rubber Balls provides an outline for this exercise and the next.

Exercise 2.13 Instead of organizing our rubber balls as both a square and a square pyramid, we could try to arrange them into both a square and a tetrahedron (triangular pyramid). For example, now $n = 4$ balls would work, arrangeable into either a 2-by-2 square ($2 \cdot 2 = 4$) or a tetrahedron with two layers ($1 + 3 = 4$). In fact, there is only one other solution.

Find a Diophantine equation that governs numbers that represent both the number of balls in a square and the number of balls in a tetrahedron. (This will involve conjecturing a formula for the number of balls in a tetrahedron and proving your formula by induction (or otherwise)). Do a computer search for solutions to your derived equation.

Exercise 2.14 Find the side and diagonal lengths of an Euler Brick with the longest face diagonal length equal to 3471. See Python worksheet "Euler Bricks."

Exercise 2.15 The process of generalizing Diophantus' chord method to a higher degree has led to some amazing mathematics. The next step up gives *elliptic curves*, equations of the form $y^2 = $ a cubic polynomial in x. The following walks you through an example of how the chord method generalizes. Use a computer algebra system to aid in your calculations. Your goal is to find a rational point on the curve $y^2 = x^3 - 2$ other than $(3, \pm 5)$. The Python worksheet "Elliptic Curves" provides an outline for this problem.

1. Find an equation of the tangent line to the curve at the point $(3, 5)$. Note this line has a rational slope. Sound familiar?
2. Solve for the x-coordinate of the other point of intersection with the tangent line.

Exercise 2.16 Let s be the sum of the Pythagorean triple $\{a, b, c\}$. When $s = 120$, there are exactly three solutions for $\{a, b, c\}$, namely $\{20, 48, 52\}$, $\{24, 45, 51\}$, and $\{30, 40, 50\}$. For which value of $s \leq 1000$, is the number of solutions maximized? This is problem 38 from the Project Euler archives. If you would like to flex your number theory and programming muscle, check out projecteuler.net. You can submit your answer to this problem, see if you got it right, and compare your approach to others'.

General Number Theory Awareness

Exercise 2.17 Who was Sophie Germain? Give a brief description of her life, accomplishments, and connection to Fermat's Last Theorem.

Exercise 2.18 Find Diophantus's solution to his *Arithmetica* Problem II.8, dividing a given square into a sum of two squares. We approached this using the chord method, but without Cartesian geometry at his disposal, Diophantus could not actually draw a pictorial representation. How does his solution proceed without such a picture? Who is responsible for the retconning of Diophantus's solution to be a graphical one?

Exercise 2.19 The text refers to the possibility of an "undecidable" Diophantine equation. Do some research. Is there an algorithm which inputs a Diophantine equation and outputs its solution? Are there explicit Diophantine equations that are undecidable?

Exercise 2.20 If Pythagorean triples are too passé for you, do some research about Pythagorean *quadruples*. What is their geometric significance? Can we parameterize them like Pythagorean triples? Find at least one unsolved problem involving them.

Exercise 2.21 Since its proof, Fermat's Last Theorem can no longer be considered the longest standing open math problem. What problem does the Guinness Book of World Records cite in its place? Do you agree with their diagnosis?

References

1. Bell, E.T.: Mathematics. Queen and Servant of Science. McGraw-Hill Book Company Inc, New York, NY (1951)
2. Yu, I.M., Alexei A.P.: Introduction to Modern Number Theory: Fundamental Problems, Ideas and Theories. Springer Science & Business Media (2006)
3. Robson, E.: Neither Sherlock Holmes nor Babylon: a Reassessment of Plimpton 322. Historia. Math. **28**(3), 167–206 (2001)
4. Andrew, R.B., Andrew, V.S.: On a question of Mordell. In: Proceedings of the National Academy of Sciences, vol. 118, pp. 153–168 (1944)
5. John, L., Simon, S., Stacy, K., Andrew, W.: The Proof. WGBH Boston Video, South Burlington, VT (1997)

Number Theory in $\mathbb{Z}$ Beginning

3

...wherein an obvious statement is rendered unobvious, and the stage is set for our journey forward.

3.1 Algebraic Structures

One of the central tenets of the last chapter is that even if we are primarily interested in a problem in $\mathbb{Z}$, there is often benefit to thinking about a closely related problem in a different number system. A flipped, and equally valuable, version of this perspective is that we understand arithmetic in $\mathbb{Z}$ well enough that we should attempt to export this mastery to other systems. What makes a number system sufficiently "$\mathbb{Z}$-like" to do algebra in? The natural numbers, for example, do not form a particularly robust algebraic system, as they lack additive inverses and hence an effective notion of subtraction. As a sample consequence of this lack of structure, we had to insist on conditions like $u > v > 0$ when parameterizing Pythagorean triples, as side lengths in Pythagorean triples must be natural numbers. Working in $\mathbb{Z}$ would have permitted a slightly crisper parameterization at the cost of allowing Pythagorean triples like $(-3, -4, 5)$ (from $u = -1$ and $v = 2$). Things go further awry if we work in systems lacking other fundamental properties of integer arithmetic: in the world of 2×2 matrices, we cannot hope for the same (u, v)-parameterization of Pythagorean triples, as even the identity

$$(U^2 - V^2)^2 + (2UV)^2 = (U^2 + V^2)^2$$

fails to hold for matrices U and V except in the rare case that they commute ($UV = VU$). On the other hand, Exploration A showed that these identities *do* hold in $\mathbb{Z}[i]$ and even $\mathbb{R}[x]$, providing (u, v)-parameterization of families of Pythagorean

Supplementary Information The online version contains supplementary material available at https://doi.org/10.1007/978-3-030-98931-6_3.

triples in these worlds as well. It behooves us, therefore, to try to write down exactly what it is about $\mathbb{Z}$ (or $\mathbb{Z}[i]$, or $\mathbb{R}[x]$, etc.) that provides such an agreeable algebraic system. The notion of a *ring* provides the baseline structure needed.

Definition 3.1.1

A **ring** is a non-empty set R with a pair of binary operations, $+$ and $\cdot$, such that the following laws hold for all elements $a, b, c \in R$:

(1) Closure: $a + b \in R$ and $a \cdot b \in R$.
(2) Associativity: $a + (b + c) = (a + b) + c$ and $a \cdot (b \cdot c) = (a \cdot b) \cdot c$.
(3) Commutativity of Addition: $a + b = b + a$.
(4) Distributivity: $a \cdot (b + c) = (a \cdot b) + (a \cdot c) = a \cdot b + a \cdot c$.
(5) Additive Identity: there exists an element $0 \in R$ such that $a + 0 = a = 0 + a$ for all $a \in R$.
(6) Additive Inverses: for each $a \in R$, there is an element "$-a$" in R such that $a + (-a) = 0 = (-a) + a$.

If multiplication is also commutative in R (i.e., $a \cdot b = b \cdot a$ for all $a, b \in R$), we say R is *commutative ring*; if there is a multiplicative identity element $1 \in R$ such that $a \cdot 1 = a = 1 \cdot a$ for all $a \in R$, we say R *is a ring with unity*. ◀

The prototypical rings, for which the above properties are all very familiar, are sets of numbers like $\mathbb{Z}$, $\mathbb{Q}$, and $\mathbb{R}$, equipped with the standard notions of addition and multiplication (see Appendix I for a more formal approach to establishing these properties). We will soon see plenty more examples, and in a course called something like *abstract algebra* one encounters a whole host of non-number-like rings and even considers non-standard operations on familiar sets (see Exercises 3.15 and 3.16). Since such enjoyable detours are not in the purview of this text, we will ignore such oddities, assuming in all cases that the implied notions of addition and multiplication are the standard ones. We will also not feel tied to the dot notation $\cdot$ for multiplication, using ab, $a \times b$, $a \cdot b$, etc., interchangeably as the context calls for.

The definition of a ring provides us with three of the four basic operations: addition, subtraction, and multiplication. Division, always the slightly estranged member of the operation family, does not come along for free with the others. Many rings, like $\mathbb{Z}$, do not even have a notion of division of arbitrary elements, and this ties ultimately to the lack of multiplicative inverses. Rings in which every non-zero element has a multiplicative inverse are special.

Definition 3.1.2

A **field** is a commutative ring F with unity in which every element except the additive identity 0 has a multiplicative inverse; i.e., for each $a \neq 0$ in F, there is an element a^{-1} such that $aa^{-1} = 1 = a^{-1}a$. ◀

In the exercises you will prove that the element a^{-1} referenced in Definition 3.1.2 is unique, allowing us to unambiguously refer to it as *the* multiplicative inverse of a. When working in general rings, it is typically safer to write a^{-1} than $\frac{1}{a}$ (assuming a happens to have an inverse), since in a non-commutative context the expression $\frac{b}{a}$ is ambiguous as to whether it refers to ba^{-1} or $a^{-1}b$, which may be different.

Note that, as we move through the terminology, we are being more and more demanding of our structure, adding commutativity, then a unity, and then the existence of multiplicative inverses. As we do, fewer and fewer of our favorite structures will satisfy all of the axioms. It is important to keep track of where each of these structures lies along this spectrum, so convince yourself of the veracity of the claims in the following examples before moving on.

▶ **Example 3.1.3** The sets $\mathbb{Z}$, $\mathbb{Z}[i]$, and $\mathbb{R}[x]$ are commutative rings with unity but are not fields.

▶ **Example 3.1.4** The set $\mathbb{N}$ is not even a ring, as elements of $\mathbb{N}$ lack additive inverses (not to mention an additive identity).

▶ **Example 3.1.5** The set of $n \times n$ matrices with real entries forms a non-commutative ring with unity. Many elements of this ring do not have multiplicative inverses.

Definition 3.1.6

Given a ring R, a subset $S \subseteq R$ forms a **subring of** R if and only if the set S forms a ring under the same operations as R. Similarly, given a subset E of a field, F, E is a **subfield of** F if and only if E forms a field under the same operations as F. ◀

▶ **Example 3.1.7** The rings $\mathbb{Q}$ and $\mathbb{R}$ are both subfields of the field $\mathbb{C}$, while $\mathbb{Z}$ and $\mathbb{Z}[i]$ are merely subrings of $\mathbb{C}$.

▶ **Example 3.1.8** The set of odd integers has additive inverses[1] but is not closed under addition. The set

$$2\mathbb{Z} = \{2n : n \in \mathbb{Z}\}$$

of even integers is a commutative ring without unity and a subring of $\mathbb{Z}$.

▶ **Remark 3.1.9** The linguistically curious will wonder about the seemingly out-of-place words "rings" and "fields" peppering the landscape of mathematical jargon. There is no universal consensus as to the etymology, though it seems likely to have come in part from unfaithful translations from French or German. For example, the French word *corps* used for fields translates most literally as "body," as in "body of water" or one's "body of knowledge" (much like, say, one's "field of

---

[1] At least, insofar as one *can* have additive inverses in the absence of an additive identity.

vision"), and so it may have been a cultural difference in the use of those words that provoked the change in metaphors[2].

Not all rings are created equal: doing algebra with real numbers is very different from doing algebra with matrices, as in the former ring we can make use of (among other things) commutativity and the fact that every non-zero element of $\mathbb{R}$ has a multiplicative inverse. As a consequence, we can solve equations like $x^2 - x = 0$ in $\mathbb{R}$ very easily:

$$x^2 - x = 0 \implies x(x - 1) = 0 \implies x = 0 \text{ or } x - 1 = 0 \implies x \in \{0, 1\}.$$

But the analogous matrix equation $M^2 - M = 0$ (where M is a square matrix) is much more challenging (and indeed has infinitely many solutions!). Why? The key step in the above calculation made use of the *zero product property*, that the only way to have a zero product, $ab = 0$, is if one of a or b itself is zero. This property is true in $\mathbb{R}$ but not true in every ring (again, matrices).

Definition 3.1.10

An **integral domain** is a commutative ring with unity in which the product of non-zero elements is always non-zero; that is, for all a and b in the ring R, $ab = 0$ if and only if $a = 0$ or $b = 0$. ◄

That is, integral domains are rings in which we can use the zero product property as needed, an important property given its pervasiveness in solving equations. One would be forgiven at this point for panicking that we might introduce new types of algebraic structures at an alarming rate, one for each algebraic property we might encounter. Fortunately, this turns out to be an unfounded fear, as the notions of commutative rings, integral domains, and fields are precisely the core structures needed to do the fascinating number theory ahead of us. Also simplifying the landscape is that not all properties are independent. For example, we can show that the existence of multiplicative inverses in fact guarantees the zero product property:

Lemma 3.1.11

Every field is an integral domain, but not every integral domain is a field. ◄

Proof Suppose R is a field, and that $a, b \in R$ satisfy $ab = 0$. We need to show that either a or b is zero, i.e., that if $a \neq 0$, then $b = 0$. If $a \neq 0$, then since R is a field, there exists $a^{-1} \in R$, and so

$$b = (a^{-1}a)b = a^{-1}(ab) = a^{-1} \cdot 0 = 0.$$

---

[2] A paper entitled "A study of bodies with no identity" would surely raise eyebrows in the English-speaking world.

Finally, $\mathbb{Z}$ is an integral domain that is not a field. □

As another example of statements of the form "if you're in a sufficiently nice ring, then algebra continues to work well," we present a couple of easy results on polynomial arithmetic. Much like $\mathbb{Z}[x]$ and $\mathbb{R}[x]$ represent rings of polynomials with respective coefficients in $\mathbb{Z}$ and $\mathbb{R}$, we can likewise make the ring of polynomials with coefficients in a general ring R.

Lemma 3.1.12

If R is an integral domain, then for non-zero polynomials $f, g \in R[x]$, we have the standard degree fact

$$\deg(fg) = \deg(f) + \deg(g).$$

◀

Note that the result still holds if f or g is zero, recalling our convention that $\deg(0) = -\infty$ and adopting the conventions $-\infty + (-\infty) = -\infty$ and $-\infty + n = -\infty$ for $n \in \mathbb{N}$.

Proof If $f = a_n x^n + \cdots + a_0$ and $g = b_m x^m + \cdots + b_0$ with $a_i, b_i \in R$, then

$$fg = a_n b_m x^{n+m} + \cdots + a_0 b_0.$$

This is a polynomial of degree $m + n$ since $a_n b_m \neq 0$ by virtue of the fact that R is an integral domain. □

Without the integral domain property on the coefficients, the result can fail. If $ab = 0$ in R, then the product of the degree 1 polynomials $f = ax + 1$ and $g = bx - 1$, for example, is $fg = abx^2 + bx - ax + 1 = (b - a)x + 1$, again of degree 1. Interesting rings in which this horror of algebra occurs naturally will appear frequently in the next chapter.

A subtle point to remark upon is that polynomials have two roles in modern mathematics. We have been treating them as elements $f \in R[x]$, i.e., as formal expressions with which we can do algebra. But as we all know from our calculus sequence, they also admit an interpretation as *functions* into which we can input an element $a \in R$ and obtain an output $f(a) \in R$ as a result. These dual perspectives frequently cross paths, e.g., when we talk about the *roots* of a polynomial.

Lemma 3.1.13

Let R be an integral domain. Every non-zero polynomial $f \in R[x]$ of degree d has at most d roots in R. ◀

Proof We proceed by induction on d, beginning with the $d = 0$ case[3], where f is constant (and non-zero) so has no roots. Now fix $d > 0$ and suppose for the sake of induction we know that all polynomials of degree $d - 1$ have at most $d - 1$ roots. Let $f = a_d x^d + \cdots + a_1 x + a_0$ be a polynomial of degree d and let r be a root of f (if no root exists, we are done). Then $f(r) = 0$, and we observe that

$$f(x) = f(x) - f(r) = a_d(x^d - r^d) + a_{d-1}(x^{d-1} - r^{d-1}) + \cdots + a_1(x - r).$$

Combined with the polynomial identity

$$x^k - r^k = (x - r)(x^{k-1} + x^{k-2}r + \cdots + xr^{k-2} + r^{k-1}),$$

we see that we can extract a factor of $x - r$ from each summand and thus conclude that

$$f(x) = (x - r)g(x)$$

for some $g \in R[x]$ (which we could write down explicitly if we wanted to). By the previous lemma, g has degree $d - 1$, so by the induction hypothesis it has at most $d - 1$ roots. Between these $d - 1$ roots and our initial root r, we see that f has at most d roots since $f(x)$ can only be zero if $x - r = 0$ or $g(x) = 0$.   □

Having dipped our toes in the waters of abstract ring theory, let us dry them off and take steps back toward the questions of number theory. Of principal importance, pervasive throughout the whole text, is the notion of divisibility.

Definition 3.1.14

Given elements a and b in a commutative ring R, we say that a **divides** b if there exists an element $c \in R$ such that $ac = b$. We write $a \mid b$ to denote that a divides b, and say that a is a **divisor** of b, or equivalently that b is a **multiple** of a. If a does not divide b, we write $a \nmid b$. ◀

To fend off a common source of confusion, note that unlike the numerical expressions $n \div d$ or $\frac{n}{d}$, the expression $d \mid n$ is either true or false, and not a numerical value: in the integers, $6 \div 3 = 2$, but $3 \mid 6$ is simply true. For the most part, when $R = \mathbb{Z}$ the definition above reflects your previously internalized grade-school notions of divisors, multiples, etc. For example, 7 is a divisor of 35 in $\mathbb{Z}$ (and 35 is a multiple of 7), since there exists an *integer* solution to the fill-in-the-blank problem $7 \cdot \_\_ = 35$. That said, there are some nuances to address: in grade school, one typically works in $\mathbb{N}$ rather than $\mathbb{Z}$, let alone rings like $\mathbb{Z}[i]$, $\mathbb{R}[x]$, etc. Particularly subtle is the prospect that we could have numbers a, b such that $a \mid b$ in one ring but $a \nmid b$ in another—a disconcerting possibility, but such is the life of the mathematician. The upcoming Exploration has us look into some of these subtleties.

---

[3] Okay, okay, maybe there is *occasionally* merit in starting the natural numbers at 0...

Exploration B: Divisibility and Linear Diophantine Equations

Divisibility in $\mathbb{Z}$:

B.1 Does 3 divide 6? Does 6 divide 3? What are all the divisors of 12? Of 13? Of -13? Of 1? Of -1? Of 0?

B.2 Which number(s) divide all other numbers? Which number(s) are divisible by all other numbers?

B.3 Our definition of $|$ implies that $0 \mid 0$. Pro: 0 is indeed a multiple of 0. Con: $\frac{0}{0}$ is indeterminate. On a scale of 0–10, how morally reprehensible is this convention? Does zero divide your score?

Divisibility in Other Rings

B.4 Explain why $3 \mid 7$ in $\mathbb{Q}$. Generally, what does divisibility look like in $\mathbb{Q}$ (or $\mathbb{R}$, or $\mathbb{C}$)?

B.5 Show that $(2 + i) \mid 10$ in $\mathbb{Z}[i]$.

B.6 What are the divisors of $2x$ in $\mathbb{Z}[x]$? In $\mathbb{R}[x]$?

B.7 In $\mathbb{R}[x]$, verify that $(x - 1) \mid (x^2 - 1)$ but $(x - 1) \nmid (x^2 - 4)$.

Linear Diophantine Equations

We have seen that solving quadratic Diophantine equations involves a rather large amount of deep number theory, so let's retreat to the linear case for now. Here is a crucially important real-world application.

B.8 Suppose that Chicken O'Nuggets can be purchased in quantities of 6 or 9. Then the possible numbers of Chicken O'Nuggets that could be purchased at once are numbers of the form

$$6x + 9y$$

where x and y are non-negative integers.

1. What can you say about the possible numbers of Chicken O'Nuggets one can buy? Can you buy 100 nuggets? 101? 102?
2. Repeat the previous problem in a store where you can *also* purchase nuggets in 20-packs. Can you now get *most* natural numbers? Half? Or only special amounts?
3. How would the answers to the previous two questions change if you were allowed to sell boxes back to the store? For example, you could now net 3 nuggets by buying a 9-nugget box and selling back a 6-nugget box. This is equivalent to taking $x, y \in \mathbb{Z}$ rather than $x, y \in \mathbb{N}$.
4. What if we allowed $x, y \in \mathbb{Q}$? Or in $\mathbb{R}$? ◄

3.2 Linear Diophantine Equations and the Euclidean Algorithm

The Chicken O'Nugget problem examined in Exploration B, along with its generalizations, encapsulate all of the central ideas behind solving linear Diophantine equations. We consider the following generalization:

▶ **Example 3.2.1** Suppose one can purchase boxes of O'Nuggets in boxes of either a or b nuggets so that by buying x boxes with a nuggets and y boxes with b nuggets, the possible numbers of Chicken O'Nuggets we can purchase are the numbers of the form

$$ax + by.$$

What are the possible numbers of nuggets we can buy? That is, for which values of c can we solve the linear equation

$$ax + by = c?$$

Perhaps the most significant takeaway from the exploration is how greatly the answer to this question depends on the world in which x and y are constrained to live. Borrowing our observations from the Exploration, taking $x, y \in \mathbb{Z}$ provides a very clean description of the solution set. When we restrict to natural or whole numbers of boxes, we lose some of this structure and get a somewhat random-feeling collection of small number of nuggets that we can buy, but then any large number we want (depending on the box sizes). At the other extreme, taking $x, y \in \mathbb{Q}$, it is reasonably easy to see that you can purchase any number of nuggets (if you want 37 nuggets, buy $\frac{37}{20}$ of a 20-pack).

An algebraic interpretation of this phenomenon is to point to the sets themselves ($\mathbb{N}$ vs. $\mathbb{Z}$ vs. $\mathbb{Q}$) having strikingly different algebraic properties. The existence of multiplicative inverses in $\mathbb{Q}$ renders the problem trivial, whereas the lack of additive inverses in $\mathbb{N}$ seems to reduce the amount of structure in the solution set. So $\mathbb{Z}$ emerges as a local maximum of interestingness in that it has enough structure for a pattern to emerge, but not so much as to make the problem trivial. We propose the following summary:

> *Unless there's an obvious reason why you*
> *can't get everything, you can get anything.*

If allowed to purchase 3-packs and 7-packs of nuggets, for example, we could buy any number of nuggets we want: To get 1 nugget, you buy a 7-pack of nuggets and sell back[4] two 3-packs. Then to buy any other number of nuggets, you repeat that exchange that many times. For contrast, when allowed to purchase 6-packs and 9-packs of nuggets, we have a problem that both 6 and 9 are multiples of 3, and adding

---

[4] The authors do not recommend attempting this in person.

and subtracting multiples of 3 seems to always leave us a multiple of 3. This gives form to the "obvious reasons" that would prevent you from achieving any number of nuggets you want: if all available box sizes are multiples of some common base, then any linear combination of those two numbers will again be a multiple of that base. Finding such obstructions is thus tantamount to determining whether there are non-trivial (i.e., greater than 1) factors in common between a and b.

Definition 3.2.2

Let a and b be integers, not both equal to 0. A **common divisor** of a and b is an integer d such that $d \mid a$ and $d \mid b$. The largest integer that divides both a and b is called the **greatest common divisor** of a and b, denoted $\gcd(a, b)$. Similarly, a **common multiple** of a and b is an integer n such that $a \mid n$ and $b \mid n$. The smallest positive integer n satisfying $a \mid n$ and $b \mid n$ is called the **least common multiple** of a and b, denoted $\operatorname{lcm}(a, b)$. We say that a and b are **relatively prime** (or **coprime**) if

$$\gcd(a, b) = 1. \qquad \blacktriangleleft$$

Again, you are likely used to interpreting these ideas in $\mathbb{N}$, and so we should pause and acclimate ourselves to interpreting these calculations in $\mathbb{Z}$ instead. Some cases of interest are dealt with in the example below.

▶ **Example 3.2.3** Be sure to convince yourself these results are what the definition above mandates:

$$\gcd(15, 10) = 5 \qquad \gcd(3, -7) = 1 \qquad \gcd(-15, -15) = 15$$

$$\gcd(0, 12) = 12 \qquad \gcd(0, 0) = \text{undefined}.$$

The remainder of the section is largely about the process of computing these gcds and the theoretical significance that accompanies this discussion.

Lemma 3.2.4 (Linear Combination Lemma)

Suppose d is a common divisor of a and b ($a, b, d \in \mathbb{Z}$). Then $d \mid (ax + by)$ for all $x, y \in \mathbb{Z}$. ◀

Proof Since $d \mid a$ and $d \mid b$, we can write $a = dm$ and $b = dn$ for some $m, n \in \mathbb{Z}$. Substituting, we find

$$ax + by = dmx + dny = d(mx + ny),$$

and thus $ax + by$ is an integral multiple of d. □

Corollary 3.2.5

If d divides any two of a, b, and $a + b$, it divides the third. ◄

It is worth pointing out that we have implicitly used algebraic properties of the integers in the previous proof, namely distributivity and the fact that the integers are closed under addition and multiplication.

Corollary 3.2.6

For $a, b \in \mathbb{Z}$, we have the identities $\gcd(a, b) = \gcd(b, a)$ and $\gcd(a, b) = \gcd(a, b - a)$. ◄

Proof The first is clear from the definition, and the second follows from the lemma as every common divisor of a and b is also a common divisor of a and $b - a$, and, conversely, a common divisor of a and $b - a$ is also a common divisor of a and $(b - a) + a = b$. Thus, the two pairs have exactly the same common divisors, so in particular, they share the same greatest common divisor. ☐

This simple corollary is the source of an amazingly efficient tool for the computation of greatest common divisors called the *Euclidean Algorithm*. While we'll formalize the process momentarily (after making an upgrade), let's see it in action first:

► **Example 3.2.7** We systematically apply the identities in Corollary 3.2.6 to compute the gcd of 52 and 91:

$$\gcd(91, 52) = \gcd(52, 91) = \gcd(52, 91 - 52) = \gcd(52, 39)$$
$$= \gcd(39, 52) = \gcd(39, 52 - 39) = \gcd(39, 13) = 13.$$

In practice, once we reach a pair where the gcd is obvious, we can stop and evaluate it. This technique has the potential for being tremendously more efficient than more naive techniques for calculating the greatest common divisor of two numbers (e.g., systematically listing all divisors, or even prime factorization—a topic that we'll be turning to shortly). There is still room for improvement, however, as repeated subtraction can often include an unreasonable number of steps:

► **Example 3.2.8** Find the gcd of 93 and 7.

$$\gcd(7, 93) = \gcd(7, 93 - 7) = \gcd(7, 86) = \gcd(7, 86 - 7) = \gcd(7, 79) = \cdots$$

Sigh.

We all see where this is heading—this process will end only after we have subtracted from 93 as many 7s as we can before the result becomes negative. "But," we hear you cry, "we have a name for the resulting quantity already – this is precisely the remainder when 93 is divided by 7." As always, beloved reader, your insight is downright Euclidean. Here are the words we'll need:

> **Theorem 3.2.9 (The Division Algorithm[5])**
> Given integers a and b with $b > 0$, there exist unique integers q and r (the *quotient* and *remainder*, respectively) such that
>
> $$a = bq + r$$
>
> and $0 \leq r < b$.

The proof unfolds quickly by invoking what's known as the *Well-Ordering Principle*, the statement that any non-empty set of non-negative integers contains a least element[6]. Note that this claim is false if we change from the set of non-negative integers to the set of all integers, or the rationals, reals, complex numbers, etc. While it isn't always called out explicitly, the Well-Ordering Principle is a foundational result in number theory, allowing the logical step "Well, if there are *some* non-negative integers with this property, there must be a *smallest* such number." Being able to name and reference that smallest number is a valuable tool in proofs. This leads to a noteworthy proof template: after picking out the smallest number with a given property, construct an even smaller number with that property, establishing a contradiction.

Proof (of Theorem 3.2.9) Let a and b be integers with $b > 0$, and consider the set of non-negative integers

$$S = \{a - bq : q \in \mathbb{Z} \text{ and } a - bq \geq 0\}.$$

We begin by observing that

(i) S consists of non-negative integers (by the $a - bq \geq 0$ clause of its definition); and
(ii) S is non-empty (since if $a \geq 0$, then $a - b(0) = a \in S$, and if $a < 0$, then $a - b(a) = -a(b - 1) \in S$).

Since S is a non-empty set of non-negative integers, the Well-Ordering Principle thus tells us that there is a least element r of S. That is, r is the least non-negative integer of the form $r = a - bq$ for some $q \in \mathbb{Z}$. Note that $r < b$ since if $r \geq b$, then $r - b = a - b(q+1)$ is an element of S smaller than r, contradicting the fact that r was the least such number. This establishes the existence of the numbers q and r in the theorem.

Finally, we check that r and q are unique. Suppose to the contrary that we simultaneously had $a = bq_1 + r_1$ and $a = bq_2 + r_2$, where q_1, q_2, r_1, r_2 are all integers and

---

[5] It is curious that this theorem is called an algorithm despite no actual algorithm in sight. Any of a multitude of actual algorithms (e.g., "long division") would supplement this result nicely, but it is also true that we will frequently just need to guarantee the *existence* of the quotient and remainder.

[6] See Appendix I for more on this fundamental principle.

$0 \leq r_1, r_2 < b$. Without loss of generality, suppose $r_1 \leq r_2$. Thus $0 \leq r_2 - r_1 < b$. Substituting, we find $0 \leq (a - bq_2) - (a - bq_1) = b(q_1 - q_2) < b$, which implies $0 \leq q_1 - q_2 < 1$ for integers q_1 and q_2. Thus, $q_1 = q_2$, and solving for r_1 and r_2 gives that $r_1 = r_2$. □

Note that uniqueness is a consequence of constraining r to live in the given range. We can, for example, write $75 = 23q + r$ with $q, r \in \mathbb{Z}$ in many ways:

$$75 = 23 \cdot 0 + 75 \qquad 75 = 23 \cdot 10 + (-155) \qquad 75 = 23 \cdot (-2) + 121 \qquad \dots$$

but only one of these ways, $75 = 23 \cdot 3 + 6$, involves a value of r in the range $0 \leq r < 23$. This specific value of r, the unique remainder r provided by the Division Algorithm, gets a special notation.

Definition 3.2.10

Let a, b, q, and r be as in the theorem. We define $a \bmod b$ by putting $r = a \bmod b$, pronounced "a modulo b" or "$a \bmod b$." Here we call the number b the **modulus**. Note that

$$a \mid b \iff b \bmod a = 0 \iff b = ak \quad \text{(for some } k \in \mathbb{Z}),$$

a handy dictionary of equivalences for translating between different types of statements in the middle of a proof. ◄

▶ **Example 3.2.11** Verify each of the following using mental arithmetic:

$$37 \bmod 5 = 2 \qquad 77 \bmod 7 = 0 \qquad 12 \bmod 15 = 12$$
$$4 \bmod 1 = 0 \qquad -3 \bmod 7 = 4 \qquad 97 \bmod 23 = 5$$

What are the quotients q in each case?

The arithmetic of adding and multiplying remainders—"modular arithmetic"—will occupy our near-exclusive attention in Chapter 4. For now, $a \bmod b$ refers to nothing more than the remainder of a upon division by b. Combining the observation made at the end of our $\gcd(93, 7)$ example with the remainder $a \bmod b$ whose existence is guaranteed by the Division Algorithm, we have the following vast improvement of Corollary 3.2.6 for use in the Euclidean algorithm.

Lemma 3.2.12 (Reduction Lemma)

For all $a, b \in \mathbb{N}$, we have

$$\gcd(a, b) = \gcd(b, a \bmod b). \qquad ◄$$

Proof Let a and b be arbitrary natural numbers and let $a = r \bmod b$, so $a = qb + r$ for some $q \in \mathbb{Z}$. By the Linear Combination Lemma (3.2.4), any common divisor of a and b is also a common divisor of a and $a - bq = r$. Similarly, any divisor of b and r is also a common divisor of b and $a = bq + r$. Thus, since the finite set of common divisors of a and b is equal to the finite set of common divisors of b and $r = a \bmod b$, the greatest elements of the two sets must be equal.   □

Of course, by symmetry, we could also have written $\gcd(a, b) = \gcd(a, b \bmod a)$. The version to apply is always clear in practice, as we only make progress if we replace the larger of the two entries with its remainder when divided by the smaller. It's worth emphasizing the mental arithmetic component to computing $a \bmod b$. In particular, note that in the Reduction Lemma, the value of the *quotient* q is completely irrelevant. For example, to compute that $119 \bmod 5 = 4$, we need only note that 119 is 4 more than some multiple of 5—we need not expend mental energy figuring out *which* multiple of 5. Similarly, to compute $93 \bmod 7$, we just have to eyeball that 91 is a multiple of 7 and that 93 is two more than that. We finish off our computation as easily as

$$\gcd(7, 93) = \gcd(7, 2) = 1.$$

This process scales quite well to larger numbers:

▶ **Example 3.2.13** Compute $\gcd(1914, 899)$.

Solution Noting that $1914 = 899 \cdot 2 + 116$, we begin with $\gcd(1914, 899) = \gcd(899, 116)$. Then continuing, we find

$$\gcd(1914, 899) = \gcd(899, 116) = \gcd(116, 87) = \gcd(87, 29) = 29$$

by repeated application of the Reduction Lemma.   □

Theorem 3.2.14 (The Euclidean Algorithm)
For natural numbers a and b, let $r_{-1} = a$, $r_0 = b$, and for $i \geq 0$ set $r_{i+1} = r_{i-1} \bmod r_i$. Then for some $n \geq 1$, the process terminates with $r_n = 0$ and $r_{n-1} = \gcd(a, b)$.

Proof To prove this process terminates, note that by the Division Algorithm we have $0 \leq r_{i+1} < r_i$ for each $i \geq 0$, and so the remainders form a strictly decreasing sequence of integers bounded below by 0:

$$0 \leq \cdots < r_{i+1} < r_i < \cdots < r_1 < r_0.$$

Thus, we must have that $r_n = 0$ for some n, and we obtain

$$\gcd(a, b) = \gcd(b, r_1) = \gcd(r_1, r_2) = \cdots = \gcd(r_{n-1}, r_n) = \gcd(r_{n-1}, 0),$$

so $\gcd(a, b) = r_{n-1}$.   □

Corollary 3.2.15

For every integer $k > 0$, $\gcd(ka, kb) = k \gcd(a, b)$. ◄

Proof The proof follows directly from the Euclidean Algorithm. Given for each $i > 0$, $r_{i-1} = r_i q_{i+1} + r_{i+1}$, it follows that $kr_{i-1} = kr_i q_{i+1} + kr_{i+1}$. Thus,

$$\gcd(ka, kb) = \gcd(kb, kr_1) = \cdots = \gcd(kr_{n-1}, 0) = kr_{n-1} = k \gcd(a, b),$$

as desired.   □

Recall that two integers are relatively prime if their gcd is 1.

Corollary 3.2.16

Suppose $g = \gcd(a, b) \neq 0$. Then $\frac{a}{g}$ and $\frac{b}{g}$ are relatively prime integers. ◄

Proof Since $g \mid a$ and $g \mid b$, both $\frac{a}{g}$ and $\frac{b}{g}$ are integers. To check that they are relatively prime, we observe that by Corollary 3.2.15, we have

$$g \gcd \left(\frac{a}{g}, \frac{b}{g} \right) = \gcd(a, b) = g,$$

and so dividing by g shows that $\frac{a}{g}$ and $\frac{b}{g}$ have a gcd of 1.   □

▶ **Remark 3.2.17** The previously mentioned efficiency of the Euclidean Algorithm can be quantified explicitly. In what might be retroactively classified as one of the first theorems in the branch of mathematics known as computational complexity theory, Gabriel Lamé proved in 1844 that the Euclidean Algorithm never takes more steps than five times the number of digits in the smaller of the two inputs (Exercise 3.41).

We draw ever nearer to the punchline of our nugget problem. We recall that a common divisor to all of the box sizes presents a restriction as to what number of nuggets we can purchase. A reasonable conjecture from here is that the greatest common divisor is indeed the smallest linear combination you can get. The hardest part remaining is to guarantee that we can actually obtain this value. Given that $\gcd(93, 7) = 1$, does this guarantee that we can purchase exactly 1 Chicken O'Nugget from boxes of size 93 and 7? The affirmative answer to this question is encapsulated in the following remarkable theorem, named after French mathematician Étienne Bézout.

Theorem 3.2.18 (Bézout's Identity)
Fix $a, b \in \mathbb{Z}$ not both zero. Then there exist integers x and y such that $\gcd(a, b) = ax + by$; i.e., $\gcd(a, b)$ can be written as a linear combination of a and b.

Proof Without loss of generality, suppose a and b are integers with $a \neq 0$. Let S be the set of positive linear combinations of a and b:

$$S = \{ax + by : x, y \in \mathbb{Z} \text{ and } ax + by > 0\}.$$

Then S is non-empty as $ax + b(0) \in S$ for $x = +1$ or $x = -1$. By the Well-Ordering Principle, since S is a non-empty subset of $\mathbb{N}$, it has a smallest element $d = ax + by$ for some $x, y \in \mathbb{Z}$. Let c be an arbitrary common divisor of a and b. By the Linear Combination Lemma, $c \mid d$.

We employ a standard technique to show that d is a common divisor of a and b: we divide each by d and show that the remainders are 0. By the Division Algorithm, there exist integers q and r such that $a = dq + r$ and $0 \leq r < d$. Since $ax + by = d$ it follows that $axq + byq = dq = a - r$ and $r = a(1 - qx) + b(-qy) \geq 0$. If r were positive, then it would be a positive linear combination of a and b less than d, contradicting our assumption that d was the least such linear combination. Thus, $r = 0$ and $d \mid a$. A similar argument shows $d \mid b$. Since d is a common divisor of a and b, and is itself divisible by every common divisor, it must be the greatest common divisor of a and b. $\qquad\square$

To consider the story complete, we should augment the existence claim of Bézout's Identity with an explicit process for writing $\gcd(a, b)$ as a linear combination of a and b. Fortunately, the Euclidean Algorithm contains all the data we need to do so. The important observation is that as we perform the Euclidean Algorithm, the entries that appear at *any* stage of the Euclidean Algorithm are linear combinations of the initial pair of numbers. Let's see a couple of small examples before we formalize:

▶ **Example 3.2.19** To purchase one O'Nugget from boxes of size 7 and 93, we apply the Euclidean algorithm and then flip it and reverse it:

$$93 = 7 \cdot 13 + 2$$
$$7 = 2 \cdot 3 + 1$$

$$1 = 7 + 2(-3)$$
$$= 7 + (93 + 7(-13))(-3)$$
$$= 7(40) + 93(-3)$$

Thus, we could buy 40 of the 7-piece boxes and sell back 3 of the 93-piece boxes to get precisely one delicious, crispy, tender Chicken O'Nugget.

▶ **Example 3.2.20** The historically literate will recall that one of the most significant applications of Bézout's Identity took place in 1995, when NYPD Detective
Lieutenant John McClane was tasked with saving New York from the machina-
tions of terrorist Simon Gruber. Here, a thoughtful application of Bézout's identity
provided the expression

$$5 - (3 - (5 - 3)) = 2 \cdot 5 - 2 \cdot 3 = 4,$$

which had profound applications to the practice of bomb defusal. It is worth
taking the time (Exercise 3.42) to fully understand this piece of Americana and
its mathematical underpinnings.

This process of performing the Euclidean Algorithm while keeping track of the
Division Algorithm steps employed is the basis of the *extended Euclidean Algorithm*,
which culminates in writing the gcd as a linear combination of the inputs. To enact
the process, we go through the Euclidean Algorithm as usual, but use a separate
column to keep track of the quotients and remainders. As a matter of practice, this is
facilitated by forgetting the values of a and b and just referring to them as placeholder
variables a and b. The example below shows a general schematic of the Extended
Euclidean Algorithm as well as how it applies to the specific example of calculating
the gcd of 273 and 429.

▶ **Example 3.2.21** Find $\gcd(429, 273)$. Let $a = 429$ and $b = 273$.

Solution We first apply the Division Algorithm to reduce the gcds.

$$
\begin{array}{ll}
429 = 273 \cdot 1 + 156 & \gcd(429, 273) = \gcd(273, 156) \\
273 = 156 \cdot 1 + 117 & \gcd(273, 156) = \gcd(156, 117) \\
156 = 117 \cdot 1 + 39 & \gcd(156, 117) = \gcd(117, 39) \\
117 = 39 \cdot 3 + 0 & \gcd(117, 39) = \gcd(39, 0) = 39
\end{array}
$$

Retracing the steps of the algorithm we can solve for the $\gcd(429, 273)$:

$$
\begin{aligned}
\gcd(429, 273) = 39 &= 156 - 117 \\
&= 156 - (273 - 156) \\
&= (429 - 273) - (273 - (429 - 273)) \\
&= 2 \cdot 429 - 3 \cdot 273.
\end{aligned}
$$

Thus, we can represent $\gcd(429, 273)$ as a linear combination of 429 and 273. Note
that we don't simplify along the way (other than combining like remainders), as that
would simply unravel all our hard work!   □

Thus, the Extended Euclidean Algorithm improves somewhat upon Bézout's
Identity, which gives only the *existence* of a linear combination of a and b equal
to their greatest common divisor as opposed to an explicit construction. The general

proof this works proceeds inductively: the algorithm terminates with r_{n-1}, where $r_{n-3} = r_{n-2}q_{n-1} + r_{n-1}$, so $r_{n-1} = r_{n-3} - r_{n-2}q_{n-1}$. Thus, we may express r_{n-1} (the gcd) in terms of the previous two remainders. In general, $r_{i-1} = r_iq_{i+1} + r_{i+1}$, so $r_{i+1} = r_{i-1} - r_iq_{i+1}$. Inductively, then, we see that we can write r_{n-1} in terms of first r_{n-2} and r_{n-3}, and then r_{n-3} and r_{n-4}, and so on, eventually arriving at an expression for r_{n-1} in terms of r_{-1} and r_0, or a and b. This construction in hand, we can return to finish off our questions about linear Diophantine equations.

> **Theorem 3.2.22**
> For $a, b, c \in \mathbb{Z}$, the equation $ax + by = c$ has an integer solution if and only if $\gcd(a, b) \mid c$.

Bézout does most of the heavy lifting for us. The principal question that remains is simple by comparison, as illustrated by the following dialogue:

Q: Okay, so I now believe that we can buy $g = \gcd(a, b)$ nuggets. But are we sure I couldn't buy *fewer* than g nuggets??

A: Yes, because g divides *every* linear combination of a and b, and g can't divide a positive number smaller than itself!

Q: Okay, so we can get g itself. How do we know we can we get *every multiple* of the gcd?

A: Because once you've figured out how to get g O'Nuggets from O'Donnell's, you can apply that algorithm n times to get ng nuggets.

Though the Socratic method may have sufficed for the ancient Greek philosophers, let's take the time to write this argument a little more carefully.

Proof We start by dispensing with the "only if" part of the statement. Let $d = \gcd(a, b)$ and suppose $ax + by = c$ for some integers x and y. Then by the Linear Combination Lemma, $d \mid c$. Conversely, suppose $d \mid c$ for some integer c, so there exists an integer q such that $dq = c$. Using the extended Euclidean Algorithm, we can find integer values for x' and y' such that $ax' + by' = d$, and so $a(x'q) + b(y'q) = dq = c$. Thus, $x = x'q$ and $y = y'q$ are integer solutions to the linear Diophantine equation $ax + by = c$. $\qquad\square$

One of the more stunning consequences of Bézout's identity is a firm foundation for more complicated divisibility statements in $\mathbb{Z}$. The following lemma is the first step along that path, and will also play an important role as we return to linear Diophantine equations.

Lemma 3.2.23 (Euclid's Lemma)

For integers a, b, and c, if $\gcd(a, b) = 1$ and $a \mid bc$, then $a \mid c$. ◄

Proof Given integers a, b with $\gcd(a, b) = 1$, we know by Bézout's Identity (Theorem 3.2.18) that there exist integers x and y such that $ax + by = 1$. Thus, $cax + cby = c$. Since a clearly divides ca, if $a \mid cb$ as well, then by the Linear Combination Lemma, $a \mid c$. □

Now, let us finish the linear Diophantine story. We've learned how to find *one* solution to an equation like

$$17x + 13y = 1$$

via the Extended Euclidean Algorithm, in this case, obtaining $x = -3$ and $y = 4$. How do we find the rest? Much like the chord method of Chapter 3, once we have one, we can find them all, and some experimenting shows how. You need only figure out the ways to buy 0 nuggets, and one cute answer emerges: buy 17 more 13-packs and sell back 13 more 17-packs. The two contributions cancel each other out, and you have the solution $x = -16$ and $y = 21$. You can repeat this process forever to get infinitely many solutions. This turns out to be *all* of them, though it's worth waiting for one more example before we write this down formally. Consider the equation

$$15x + 33y = 48.$$

This has the trivial solution $x = 1$ and $y = 1$. Any more? Well, we can play the same trick as before: increase x by 33 and decrease y by 15...but now there's a little more. Since 33 and 15 have a factor in common, there's a smaller combination of 15 and 33 we can use to cancel out (a smaller lcm, if you will); namely, we can buy eleven 15-packs and sell back five 33-packs of O'Nuggets.

Theorem 3.2.24

If (x_0, y_0) is a solution to the linear Diophantine equation $ax + by = c$, then for all integers t, another solution is given by

$$x = x_0 - t\left(\frac{b}{\gcd(a, b)}\right) \qquad y = y_0 + t\left(\frac{a}{\gcd(a, b)}\right)$$

Furthermore, these are all of the solutions.

Proof We leave to the reader the verification that if (x_0, y_0) is a solution, so is the (x, y) pair described in the theorem. To prove that this collection forms *all* solutions, suppose (x, y) is an arbitrary second solution, so that $ax + by = c$. Subtracting from this the equation $ax_0 + by_0 = c$ and then dividing by $g = \gcd(a, b)$ gives

$$\frac{a}{g}(x_0 - x) = \frac{b}{g}(y - y_0).$$

This last equality above shows that $\frac{a}{g} \mid \frac{b}{g}(y - y_0)$, and since $\frac{a}{g}$ and $\frac{b}{g}$ are relatively prime by Corollary 3.2.16, Euclid's Lemma shows that $\frac{a}{g} \mid (y - y_0)$. Writing $y - y_0 = t\left(\frac{a}{g}\right)$ (which gives the desired parameterizaton of y), substitution into that last equality gives

$$\frac{a}{g}(x_0 - x) = \frac{b}{g}\left(\frac{ta}{g}\right),$$

which provides

$$x = x_0 - t\left(\frac{b}{g}\right) = x_0 - t\left(\frac{b}{\gcd(a, b)}\right)$$

after multiplying by $\frac{g}{a}$ and then rearranging to solve for x. $\square$

Depending on whom you ask[7], the actual process of solving linear Diophantine equations is less exciting than the structural properties of *integers* we discover as a byproduct of this discussion. Namely, we've learned that the gcd of any two integers, which is defined completely in terms of *multiplicative* properties, also admits an interpretation in terms of *additive* properties, namely linear combinations. Further, this translation of language has a very curious effect: the word "greatest" in "greatest common divisor" gets translated into "smallest" as in "smallest linear combination." This duality is remarkable, so let's make it a theorem – nay, a *Fundamental* Theorem:

Theorem 3.2.25 (Fundamental Theorem of GCDs)
For integers a and b not both 0:

The smallest positive linear combination of a and b	$=$	The largest integer dividing both a and b

---

We concede that the name given is sadly not a standard one and that it is essentially equivalent to Bézout's Identity, but its memorable reformulation above deserves a special place in your heart. It also plays a crucial role in the development of our big story: we are quietly building toward the single most important structural result of the integers, and the upcoming Exploration provides a segue, dissecting precisely what makes the set $\{\pm 1\}$ such a special subset of the integers.

---

[7] ...and we guess you've implicitly asked *us*, so...

Definition. Let R be a commutative ring with unity. An element $u \in R$ is called a **unit** if $u \mid 1$, i.e., if u has a multiplicative inverse in R.

C.1 What are the units of $\mathbb{Z}$? Of $\mathbb{Q}$? Of $\mathbb{R}[x]$?

C.2 Prove that if $a, b \in R$ are units, then so is ab.

Gaussian Units

The units of $\mathbb{Z}[i]$ are less quickly addressed. Recall that for $z = a + bi \in \mathbb{Z}[i]$, we define $\overline{z} = a - bi$ and $N(z) = z\overline{z} = (a + bi)(a - bi) = a^2 + b^2$.

C.3 Verify to your satisfaction the identities

$$\overline{z_1 + z_2} = \overline{z_1} + \overline{z_2} \qquad \overline{z_1 z_2} = \overline{z_1} \cdot \overline{z_2} \qquad N(z_1 z_2) = N(z_1)N(z_2)$$

for $z_1, z_2 \in \mathbb{Z}[i]$ which will be of use below.

C.4 Prove that if z is a unit of $\mathbb{Z}[i]$, then $\overline{z}$ is also a unit. Conclude that z is a unit if and only if $N(z) = 1$.

C.5 What are all the units of $\mathbb{Z}[i]$?

To mirror some questions from the previous Exploration on divisibility in $\mathbb{Z}$:

C.6 Which Gaussian integer(s) divide all other Gaussian integers? Which Gaussian integer(s) are divisible by all other Gaussian integers?

C.7 What is the smallest number of divisors a non-unit $z \in \mathbb{Z}[i]$ can have?

Gaussian GCDs

To foreshadow some of the nuances of gcds in $\mathbb{Z}[i]$ that we will take up in Chapter 5:

C.8 Propose a definition of "greatest common divisor" in $\mathbb{Z}[i]$. What should "greatest" mean here?

C.9 Check that $2 + i$ is a common divisor of $3 - i$ and 5. Find the 7 other common divisors. Which is the "greatest"?

C.10 An aside on the necessity of unity: in the commutative ring $2\mathbb{Z}$ of even integers, compute:

$$\gcd(2, 4) \qquad \gcd(8, 12) \qquad \gcd(12, 16) \qquad \gcd(6, 24)$$

Be sure to check divisibility in $2\mathbb{Z}$ carefully!   ◄

3.3   The Fundamental Theorem of Arithmetic

> The history of the Fundamental Theorem of Arithmetic is strangely obscure. It is not too much of an exaggeration to say that the result passed from being unknown to being obvious without a proof passing through the head of any mathematician.
>
> Agargün and Fletcher,
> *The Fundamental Theorem of Arithmetic Dissected*

Unlike our typical process of discovering a result by systematic exploration, let's start this section with one of the fundamental structural results of the natural numbers. At some point in your mathematical career thus far, you've probably had someone tell you that a *prime number* is a number whose divisors are only 1 and itself, and explain to you that these numbers are used to find the factorization of any number, culminating in a statement something like the following:

Non-Theorem 3.3.1 For every integer n, there is one and only way to write an equation of the form

$$n = p_1 p_2 p_3 \cdots p_k,$$

where each p_k is prime.

This statement is eminently plausible at first glance. For example, $8 = 2 \cdot 2 \cdot 2$, $84 = 2 \cdot 2 \cdot 3 \cdot 7$, and $5 = 5$. You may even have internalized a procedure for coming up with prime factorization (e.g., factor trees). But boy howdy is that a false statement, and for quite a few reasons. Let's collect some:

- The "one and only one" clause is a little ambiguous. For example, $15 = 3 \cdot 5$ and $15 = 5 \cdot 3$ are both factorizations of 15, and the theorem as written fails to account for this.
- The theorem's statement includes negative numbers, and it's a little unclear what to do with these. What do we do about $-6 = 2 \cdot (-3) = -2 \cdot 3$? Or $15 = (-3)(-5)$?
- The case of $n = 0$ seems problematic. Likewise, for the cases $n = \pm 1$.
- There is some ambiguity as to the definition of prime numbers—is 1 a prime number? After all, its positive divisors are merely 1 and itself....

The easiest and most common way to fix Theorem 3.3.1 is simply to throw out all the problematic cases, stating the theorem only for natural numbers at least 2 and insisting on prime factorizations being written, for example, in increasing order of primes. But it's more instructive to think about the structural implications at stake here. Let's begin at the bottom of the list, with the role of 1, where we note first and foremost that this is solely a matter of definition. We can choose to either agree that 1 is a prime by sneaking its way through the definition on a technicality or to explicitly forbid it by insisting that to be prime you need exactly two distinct positive divisors. But second, and even foremost, we stress that choosing the "right" definition is one of the most important parts of setting up a field of study. *If*, for example, we choose a definition which admits 1 as a prime number, then we have to deal with the consequences of our choice, and concede that we have the following infinitely many violations of the "only one way" claim for prime factorization:

$$6 = 3 \cdot 2 = 3 \cdot 2 \cdot 1 = 3 \cdot 2 \cdot 1 \cdot 1 = 3 \cdot 2 \cdot 1 \cdot 1 \cdot 1 = \cdots$$

. Of course, we could change the theorem to say that every integer has infinitely many prime factorizations, all differing in the number of 1's tacked on at the end, but I think we can all agree that the statement of this version lacks some of the punch of the original. Nothing fundamentally breaks[8] by the inclusion of 1 as a prime number, but everything gets a little more cumbersome to state. A good way to proceed is to see which option generalizes more nicely to other rings. To this end, let us consider once more the Gaussian integers.

Mirroring the standard definition in $\mathbb{N}$, we could try defining $\alpha \in \mathbb{Z}[i]$ to be prime if its only divisors are 1 and itself. As with $\mathbb{Z}$, this still leaves 1 ambiguous, but it also makes everything not prime, for the silly reason that, e.g., $-3 \mid 3$ and $-1 \mid 3$ as well. Even taking into account negatives, we would have issues for Gaussian integers: For any $\alpha \in \mathbb{Z}[i]$, the identity $i^2 = -1$ provides a multitude of factorizations of α:

$$\alpha = \alpha \cdot 1 = (-\alpha) \cdot (-1) = (i\alpha) \cdot (-i) = (-i\alpha) \cdot i.$$

We conclude that any $\alpha \in \mathbb{Z}[i]$ has at the very least the eight divisors $\pm\alpha$, $\pm i\alpha$, ± 1, $\pm i$. Measuring primeness by counting divisors seems arbitrary and obfuscating. The problem seems to be that we have one trivial factorization, $\alpha = \alpha \cdot 1$, and then all the rest are obtained by messing around with units—we can remove or insert from the factorization any factorization of 1. Numbers that divide 1 are thus a key set of elements to identify in any ring.

Definition 3.3.2

In a ring R with unity, an element $a \in R$ is a **unit** if $a \mid 1$; that is, if a has a multiplicative inverse a^{-1} in R. Two non-zero elements $c, d \in R$ are **associates** if each is a unit times the other. ◄

▶ **Example 3.3.3** As you may have discovered in Exploration C,

- the units of $\mathbb{Z}$ are precisely ± 1,
- the units of $\mathbb{Z}[i]$ are $\{\pm 1, \pm i\}$,
- the units of $\mathbb{R}$ and $\mathbb{Q}$ (and any field) are everything except 0, and
- the units of $\mathbb{R}[x]$ are all non-zero constant polynomials

The units of a ring are of crucial importance in understanding its structure, especially its divisibility and factorization properties. In particular, the general phenomenon is that units aren't particularly well classified as *either* prime or composite.

---

[8] passionate internet flame wars to the contrary notwithstanding

---

| Definition 3.3.4 |

Let p be an integer that is neither zero nor a unit. We say that p is **prime** if its only divisors are units and associates of p; otherwise, p is **composite**. ◄

This definition of prime is more structural than the one given at the start of the section and differs in how it treats negative numbers: 7 is prime since its divisors (± 1 and ± 7) are all units or associates of 7, and -7 is prime for the same reason.

This presents an explicit partition of the integers into four categories, with every integer belonging to one and only one category in this list:

- **Zero**: Unsurprisingly, this consists only of the integer 0.
- **Units**: The units are precisely ± 1.
- **Prime numbers**: ± 2, ± 3, ± 5, ± 7, ± 11, etc[9], appearing in associate ($\pm$) pairs.
- **Composite numbers**: By definition everything not yet addressed. This includes ± 4, ± 6, ± 8, ± 9, etc.

More importantly, we can now state a careful and explicit (and even better, *true!*) version of Theorem 3.3.1:

Theorem 3.3.5 (Fundamental Theorem of Arithmetic)
Every non-zero integer $n \in \mathbb{Z}$ can be written in the form

$$n = u p_1 p_2 \dots p_k,$$

where k is a non-negative integer, u is a unit of $\mathbb{Z}$, and each p_k is a prime of $\mathbb{Z}$; moreover, this form is unique up to reordering and associates. Note that the $k = 0$ case corresponds precisely to the units.

Now *that's* a theorem! It takes care of all of our problem cases above, in that it recognizes

$$15 = 3 \cdot 5 = 5 \cdot 3 = (-3) \cdot (-5) = -1(3 \cdot (-5))$$

all as factorizations of 15, and then dictates what flexibility we have in finding other factorizations: we can only rearrange the placement of the units (in this case, just ± 1) and the order of the factors. The language "unique up to order and units" for this phenomenon will become a familiar refrain as we move forward. The theorem also address the fringe cases we ran into before. For $n = 1$, we have the prime factorization with $k = 0$ and $u = 1$, and likewise for $n = -1$ and $u = -1$. When

---

[9] The contents of this "etc" and the patterns in the continuation of this list will be a motivating source of mystery and a mysterious source of motivation.

$n = 0$, the theorem says nothing, implicitly declaring that we will not attempt to ascribe a prime factorization to zero.

Now, to return to the quote at the start of this section: isn't this theorem just...*obvious*[10]? After all, you've likely been using this result since grade school—don't you just keep pulling out prime factors until there's nothing left? We argue to the contrary that there's quite a bit of meat in this result. First, it does not follow simply from the ring axioms, as there are rings where unique factorization does not hold!

▶ **Example 3.3.6** Consider again the ring of even integers, $2\mathbb{Z}$. We can still talk about divisibility since, e.g., $10 \mid 20$ (because $20 = 2 \cdot 10$). On the other hand, we now have $2 \nmid 6$ since there is no $k \in 2\mathbb{Z}$ such that $2k = 6$. In fact, 6 has *no* divisors in this ring (not even itself!), making it vacuously prime, and likewise, for any even number that is not a multiple of 4. But now consider the identity

$$2 \cdot 30 = 60 = 6 \cdot 10.$$

This provides two distinct factorizations of 60 as the product of two prime elements of this ring!

So if unique factorization does not follow "for free" in an arbitrary ring, it must be that there is some property of $\mathbb{Z}$ that provides it that not all rings have. Even in the context of $\mathbb{Z}$, the result has profound implications. As illustrative examples, ponder for a moment whether or not the following equations are true:

$$2017 \cdot 4973 \overset{?}{=} 2687 \cdot 3733$$

$$1517 \cdot 2021 \overset{?}{=} 1591 \cdot 1927$$

There are certainly some tests to run before simply doing the multiplication, e.g., checking the last digit, or checking that they should have about the same size, etc., though neither of these attempts resolves these particular equations. Suppose we told you that the numbers 2017, 4973, 2687, and 3733 were all prime. Would that help resolve the first equation? Certainly! It tells us that if the first equation were true, then this common product would have two distinct prime factorizations, contradicting the Fundamental Theorem of Arithmetic.

For the second equation, note how quickly prime factorizations resolve the question:

$$1517 \cdot 2021 = (37 \cdot 41) \cdot (43 \cdot 47) = (37 \cdot 43) \cdot (41 \cdot 47) = 1591 \cdot 1927.$$

The Fundamental Theorem of Arithmetic tells us this *had* to be true for the two sides of this to be equal. The two factorizations of their product (3,065,857) must necessarily come from two different partitions of its prime divisors into factors.

---

[10] "Obvious" is the most dangerous word in mathematics. - E.T. Bell

Returning to the first equation, suppose we knew *only* that 2017 were prime—would this be sufficient to conclude that the equality was false? Certainly, neither 2687 nor 3733 is a multiple of 2017, so it's hard to imagine how the prime factorizations could then work out to be the same. This application of the Fundamental Theorem turns out to be the key insight in its proof. The first step along this path is the Prime Divisor Property, whose proof, by way of Euclid's Lemma, is a direct consequence of Bézout's identity.

Corollary 3.3.7 (The Prime Divisor Property)

For $b, c \in \mathbb{Z}$, if p is a prime with $p \mid bc$, then $p \mid b$ or $p \mid c$. ◄

Proof Suppose $p \mid bc$. If $p \mid b$ we are done. If $p \nmid b$, then $\gcd(p, b) = 1$, so by Euclid's Lemma (Lemma 3.2.23), $p \mid c$. □

▶ **Remark 3.3.8** In fact, the Prime Divisor Property provides a property of primes that could have been used as the *definition* of a prime number. As it turns out, this is precisely the definition we will take of a prime element in a more general ring when we come to that stage in Chapter 6.

The contrapositive of this result is applied just as frequently: you can't get a multiple of a prime p by multiplying two non-multiples of p. The generalization to n factors is a standard proof by induction.

Corollary 3.3.9

If $a_1, \ldots, a_n \in \mathbb{Z}$ and p is a prime with $p \mid a_1 \cdots a_n$, then $p \mid a_i$ for some $1 \le i \le n$. ◄

Proof We induct on n, the number of factors. The base case of $n = 2$ is the previous lemma. Now we assume that whenever p divides a product of $n - 1$ factors for $n > 2$, it divides one of the $n - 1$ individual factors, and prove the analogous statement for n. Suppose $p \mid a_1 \cdots a_n = a_1(a_2 \cdots a_n)$. By the previous lemma, $p \mid a_1$ or $p \mid a_2 \cdots a_n$. But by the induction hypothesis, the latter case implies that $p \mid a_i$ for some $2 \le i \le n$. In either case, p divides one of the a_i and we are done. □

We will make use of this handy series of results extensively for the remainder of the book. Here are several quick applications to get us started. For example, taking all a_i to be the same constant a, we arrive at the following:

Corollary 3.3.10

Let p be prime in $\mathbb{Z}$. For $n \in \mathbb{N}$, if $p \mid a^n$, then $p \mid a$. ◄

In particular, we get a slick generalization of the standard even/odd argument that $\sqrt{2}$ is irrational, now applying to any prime.

Corollary 3.3.11

If p is prime, then $\sqrt{p}$ is irrational. ◀

Proof Suppose for the sake of contradiction that $\sqrt{p} = \frac{a}{b}$ with $\gcd(a, b) = 1$. Then $pb^2 = a^2$, so $p \mid a^2$, and thus $p \mid a$ by the previous corollary. Write $a = pc$ for some $c \in \mathbb{Z}$ and substitute to get $pb^2 = p^2c^2$, which gives $b^2 = pc^2$. From this, we conclude that $p \mid b^2$, so $p \mid b$. But now p is a common divisor of a and b, contradicting that $\gcd(a, b) = 1$.   □

And, finally, the big one.

Theorem 3.3.12 (Fundamental Theorem of Arithmetic)
Every non-zero integer can be written in the form

$$n = up_1 p_2 \ldots p_k,$$

where u is a unit, $k \geq 0$, and each p_k is prime. Moreover, this form is unique up to reordering and associates.

Proof Let n be a non-zero integer. First, let's dispense with the case that n is a unit (i.e., $n = \pm 1$). In this case, $n = n$ is an expression for n in the form $n = up_1 p_2 \ldots p_k$, where $k = 0$.

Notice that any non-zero integer n has a unique prime factorization up to re-ordering and associates if and only if $-n$ does, so for the remainder of the proof, we consider the case n is positive, assume $u = 1$, and suppose that any prime p appearing in the factorization is positive. We break the proof into the existence and uniqueness of a factorization for such an integer.

First, we argue existence by strong induction. If $n = 2$, then n is prime, and taking $u = 1$, $k = 1$, $p_1 = 2$ provides a prime factorization. Now suppose t can be written as a product of primes for all natural numbers $2 \leq t < n$, and consider n itself. If n is prime we are done. If not, $n = ab$ for some $2 \leq a, b < n$ in $\mathbb{N}$. By our strong induction hypothesis, a and b can each be expressed as a product of primes, $a = p_1 p_2 \cdots p_l$ and $b = q_1 q_2 \cdots q_m$. Thus, $n = p_1 p_2 \cdots p_l q_1 q_2 \cdots q_m$ is an expression for n as a product of primes.

It remains to prove uniqueness. We show that any time we have two prime factorizations of an integer $n \geq 2$ into a product of positive primes, then those two products are in fact simply rearrangements of one another. That is, suppose that, for some positive primes p_i and q_i, we have

$$p_1 \cdots p_k = q_1 \cdots q_\ell.$$

After canceling off any factors in common to both sides, we can assume that no p_i is equal to any q_j. Now pick any prime p_i you want. The equality of the two products implies $p_i \mid q_1 \cdots q_\ell$. Thus, by Corollary 3.3.9, we have $p_i \mid q_j$ for some j. However, the only divisors of q_j are ± 1 and $\pm q_j$, and we reach the contradiction $p_i = q_j$ for some j. $\qquad\square$

3.4 Factors and Factorials

> Teacher: How many times does 5 go into 75?
> Student: Every single time I do it!
> You: Lol.
> - Anonymous Jokester

One of the most concise ways of seeing the power of the Fundamental Theorem of Arithmetic is that we can unambiguously make sense of "the power of p that goes into n." This phrase is simultaneously bulky enough and important enough that it merits some terminology and notation.

Definition 3.4.1

For a non-zero integer n and prime p, define the **p-adic valuation** of n, denoted $v_p(n)$, to be the power of p appearing in the prime factorization of n. In other words, $v_p(n)$ is the unique integer such that we can write

$$n = p^{v_p(n)} n'$$

for some integer n' with $p \nmid n'$. ◀

For $n = 3500 = 2^2 \cdot 5^3 \cdot 7$, we have $v_2(n) = 2$, $v_5(n) = 3$, and $v_7(n) = 1$, since these are precisely the powers of each of those primes showing up in its prime factorization. The n' in the definition is a catch-all for "the rest of the factorization," i.e., the part not involving the specific prime p (so for a given n, the value of n' depends on p). The definition also makes sense for primes p not dividing n: since $3500 = 11^0 \cdot 3500$ and $11 \nmid 3500$, we conclude that $v_{11}(3500) = 0$, and likewise for every prime other than 2, 5, and 7.

Many of the observations about prime factorizations, relatively prime numbers, gcds, perfect squares, etc., can be neatly encoded in terms of these valuations, and often more easily proved in this language as well. Specifically, many are direct consequences of the following rather trivial lemma.

Lemma 3.4.2

For non-zero integers a and b, we have

$$v_p(ab) = v_p(a) + v_p(b).$$ ◀

Proof Let $k = v_p(a)$ and $\ell = v_p(b)$ so we can write $a = p^k a'$ and $b = p^\ell b'$ with $p \nmid a'$ and $p \nmid b'$. Then $ab = p^{k+\ell} a'b'$, and by the contrapositive of the Prime Divisor Lemma, $p \nmid a'b'$, so $v_p(ab) = k + \ell = v_p(a) + v_p(b)$. □

The identity shows how similar v_p is to the base-p logarithm function $\log_p$, and in fact, the two functions give the same result on an input of the form p^k. For other integer inputs, they differ in that v_p simply ignores the part of the input relatively prime to p. As always, we take our new definition out for a test run before engaging in any life-or-death street racing with it. How, for example, does it handle units? Once you answer that question, it's easy to see that the sign of n is inconsequential: since $-3500 = -1 \cdot 2^2 \cdot 5^3 \cdot 7$, the valuations $v_p(3500)$ and $v_p(-3500)$ are the same for all primes p. Alternatively, since $v_p(-1) = 0$ for all primes p, the result follows from the lemma: $v_p(-n) = v_p(-1) + v_p(n) = v_p(n)$. Next, what about 0? Since we have not defined the prime factorization of 0, the official definition provides no guidance, but if we let Lemma 3.4.2 lead the way, we find our hand forced! That is, if we were to insist that

$$v_p(0) = v_p(0 \cdot n) = v_p(0) + v_p(n)$$

hold for all n, then we are honor-bound to accept that $v_p(0)$ has the property that it remains unchanged upon addition of any natural number! There is but one solution, which is to adopt the convention that $v_p(0) = \infty$! This rather unorthodox convention protects the legitimacy of the lemma, but also bears some semblance of plausibility— it is indeed the case that $p^k \mid 0$ for every integer k, so if we adopt the stance that $v_p(0)$ should be the highest power of p that divides zero, we are also led to the convention of setting $v_p(0) = \infty$. (This latter piece of reasoning also explains why we do not choose $v_p(0) = -\infty$.)

Let us put these valuations to use. First, we spend some time rephrasing key arithmetic concepts (divisibility, gcds, etc.) in the language of valuations, proofs of which typically follow immediately from prime factorizations.

Lemma 3.4.3

Given non-zero integers a and b, we have $a \mid b$ if and only if $v_p(a) \le v_p(b)$ for *all* primes p. In particular, we have $v_p(a) = v_p(b)$ for all p if and only if $a \mid b$ and $b \mid a$; that is, $a = \pm b$. ◀

This seemingly trivial lemma greatly facilitates computations of important number-theoretic quantities in the case that the prime factorizations are available to us. For example, finding $\gcd(378675, 199875)$ would be tiresome by hand, even with the Euclidean algorithm. But if we were in a context where we already knew the prime factorizations $378675 = 3^4 \cdot 5^2 \cdot 11 \cdot 17$ and $199875 = 3 \cdot 5^3 \cdot 13 \cdot 41$, computing their gcd is greatly simplified: we simply reason via the prime factorization of the gcd of these two numbers. In particular, the gcd can't have a 17 in its prime factorization, since then it wouldn't divide 199875 by the lemma. Similarly, it can't have a 13, lest it fails to divide 378675. More interestingly, we can also reason that $v_5(\gcd(378675, 199875)) = 2$. If it were any smaller, we could multiply by 5 to find a larger common divisor, and if it were any larger, it would fail to divide 378675. Now we see that this value of 2 came from the smaller of the two exponents of 5 in the two respective prime factorizations. With this observation, we arrive more generally at the following lemma.

Lemma 3.4.4

For each prime p we have

$$v_p(\gcd(a, b)) = \min\{v_p(a), v_p(b)\} \quad \text{and} \quad v_p(\operatorname{lcm}(a, b)) = \max\{v_p(a), v_p(b)\}.$$

In particular, a and b are relatively prime if and only if for each prime p, either $v_p(a) = 0$ or $v_p(b) = 0$ (or both). ◀

As an interesting consequence, we have the following algebraic identity:

Lemma 3.4.5

For all natural numbers a and b, we have

$$\gcd(a, b) \cdot \operatorname{lcm}(a, b) = ab. \qquad \blacktriangleleft$$

Proof By Lemma 3.4.4, we have

$$v_p(\gcd(a, b) \cdot \operatorname{lcm}(a, b)) = \min(v_p(a), v_p(b)) + \max(v_p(a), v_p(b))$$
$$= v_p(a) + v_p(b) = v_p(ab)$$

for all primes p, and so the result follows from Lemma 3.4.3. □

We note that above and for the next few lemmas we restrict our attention to natural numbers rather than arbitrary integers. This stems from the asymmetry implicit in the definition of gcd and lcm whereby they were taken by fiat to be positive. The previous result would necessarily fail, for example, if ab were negative. Still, in some sense, the spirit of the result is still true as all of the valuations work out correctly—so while we skirt the issue, for now, addressing the matter of units in questions of this type will require more care in upcoming chapters.

From here on, we will drop the word "perfect" from "perfect square," making the assumption that, like our readers, squares are inherently perfect.

Lemma 3.4.6

A natural number n is a square if and only if $v_p(n)$ is even for all primes p. ◄

Proof If $n = k^2$, then for any prime p, we have by Lemma 3.4.4

$$v_p(n) = v_p(k^2) = v_p(k) + v_p(k) = 2v_p(k)$$

which is even. Conversely, if each exponent in the prime factorization of n is even, then we can write

$$n = p_1^{2a_1} p_2^{2a_2} \cdots p_k^{2a_k},$$

and so evidently $n = k^2$ for $k = p_1^{a_1} p_2^{a_2} \cdots p_k^{a_k}$. ☐

Compare the following to Corollary 3.3.11.

Corollary 3.4.7

If a natural number n is not a square, then $\sqrt{n}$ is irrational. ◄

Proof Since n is not a square, there exists a prime $p \mid n$ such that $v_p(n)$ is odd. Now suppose $\sqrt{n} = \frac{a}{b}$ were rational. Then $nb^2 = a^2$, but $v_p(a^2) = 2v_p(a)$ is even, whereas $v_p(nb^2) = v_p(n) + 2v_p(b)$ is odd, giving a contradiction. ☐

This result admits a vast generalization, the so-called *Rational Root Test*.

Theorem 3.4.8 (Rational Root Test)
If $f(x) = a_n x^n + a_{n-1} x^{n-1} + \cdots + a_0$ has a rational root $x = \frac{b}{c}$, with $\gcd(b, c) = 1$, then $b \mid a_0$ and $c \mid a_n$.

Proof Suppose $x = \frac{b}{c}$, with $\gcd(b, c) = 1$, is a rational root of $f(x)$. Then the equation $f\left(\frac{b}{c}\right) = 0$ gives, after multiplying by c^n, that

$$a_n b^n + a_{n-1} b^{n-1} c + \cdots + a_0 c^n = 0.$$

Organizing this equation as

$$-a_0 c^n = b(a_n b^{n-1} + \cdots + a_1 c^{n-1}),$$

we see that $b \mid a_0$ (since $\gcd(b, c) = 1$ implies $\gcd(b, c^n) = 1$). Similarly, organizing the equation as

$$a_n b^n = -c(a_{n-1} b^{n-1} + \cdots + a_0 c^{n-1}),$$

we see that $c \mid a_n$. $\square$

Corollary 3.4.9

In particular, if f is *monic* ($a_n = 1$), then any root of f is either an integer or irrational. ◄

Applied to the polynomial $f(x) = x^2 - n$, we see that any root (i.e., $\pm\sqrt{n}$) is either an integer (if n is a square) or irrational (if not). This is where the "vast generalization" we mentioned above comes into play. We can generalize easily to k-th powers like $\sqrt[5]{2}$ or $\sqrt[3]{17}$, and also to roots of other polynomials which are harder to write down. For example, all roots to the polynomial

$$f(x) = x^7 + 23x^6 - 19x^5 + 24x^2 - 37x + 1$$

are irrational, since by the corollary the only possible rational numbers that could be roots, ± 1, are not (as neither $f(1)$ nor $f(-1)$ is zero).

Finally, we mention one more key result that follows from unique factorization and admits a nice proof using valuations.

Lemma 3.4.10 (Power Lemma)

Let a and b be relatively prime natural numbers. If ab is a square, then a and b are themselves squares. More generally, if ab is an n-th power, then so are a and b. ◄

Proof Suppose $ab = k^2$ with $\gcd(a, b) = 1$. Then for any prime p, $v_p(a) + v_p(b) = v_p(ab) = 2v_p(k)$ is even, but either $v_p(a) = 0$ or $v_p(b) = 0$ since $\gcd(a, b) = 1$, so both must be even. That is, for each prime p, both $v_p(a)$ and $v_p(b)$ are even, so a and b are squares by Lemma 3.4.6. For the general argument, we need only replace "even" with "multiple of n." $\square$

We close the section with a return to Pythagorean triples, highlighting both how far we've come and some tantalizing prospects for what lies ahead. When we last left our heroes, we had observed that Pythagorean triples come in infinite families: $(3, 4, 5)$ and its multiples, $(5, 12, 13)$ and its multiples, etc., and that (Theorem 2.2.4) every Pythagorean triple was a multiple of one of the form

$$(a, b, c) = (u^2 - v^2, 2uv, u^2 + v^2)$$

for some $u > v > 0$. The mapping between pairs (u, v) and triples (a, b, c), along with their multiples, is not particularly crisp. Not every Pythagorean triple can be written in the form $(u^2 - v^2, 2uv, u^2 + v^2)$, and not every (u, v) pair generates a triple that's not a multiple of some smaller triple. The literature calls a Pythagorean triple (a, b, c) **primitive** if no prime divides all three side lengths. We did not at the time have the tools to precisely parameterize the primitive triples, but we're much better equipped now:

Corollary 3.4.11

The primitive Pythagorean triples are precisely those coming from pairs (u, v) with $u > v > 0$, $\gcd(u, v) = 1$, and u, v of opposite parity (that is, one odd and one even). ◄

Proof First, suppose u and v satisfy all of the conditions. We show that $(u^2 - v^2, 2uv, u^2 + v^2)$ is primitive; i.e., no prime p divides all three side lengths. We can rule out $p = 2$ since the even/odd hypothesis on u and v forces both of $u^2 \pm v^2$ to be odd. Next, suppose that an odd prime p divides $u^2 - v^2$ and $u^2 + v^2$, and thus their sum and difference, $2u^2$ and $2v^2$, respectively. By the prime divisor property and Corollary 3.3.10, we see that p also divides both u and v, contradicting that $\gcd(u, v) = 1$. For the converse, suppose $(u^2 - v^2, 2uv, u^2 + v^2)$ is a primitive Pythagorean triple. The only substantive check is that u and v are relatively prime. If any prime p were to divide both u and v, it would also divide each of $u^2 - v^2$, $2uv$, and $u^2 + v^2$, culminating in a non-primitive triple.   □

Finally, we note that Pythagorean triples provide one of the more compelling justifications for trying to develop number theory (and in particular, analogs of the Fundamental Theorem of Arithmetic) in other rings. As a quick foreshadowing, let's revisit the Pythagorean Theorem yet again, but this time in a marvelous future where we have developed an analog of gcds, Bézout's Identity, etc., all the way up to the Power Lemma for $\mathbb{Z}[i]$, the ring of Gaussian integers. In this ring, we can re-write the Pythagorean identity as

$$c^2 = a^2 + b^2 = a^2 - (-bi)^2 = (a + bi)(a - bi).$$

If we could assume that the Power Lemma (3.4.10) continues to hold in $\mathbb{Z}[i]$ and that $a + bi$ and $a - bi$ are relatively prime (whatever *that* means...), then the lemma guarantees that $a + bi$ and $a - bi$ are themselves squares:

$$a + bi = (u + vi)^2 = (u^2 - v^2) + 2uvi$$

By equating real and imaginary parts, we get $a = u^2 - v^2$, $b = 2uv$, and from $a^2 + b^2 = c^2$, we get $c = u^2 + v^2$. This is the complete classification of Pythagorean triples we found in Theorem 2.2.4 from the chord method, but from purely number-theoretic techniques.

Exploration D: Consequences of the Fundamental Theorem

Factorials

D.1 How do the prime factorizations of $n!$ and $(n + 1)!$ relate? Experiment by writing down the prime factorizations of the first few factorials.

D.2 Compute $v_2(100!)$, $v_3(100!)$ and $v_{53}(100!)$.

D.3 Find the number of zeroes at the end of the decimal expansion of $125!$.

D.4 Explain the following two important and pleasantly opposite facts:

- All prime divisors p of $n!$ must satisfy $p \leq n$.
- All prime divisors p of $n! + 1$ must satisfy $p > n$.

D.5 Use the previous problems to describe an algorithm for writing down the prime factorization of $n!$. For those who prefer formulas to algorithms, you might find the expression $\lfloor \frac{n}{p^i} \rfloor$ helpful (here $\lfloor x \rfloor$ denotes the *floor* of x, the greatest integer less than or equal to x).

Divisor-Counting

While working in $\mathbb{Z}$ (allowing negative primes and whatnot) has some clear structural benefits, it does render some aspects of number theory cumbersome. For example, each prime p now has four divisors (± 1 and $\pm p$) instead of just "1 and itself." For the following counting problems, work in $\mathbb{N}$ and take all integers, primes, divisors, etc., under consideration to be positive.

D.6 Suppose p, q, and r are primes. How many divisors does p^3 have? p^6? How about $p^2 q^3$? pqr?

D.7 Generalize: Find a formula for the number of divisors of $n = p_1^{e_1} \cdots p_k^{e_k}$, where $k \geq 1$, each p_i is prime, and each e_i is a non-negative integer.

D.8 It is a little-known fact that math professors do not actually read homeworks in order to grade them. Given, say, 600 calculus problems to grade, we begin by marking them all correct. Then, fearing this is too obvious, we go through and change every second problem (all the even-numbered problems) to incorrect. Of course, this is a pretty recognizable pattern as well, so we go through and change every third problem (changing problems 3, 6, 9, etc. to correct if it was marked incorrect, and vice versa), then change every 4th problem, then every 5th problem, etc. On the 600th and last run-through, only the 600th problem is changed one last time. How many problems ended up being marked correct? The previous problem is relevant.   ◀

3.5 The Prime Archipelago

The topic of valuations shepherds in with it an important philosophical perspective: to understand one global question about integers ("Does a divide b?") it often suffices to resolve related questions one prime at a time ("For each prime p, is $v_p(a) \leq v_p(b)$?"). The significance of this perspective will increase over time and is the explicit focus of Chapter 8, but for now, we use it as motivation for moving away from how arithmetic is governed by primes to a closer look at the set of primes itself.

We have thus far cavalierly thrown around numbers like 2, 3, 5, 7, 11, etc., as examples of primes. How do we know these are primes? Presumably, because we're so well versed in our small-integer multiplication tables that we're convinced that no two natural numbers greater than 1 multiply together to give, for example, 11. The mental hard-wiring of multiplication tables can only take us so far, though—for 4-digit primes, one would need to know the 100×100 multiplication table. But even this observation, that $100 \times 100 = 10000$ is bigger than any 4-digit number, holds some insight:

Lemma 3.5.1

Every composite $n \in \mathbb{N}$ has a prime divisor p with $p \leq \sqrt{n}$. ◄

Proof Since n is composite, we can write $n = ab$ with $a, b > 1$. Now we couldn't have both $a, b > \sqrt{n}$, as this would imply $ab > \sqrt{n} \cdot \sqrt{n} = n$. Choose p to be any prime divisor of whichever factor is at most $\sqrt{n}$.   □

Most directly, this simplifies the brute-force test for primeness: we need not test n for divisibility by *all* integers up to n, but instead only all primes up to $\sqrt{n}$. For $n = 119$, for example, we can use the lemma to conclude primeness after testing divisibility only by 2, 3, 5, and 7 (the primes less than $\sqrt{119}$). These four divisions represent amazing time savings over dividing by all 117 integers from 2 up to 118. But there is a second boon of the lemma which comes into play if we wish to enumerate *all* primes up to n. The algorithm is known as the *Sieve of Eratosthenes*:

▶ **Sieve 3.5.2 (Sieve of Eratosthenes)** Fix $n > 2$. To find all primes $p \leq n$:

- Begin with a list of the integers $2 \leq k \leq n$. Initially, all numbers are neither circled nor crossed out.
- Circle the smallest remaining number (that is, not circled and not crossed out) and cross off all of its multiples in the list.
- Repeat the previous step until all numbers less than or equal to $\sqrt{n}$ have been circled or crossed out.
- Circle every remaining number on the list not yet crossed out. Now the circled numbers are all prime, and all crossed-out numbers are composite.

The argument that this procedure produces the primes up to n is simply the lemma: the elements of the list that get crossed off are precisely the composite numbers up to n with a prime factor at most $\sqrt{n}$. The sieve is vastly more efficient for producing all primes up to n than simply testing each one in that range individually for primeness, and so with very little work we can extend our initial list of primes:

$$2, \ 3, \ 5, \ 7, \ 11, \ 13, \ 17, \ 19, \ 23, \ 29, \ 31, \ 37, \ 41, \ 43, \ 47, \ 53, \ 59, \ \text{etc.}$$

From a list of the integers up to 100, the 25 primes up to 100 can be enumerated in a matter of seconds—we need only cross out the multiples of 2, 3, 5, and 7.

Still, the "etc" in that list of primes quietly conceals a profound mystery—looking out over the ocean of integers, islands of primes speckle the landscape. As we strain to look further and further out, does this archipelago eventually end? Are there finitely many primes or infinitely many? Is there a formula for the n-th term in that list, like there is for the n-th triangular number, or a recursive formula like for the n-th Fibonacci number? Is it true that the primes get sparser as we go further down the number line? Could we even make a statement like that precise?

None of these questions are trivial to resolve. At first glance, it's quite plausible that the list of primes might contain only finitely many terms. As a thought experiment, consider that at least half the natural numbers are composite simply because they're even, and then a third of the remaining odd numbers are composite since they're multiples of three, then an extra fifth are multiples of 5, etc. For any given 100-digit number to be prime, it would have to simultaneously avoid being a multiple of 2, *and* avoid being a multiple of 3, *and* avoid being a multiple of 5, etc. So it is conceivable there are only, say, fourteen trillion numbers which manage to beat the odds and avoid being multiples of each smaller number. As it turns out, this is not the case: one of the most celebrated pieces of mathematics from antiquity, with one of the most elegant proofs ever written, is the denial of that possibility.

> **Theorem 3.5.3 (Euclid, 300 BCE)**
> There are infinitely many primes.

Proof We will show that given any finite set of primes, $\{p_1, \ldots, p_k\}$, we can always find another prime p not in the set. To do this, consider the number

$$N = p_1 p_2 \cdots p_k + 1,$$

and let p be the smallest prime divisor of N. Then none of the p_i divide N (by its form above, we can see that the remainder when N is divided by any of the p_i is 1) but p *does*, so p is distinct from each of the primes p_i in the initial set. We conclude that no finite set of primes can contain them all. $\square$

The proof is almost *too* elegant, in that it so quickly disposes of a profoundly interesting question. We encourage the reader to ponder the approach carefully and consider any number of follow-up questions that arise: is the number N constructed in the proof always prime? Is the proof "constructive" in the sense that it provides an explicit new prime? What if we didn't choose the smallest prime divisor of N but the largest? Could you generate *all* primes by repeating the procedure in the proof over and over?

Knowing that there is an infinitude of primes is just a first step toward understanding how the primes are distributed along the natural number line. In particular, a belief that there are only finitely many primes can be defended with plausible reasoning, even if it ultimately wrong. It *is* "harder" for a random large number to be prime than a random small one, for the simple reason that there are more potential prime divisors it has to dodge. Making precise claims of this type falls most cleanly in a branch of number theory called *analytic number theory*, as opposed to the *algebraic number theory* that this book fawns over. We do not have space to do the topic justice here, but nor could we neglect to at least inform you of the story. One way of attempting to make precise claims about the decreasing density of primes as we progress down the number line is to count all the primes up to an indeterminate value x, and then see how that count evolves as x increases. This leads to the following definition.

Definition 3.5.4

For a positive real number n, the prime counting function $\pi(n)$ is defined to be the number of prime natural numbers $p \leq n$. ◀

For example, since there are precisely four prime numbers less than 10 (namely, 2, 3, 5, and 7), we have $\pi(10) = 4$, and with the aid of a computer[11], we can tabulate that $\pi(1,000,000) = 78,498$. One of the success stories of nineteenth-century number theory was a proof, independently completed by Jacques Hadamard and Charles Jean de la Vallée Poussin, of an asymptotic growth rate for $\pi(n)$. One need only look at the name of the result, and the absence of words in its statement, to get a sense of its significance.

Theorem 3.5.5 (The Prime Number Theorem)

$$\lim_{n \to \infty} \frac{\pi(n)}{n / \ln(n)} = 1.$$

---

[11] or a bevy of very well-trained chimpanzees

The theorem presents some compelling information on how the primes are distributed. Namely, we should expect that for large n, we have $\pi(n) \approx \frac{n}{\ln(n)}$, approximate in the sense that the ratio between the two sides should be close to 1, and gets closer to 1 as n increases. For one, the prediction

$$\pi(1,000,000) \approx \frac{1,000,000}{\ln(1,000,000)} \approx 72,382$$

is about 8% off from the correct answer given above, but much, much easier to evaluate. Further, since there are n natural numbers up to n, dividing the approximation by n describes the *proportion* of numbers in this range which are prime. Loosely re-interpreting this as a probability, we conclude that the probability that a number up to n is prime is about $\frac{1}{\ln(n)}$. Since this goes to 0 as $n \to \infty$, we have successfully put form to our prediction that the primes become sparser and sparser as we go. Still, loads of questions remain: Are there *exact* formulas for $\pi(n)$? An error of 8% is still reasonably large—can we find better approximations? Analytic number theory focuses around questions of this type, and while not the focus of the current text, we will endeavor to make references to these questions throughout.

To wrap up, let's take a moment (or ten) of quiet introspection on the major revelations which allowed us to reduce so much of the structure of $\mathbb{Z}$ to the study of prime numbers. Especially relevant is diagnosing what obstacles lie ahead as we try to replicate these arguments in other rings (e.g. $\mathbb{Z}[i]$, $\mathbb{R}[x]$). If we trace backward, we see the necessity of the Fundamental Theorem of Arithmetic, which in turn rested upon The Fundamental Theorem of GCDs, which in turn rested on the notion of divisibility, and in particular on the Division Algorithm. In a very real sense, which we will formalize in the next chapter, the starting point of much of number theory is trying to make sense of remainders upon division in some arbitrary ring. We've already made reference to the notation of $a \bmod n$ for these remainders, and we will need to develop tools for doing arithmetic with them ("modular arithmetic"). As per the central themes of this book, our approach to developing this arithmetic takes a very algebraic bent—while we currently lack a ring structure on the set of remainders, we will rectify this deficiency in the next chapter.

3.6 Exercises

Calculation & Short Answer

Exercise 3.1 Let G denote one googol: $G = 10^{100}$. Compute $\gcd(G-1, G+1)$. Repeat for $G = 11^{111}$.

Exercise 3.2 Write down the prime factorization of 10!. Use the factorization to explain why there are precisely 10! seconds in 6 weeks.

Exercise 3.3 Give the complete prime factorization of 20!.

Exercise 3.4 How many zeroes are at the end of 117!?

Exercise 3.5 Show that there is no n such that $n!$ ends with exactly 5 zeroes.

Exercise 3.6 What are the possible values of

$$\gcd(7a + 12, 4a - 3)$$

for an integer a? Explain your reasoning and find a value of a that gives each of these possible values.

Exercise 3.7 Perform the Extended Euclidean Algorithm to find a linear combination of the 10th and 11th Fibonacci numbers (55 and 89 respectively) equaling 1, and then repeat for $F_9 = 34$ and $F_{10} = 55$. Make some observations.

Exercise 3.8 Recall that for α, β in $\mathbb{Z}[i]$, we say α divides β (denoted $\alpha \mid \beta$) if there exists a Gaussian integer $\gamma \in \mathbb{Z}[i]$ such that $\alpha\gamma = \beta$. Using this definition, decide which of the following are true:

(a) $2 + 3i \mid 5 + 7i$ (c) $2 + i \mid 7 + i$.
(b) $3 + 0i \mid 5 + 0i$.   (d) $2 + i \mid 10$.

Exercise 3.9 Recall that for f, g in $\mathbb{R}[x]$, we say f divides g (denoted $f \mid g$) if there exists $h \in \mathbb{R}[x]$ such that $fh = g$. Using this definition, decide which of the following are true:

(a) $x \mid x^3 - 2x^2 + 4x - 1$ (c) $x - 1 \mid x^3 + 7x^2 - 13x + 5$
(b) $x \mid x^5 + x^4 - x$ (d) $2x^2 + 6x \mid 3x^3 + 2x^2 - 6x$.

Then answer these same four divisibility questions in $\mathbb{Z}[x]$.

Exercise 3.10 What are the divisors of $6x^2 + 24$ in $\mathbb{R}[x]$? in $\mathbb{Z}[x]$? Repeat both questions for $6x^2 - 24$.

Exercise 3.11 Parameterize all integer solutions to each equation.

(a) $24x + 40y = 16$
(b) $13x + 12y = 1$
(c) $23x - 41y = 1$.

Exercise 3.12 Find all rational roots of $x^7 - 3x^5 - 4x^4 + 2x^3 + 12x^2 - 8$.

Exercise 3.13 Find all real roots of the polynomial $3x^3 + 11x^2 - x - 1$.

Exercise 3.14 Begin with the set of primes $S = \{2\}$ and enact the construction of Euclid's Proof in Theorem 3.5.3 to generate a new prime (the smallest divisor of $1 + \prod_{p \in S} p$). Add this prime to S and repeat, recording the sequences of primes generated, until you get the prime 5.

Formal Proofs

Exercise 3.15 Let S be a set, and let P be the **power set** of S: $P = \{A : A \subseteq S\}$ (the set of all subsets of S). Define the **symmetric difference** Δ on P by $A \Delta B = (A \cup B) - (A \cap B) = (A - B) \cup (B - A)$. Prove that P is a commutative ring with unity with "addition" given by Δ and "multiplication" given by $\cap$.

Exercise 3.16 Define new operations $\oplus$ and $\otimes$ on $\mathbb{Z}$ by $a \oplus b = a + b + 1$ and $a \otimes b = ab + a + b$. Prove that $\mathbb{Z}$ is a ring under these operations.

Exercise 3.17 What goes wrong in the proof of Corollary 3.3.11 if $p = 6$? How could the argument be modified to make it work?

Exercise 3.18 Let R be a commutative ring and let $a, b, c \in R$. Prove that if $a \mid b$ and $b \mid c$, then $a \mid c$.

Exercise 3.19 Let R be a commutative ring and let $a, b, c \in R$. Prove that if $a \mid b$ and $a \mid c$, then $a \mid (b + c)$ and $a \mid (b - c)$.

Exercise 3.20 Let R be a commutative ring. Prove that for all $a, b \in R$, we have the following three identities:

$$(-1) \cdot a = -a \qquad (-a) \cdot b = -(ab) \qquad (-a) \cdot (-b) = ab.$$

Exercise 3.21 Prove that in any ring with unity:

- the product of two units is a unit
- the multiplicative inverse of a unit is a unit
- the additive inverse of a unit is a unit

Exercise 3.22 Prove that the multiplicative inverse of a unit u in a commutative ring R is unique.

Exercise 3.23 Let R be a ring and 0 its additive identity. Prove that 0 cannot have a multiplicative inverse.

Exercise 3.24 Prove that there are no integers between 0 and 1. [Hint: use the Well-Ordering Principle.]

Exercise 3.25 We can extend the definition of p-adic valuations to rational numbers by

$$v_p \left(\frac{a}{b} \right) = v_p(a) - v_p(b).$$

Prove that Lemma 3.4.2 continues to hold. What happens to Lemmas 3.4.3 and 3.4.4?

Exercise 3.26 Prove Lemma 3.4.4.

Exercise 3.27 Euclid's proof gives us our first analytic estimate for how small the n-th prime p_n could be. Deduce from Euclid's proof that

$$p_n \leq p_1 \cdots p_{n-1} + 1 < 2^{2^n}.$$

Use this to deduce a lower bound for $\pi(n)$.

Exercise 3.28 Prove that if (a, b, c) is a Pythagorean triple and p is a prime dividing any two of $a, b,$ or c, then p must divide the third. Does the statement hold if we replace p with an arbitrary integer n?

Exercise 3.29 Show, via a formal induction proof, that every pair of consecutive Fibonacci numbers are relatively prime.

Exercise 3.30 Find and prove a formula for $\mathrm{lcm}(n, n + 1)$ for a natural number n. Repeat for $\mathrm{lcm}(n, n + 3)$ (consider breaking into cases).

Exercise 3.31 For a natural number n, let $\sigma_0(n)$ be the number of natural number divisors of n.

- Develop a formula for $\sigma_0(n)$ in terms of the p-adic valuations of n.
- Prove that if $\gcd(m, n) = 1$, then $\sigma_0(mn) = \sigma_0(m)\sigma_0(n)$.

Exercise 3.32 Use Bézout's identity to show that if $a, b,$ and c are integers such that $c \mid ab$, then $c \mid \gcd(a, c) \gcd(b, c)$.

Exercise 3.33 A *Mersenne prime* is a prime number that is one less than a power of 2. A *perfect number* is a positive integer that is the sum of its positive proper divisors (i.e., the positive divisors strictly less than the number itself). For example, 6 is a perfect number since $6 = 1 + 2 + 3$.

1. Find the first four Mersenne primes.
2. Given a prime p, show that if $2^p - 1 = q$ is prime (and hence a Mersenne prime), then $2^{p-1}q$ is perfect.
3. Show that $2^n - 1$ is never prime if n is composite.

Exercise 3.34 Let $n \in \mathbb{Z}$. Use properties of the 2-adic valuation to prove that if $16 \mid n^3$ then $16 \mid n^2$.

Exercise 3.35 Let a and b be positive even integers. Prove that the gcd of a and b in $\mathbb{Z}$ is never equal to the gcd of a and b in $2\mathbb{Z}$.

Exercise 3.36 Prove that $\log_2(7)$ is irrational.

Exercise 3.37 Find all rational numbers that are 30 less than their 5th power. Repeat for 31.

Computation and Experimentation

Exercise 3.38 Use a computer algebra system to build the functions below. See the Python worksheet "Chapter 3 Tools."

1. Write a "prime checker"—a function that tests an input to see whether it is prime.
2. Write a function that implements the Euclidean Algorithm to find the gcd of two inputs.
3. Write a function that factors a given integer into prime factors.
4. Write a function that counts the number of primes that are less than or equal to a given integer.
5. Write a function that finds the p-adic valuation of the input n.

Exercise 3.39 Here we explore the *Fermat numbers*.

1. Write a function that outputs the nth Fermat number, $F_n = 2^{2^n} + 1$, given the input n. Use it to verify that the third Fermat number is 257.
2. Find the first few Fermat numbers along with their prime factorizations.
3. Compute $\gcd(F_m, F_n)$ for all $m, n \leq 20$ with $m \neq n$. Then write down a clear conjecture based on your observations.
4. Prove your conjecture and deduce as a corollary another proof that there are infinitely many prime numbers.

Fermat conjectured that *every* Fermat number was prime. Research the current status of this conjecture. Between this and his "proof" of Fermat's Last Theorem... sheesh!

General Number Theory Awareness

Exercise 3.40 What's the deal with "twin primes"? Give a brief status report on what's known about them.

Exercise 3.41 Do some research on the runtime and efficiency of the Euclidean Algorithm. Just how good *is* it? Why do Fibonacci numbers make an appearance?

Exercise 3.42 The text makes reference (Example 3.2.20) of an application of the Euclidean Algorithm (and/or Bézout's Identity) to the genre of riddles in which one uses jugs of a fixed size to measure out a specific volume. Sort out this general connection and explain the equation given in that Example.

Exercise 3.43 Complete Exercise 3.14 first. Enter the first few primes you encounter into the Online Encyclopedia of Integer Sequences, and use the information there as a springboard to investigate what is known and not known about this and/or similar sequence(s).

Exercise 3.44 Learn about the difference in efficiency between checking that an integer is prime and factoring it. About how long would it take on the fastest computers in existence at the moment to:

- Check if a 400-digit number is prime?
- Factor a 400-digit number?

Exercise 3.45 The ABC conjecture is one of the most tantalizing open conjectures in mathematics, relating the additive and multiplicative aspects of number theory. Learn and understand what the conjecture says, and what some of the corollaries to the conjecture would be. Without getting too bogged down, see what you can figure out about the current status of a proof.

Exercise 3.46 Speaking of tantalizing conjectures, it's hard to get more tantalizing than the Riemann hypothesis, considered by many to be one of the most important unproven propositions in mathematics. Look into this hypothesis and what it says about the distribution of primes. Get a sense of why it inspires quotes like the following from eminent mathematicians:

> *...that the distribution of prime numbers can be so accurately represented in a harmonic analysis is absolutely amazing and incredibly beautiful. It tells of an arcane music and a secret harmony composed by the prime numbers.*
> – Enrico Bombieri

> *...that, despite their simple definition and role as the building blocks of the natural numbers, the prime numbers belong to the most arbitrary and ornery objects studied by mathematicians: they grow like weeds among the natural numbers, seeming to obey no other law than that of chance, and nobody can predict where the next one will sprout.*
> *The second fact is even more astonishing, for it states just the opposite: that the prime numbers exhibit stunning regularity, that there are laws governing their behaviour, and that they obey these laws with almost military precision.*
> – Don Zagier

Number Theory in the Mod-n Era

4

...wherein Carl Friedrich Gauss's theory of congruence sets into motion a sequence of discoveries that revolutionizes mathematics and leaves the world a slightly better place.

4.1 Equivalence Relations and the Binary World

Consciously or not, many of us had our nascent interest in number theory ignited at an early stage not by Fermat's Last Theorem or the Fundamental Theorem of Arithmetic, but by patterns in multiplication tables, divisibility tests, and even odd observations like

the sum of two odd natural numbers is even,

a fact which one often casually abbreviates as "Odd + Odd = Even." As it turns out, almost all of these early patterns, tests, and observations trace their source back to the Division Algorithm and reasoning with remainders. This chapter develops this type of reasoning, moving from understanding "mod" as an operation on integers to the bedrock of new number systems. The algebra of those number systems arises from continuing the shorthand above, where we can collect a number of similar easily verified observations:

Even + Even = Even	Even $\times$ Even = Even
Even + Odd = Odd	Even $\times$ Odd = Even
Odd + Even = Odd	Odd $\times$ Even = Even
Odd + Odd = Even	Odd $\times$ Odd = Odd

Supplementary Information The online version contains supplementary material available at https://doi.org/10.1007/978-3-030-98931-6_4.

In particular, note that for any $x \in \{\text{Odd}, \text{Even}\}$, we have the identities

$$\text{Even} + x = x \qquad \text{and} \qquad \text{Odd} \times x = x.$$

That is, on the set {Odd, Even}, we have a notion of addition and multiplication, with the element "Even" serving as an additive identity and "Odd" as a multiplicative identity. We have likely already hinted away the punchline: though we will temporarily take these facts for granted, we will soon see that these new operations inherit from the integers many basic algebraic properties (associativity, commutativity, distributivity, etc.) and thus this very small set of two elements actually forms a ring. For the sake of foreshadowing and expedience of notation, let's relabel "Even" as "[0]" and "Odd" as "[1]," a nod to the fact that, when divided by 2, all even numbers leave a remainder of 0 and all odd numbers leave a remainder of 1. The brackets are used here to distinguish [0], the set of all even numbers, from the literal number 0.

Definition 4.1.1

The ring $\mathbb{Z}/(2)$ (pronounced "zee mod 2") is the set $\{[0], [1]\}$ with addition and multiplication defined as follows:

+	[0]	[1]
[0]	[0]	[1]
[1]	[1]	[0]

and

×	[0]	[1]
[0]	[0]	[0]
[1]	[0]	[1]

◄

This is precisely the notion of binary arithmetic that permeates computer science. As an intriguing aside, note that if we chose to identify [0] and [1] with "False" and "True" rather than "Even" and "Odd," then the two tables above are also the respective truth tables for *exclusive or* and *and*, so in some sense the entirety of manipulating logical expressions can be thought of as doing arithmetic in $\mathbb{Z}/(2)$.

Now, hoping that what you can do for one prime you can automatically do for the rest, we might try to replace even/odd with, say, multiple of 3 vs. non-multiple of 3. But this naive analogy ultimately fails, for if we add two "non-multiples of 2" we *always* get a multiple of 2, whereas if we add two "non-multiples of 3" we may or may not get a multiple of 3. Instead, an appeal to the Division Algorithm provides a more compelling analogy than the dichotomy $3 \mid n$ vs. $3 \nmid n$: We introduce a *tri*chotomy of possibilities corresponding to whether an integer is 0, 1, or 2 more than a multiple of 3. Lacking English words for each of these three terms[1], let us temporarily encode these classes with shapes and colors:

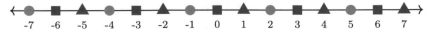

Integers that are multiples of 3 ($n \bmod 3 = 0$) are magenta squares, integers with $n \bmod 3 = 1$ are blue triangles, and integers with $n \bmod 3 = 2$ are orange circles. We

---

[1] Though there is a certain ring to *threven*, *thrwon*, and *thrwoo*...

mirror our even/odd discussion by considering magenta, blue, and orange, interpreted as the sets of integers they represent, as objects in and of themselves. Again we can convince ourselves (informally, for now) that we have a well-defined notion of addition and multiplication here as well. For example, we have "Blue + Orange = Magenta" in the same sense as before, in that if you add a blue number and an orange number, you *always* get a magenta number. Why? Because blue numbers are numbers one more than a multiple of 3 (i.e., expressible in the form $3k + 1$), and likewise orange numbers are expressible in the form $3\ell + 2$, so the sum of a blue and an orange is always of the form of a magenta number:

$$(3k + 1) + (3\ell + 2) = 3k + 3\ell + 1 + 2 = 3(k + \ell + 1) + 0.$$

Multiplication similarly depends not on the specific numbers being multiplied, but only their color. For example, we see "Orange × Orange = Blue," since

$$(3k + 2)(3\ell + 2) = 9k\ell + 6k + 6\ell + 4 = 3(3k\ell + 2k + 2\ell + 1) + 1$$

is one more than a multiple of 3 and hence blue. In general, the color of the product depends only on the color of the factors, not on the numbers themselves. This observation leads us to another small new ring. While it is tempting to develop a full range of chromatic number theory, it is perhaps more sensible to abbreviate as before, each color class being denoted by the mod-3 remainders they represent. If we denote by [0] the set of all magenta numbers, and similarly [1] for blue and [2] for orange, then we have the following ring.

Definition 4.1.2

The ring $\mathbb{Z}/(3)$ is the set $\{[0], [1], [2]\}$ with addition and multiplication defined by

+	[0] [1] [2]
[0]	[0] [1] [2]
[1]	[1] [2] [0]
[2]	[2] [0] [1]

and

×	[0] [1] [2]
[0]	[0] [0] [0]
[1]	[0] [1] [2]
[2]	[0] [2] [1].

◀

Filling in these tables can be done in a number of ways. One is to reason as we did above, using the general form of the elements in each set. Another technique, somewhat easier and certainly worth mastering, is to "randomly" choose an element of each type, add or multiply them in the integers, and then determine the type of the result. For example, to compute $[2] + [1]$ in $\mathbb{Z}/(3)$, we might randomly choose $8 \in [2]$ and $31 \in [1]$, and compute

$$8 + 31 = 39 \in [0] \quad \text{and} \quad 8 \cdot 31 = 248 \in [2],$$

so $[2] + [1] = [0]$ and $[2] \cdot [1] = [2]$. For this approach to work, a critical point to be checked in the next section is that the choice of the representatives 8 and 31 must be completely immaterial to the final result—having instead chosen $5 \in [2]$

and $7 \in [1]$ (or most easily, $2 \in [2]$ and $1 \in [1]$) would have resulted in the same sum and product types.

It is reasonably clear what we need to do to formalize and generalize these ideas to an arbitrary modulus n. We will partition the integers into sets based upon each integer's remainder mod n, and then form the "ring of integers mod n" out of these sets. This construction, making an algebraic structure out of subsets of another structure, is an odd mix of something completely familiar (e.g., the even/odd arithmetic) and something bizarrely abstract (adding and multiplying subsets of the integers). The formal version of this is the language of equivalence relations and equivalence classes and will be of repeated interest both in this book and in algebra as a whole.

Definition 4.1.3 (Equivalence Relation)

Given a set S, an **equivalence relation** $\mathcal{R}$ on S is a set of ordered pairs of elements of S having the following properties:

  (i) (Reflexivity) For all $a \in S$, $(a, a) \in \mathcal{R}$.
  (ii) (Symmetry) If $(a, b) \in \mathcal{R}$ then $(b, a) \in \mathcal{R}$.
  (iii) (Transitivity) If $(a, b) \in \mathcal{R}$ and $(b, c) \in \mathcal{R}$, then $(a, c) \in \mathcal{R}$.

Given such an equivalence relation $\mathcal{R}$, we say a **is equivalent to** b if and only if $(a, b) \in \mathcal{R}$. ◄

Equivalence relations are the quiet workhorses of the mathematics industry. Every time we encounter a new type of object, we invariably have a notion of when two of those objects are considered "essentially the same," and the formal notion of an equivalence relation encodes this. That is, $\mathcal{R}$ consists of the pairs of elements of S we want to consider equivalent. This idea subsumes many concepts in modern mathematics, under a wide variety of names:

- Two Gaussian integers $a + bi$ and $c + di$ are *equal* if $a = c$ and $b = d$.
- Two vectors v and w are *equal* if they have the same direction and length.
- Two polynomials f and g are *equal* if they have the same coefficients.
- Two fractions $\frac{a}{b}$ and $\frac{c}{d}$ are *equivalent* if $ad = bc$.
- Two triangles T and T' are *congruent* if there exists a rigid motion of the plane taking T to T'.
- Two square matrices A and B are *similar* if there exists an invertible matrix S such that $SAS^{-1} = B$.
- Two square matrices A and B are *congruent* if there exists an invertible matrix S such that $SAS^{T} = B$.

In each bullet, the italicized word defines an equivalence relation, and it's worth verifying that the three defining conditions in Definition 4.1.3 are satisfied. For example, one checks that (i) every square matrix A is similar to itself; (ii) if A is similar to B, then B is similar to A; and (iii) that if A is similar to B and B is similar to C, then A is similar to C.

Notions of equivalence greatly streamline our thinking about objects by lumping together things we think of as being the same anyway. In such contexts, it becomes useful to think of such lumps, the class of all objects equivalent to a given one (and hence to each other), as objects in their own right.

Definition 4.1.4 (Equivalence Class)

Let $\mathcal{R}$ be an equivalence relation on a set S. Then for an element $a \in S$, the **equivalence class of a under the relation $\mathcal{R}$** is the set

$$[a] = \{b \in S : (a, b) \in \mathcal{R}\},$$

that is, the set of elements to which a is equivalent under $\mathcal{R}$. (In contexts with multiple equivalence relations around, one might use $[a]_{\mathcal{R}}$ to emphasize equivalence under $\mathcal{R}$ specifically.) Any element of an equivalence class is called a **representative** of that class. ◄

Continuing the examples above, the equivalence class of a vector consists of itself and all of its translates, the equivalence class of a matrix consists of all matrices similar to it or all matrices congruent to it, the equivalence class of a Gaussian integer consists of only itself, and the equivalence class of $\frac{2}{3}$ consists of all fractions that reduce to it (e.g., $\frac{4}{6}$, $\frac{-6}{-9}$, etc.). This last example is particularly compelling, as the notion of a rational number can be (perhaps overly) formalized as an equivalence class of fractions: the *rational number* $\frac{2}{3} \in \mathbb{Q}$ is the equivalence class of the *fraction* $\frac{2}{3}$. This perspective, making the objects of interest the equivalence classes of some other objects, is what we did above when we partitioned the integers into the two sets $[0]$ and $[1]$, based on a notion of equivalence determined by their remainders mod 2. Eager to return to this context, we mention only one critical result about equivalence classes in general, showing that this partitioning goes hand-in-hand with the equivalence relation.

Lemma 4.1.5

Let $\mathcal{R}$ be an equivalence relation on a set S. Then:

(1) Any two equivalence classes of $\mathcal{R}$ are either equal or disjoint. Specifically, for $a, b \in S$, we have $[a] = [b]$ if $(a, b) \in \mathcal{R}$, and $[a] \cap [b] = \emptyset$ otherwise.
(2) The union of all of the equivalence classes of the relation $\mathcal{R}$ on S is S.

That is, the various equivalence classes form a *partition* of S, a collection of non-empty mutually disjoint subsets of S whose union is equal to S. ◄

Proof If $[a] \cap [b] \neq \emptyset$, choose any $c \in [a] \cap [b]$. Then $(a, c), (b, c) \in \mathcal{R}$ and thus $(c, b) \in \mathcal{R}$ by symmetry, and finally $(a, b) \in \mathcal{R}$ by transitivity. Transitivity further implies that anything equivalent to a is equivalent to anything equivalent to b, and vice versa, so $[a] = [b]$. For the second claim, we note that each element $a \in S$ is an element of its own equivalence class $[a]$ (by reflexivity), so the union of all the equivalence classes contains all elements of S.   □

▶ **Remark 4.1.6** It is also true that any partition of S can be used to define an equivalence relation, namely, the equivalence relation which declares two elements to be equivalent if and only if they are in the same part of the partition. Thus, the two notions are roughly synonymous concepts, and we speak freely of taking a set and "partitioning it into equivalence classes."

4.2 The Ring of Integers Modulo n

In 1798, 21-year-old Carl Friedrich Gauss completed his landmark text *Disquisitiones Arithmeticae*[2] , revolutionizing the field of number theory by synthesizing the work of Fermat, Euler, and others, and laying the groundwork for algebraic number theory. Of particular interest here is that contained in *Disquisitiones* is an invitation to explore the world of modular remainders. The story begins with Gauss's notion of modular congruence:

Definition 4.2.1 (Modular Congruence)

Given a natural number n, we say integers a and b are **congruent modulo** n, denoted $a \equiv b \pmod{n}$, if $a \bmod n = b \bmod n$, i.e., they have the same remainder when divided by n. The equivalence classes with respect to this relation are called **congruence classes** and denoted

$$[a] = \{b \in \mathbb{Z} : a \equiv b \pmod{n}\}.$$

It is common to use the notation $[a]_n$ in place of simply $[a]$ when the modulus n needs emphasizing, but the subscript is otherwise typically left off. We note the equivalent statements

$$[a]_n = [b]_n \iff a \equiv b \pmod{n} \iff n \mid a - b. \qquad ◀$$

Gauss chose the symbol $\equiv$ for modular congruence to emphasize the similarity between modular congruence and equality, and modern mathematics formalizes this

---

[2] Pronounced "Gauss's book"

as the statement that congruence, like equality, defines an equivalence relation on $\mathbb{Z}$. It is fairly straightforward to check the details: every number is congruent to itself modulo n, congruence is clearly symmetric, and transitivity follows from a brief calculation: if $a \equiv b \pmod{n}$ and $b \equiv c \pmod{n}$, then $n \mid a - b$ and $n \mid b - c$, so $n \mid (a - b) + (b - c) = a - c$. It was Gauss's student Richard Dedekind who embedded congruence into the language of equivalence relations and equivalence classes. Having done so, the definitions and results of the last section thus come into play, e.g., the results of Lemma 4.1.5 that these congruence classes are disjoint and make up all of $\mathbb{Z}$.

▶ **Example 4.2.2** Consider congruence modulo 7 on $\mathbb{Z}$, where two integers are congruent if they have the same remainder when divided by 7. Since there are only 7 possible remainders when divided by 7, there are 7 disjoint equivalence classes of $\mathbb{Z}$ under this equivalence relation: $[0], [1], [2], [3], [4], [5]$, and $[6]$, where $[a]$ consists of those integers n such that $n \bmod 7 = a$. The Division Algorithm confirms the results of Lemma 4.1.5 in this case: every integer is in exactly one of these 7 classes. Of course, we can refer to an equivalence class by any of its other elements as well (other *representatives* of the equivalence class). For example, we have

$$\cdots = [-9] = [-2] = [5] = [12] = [19] = \cdots .$$

Dedekind's contribution of considering these equivalence classes as the appropriate objects of study marks an important step in the development of the subject and suggests that we should develop notions of arithmetic with them. Enter, once again, the language of rings.

Definition 4.2.3

For an integer $n > 1$, the **ring of integers modulo** n, $\mathbb{Z}/(n)$, is defined[3] by

$$\mathbb{Z}/(n) = \{[0], [1], \ldots, [n - 1]\},$$

where $[a]$ denotes the equivalence class of a modulo n, i.e., the set of integers that have the same remainder as a when divided by n. We define addition and multiplication in this ring by setting

$$[a] + [b] = [(a + b) \bmod n]$$

and

$$[a] \cdot [b] = [ab \bmod n]. \qquad \blacktriangleleft$$

---

[3] Other notations for $\mathbb{Z}/(n)$ include $\mathbb{Z}/n\mathbb{Z}$, $\mathbb{Z}/n$, and $\mathbb{Z}_n$. You may encounter these in other books, or perhaps even in the current one, depending on our proof-reading skills.

Before getting too comfortable, the cautious reader will note we have been presumptuous in calling this a ring, as there remain some details to check, and so we turn to this now. First, a very important habit to develop when working with equivalence classes is to make sure that any time you define something by choosing a representative from an equivalence class, you wouldn't have gotten a different answer by choosing a different representative. Here, we need to check that the notions of addition and multiplication defined above are *well-defined*, that is, when evaluating $[a] \cdot [b]$ in $\mathbb{Z}/(n)$ by computing ab mod n, as directed by Definition 4.2.3, we could have instead chosen any other other integer $c \in [a]$ and $d \in [b]$ and computed cd mod n and obtained the same result (since in this case we have $[a] = [c]$ and $[b] = [d]$).

Lemma 4.2.4 (Well-definedness of Modular Arithmetic)

Let a, b, c, d be integers. If $a \equiv c \pmod{n}$ and $b \equiv d \pmod{n}$, then $a + b \equiv c + d \pmod{n}$, and $ab \equiv cd \pmod{n}$. Equivalently, in the language of congruence classes, if $[a] = [c]$ and $[b] = [d]$ in $\mathbb{Z}/(n)$, then

$$[a] + [b] = [c] + [d] \qquad \text{and} \qquad [a] \cdot [b] = [c] \cdot [d]. \qquad \blacktriangleleft$$

Proof If $a \equiv c \pmod{n}$ and $b \equiv d \pmod{n}$, then $n \mid (c - a)$ and $n \mid (d - b)$, so

$$n \mid (c - a) + (d - b) = (c + d) - (a + b),$$

showing that $a + b \equiv c + d \pmod{n}$. Multiplication is similar (Exercise 4.30). $\square$

Finally, it remains to check the ring axioms for $\mathbb{Z}/(n)$, and we deliberately do not belabor the point here. We have the additive identity $[0]$; the multiplicative identity $[1]$; additive inverses via $-[a] = [-a]$; and the associative, commutative, and distributive laws all follow directly from the corresponding laws in $\mathbb{Z}$, e.g.,

$$([a] + [b]) + [c] = [(a + b + c) \bmod n] = [a] + ([b] + [c]).$$

We will soon tackle the problem of doing algebra in these brand new rings, but first a quick summary of where we stand. We have at least three equivalent ways of saying that two integers a and b are "the same mod n":

$$a \bmod n = b \bmod n \iff a \equiv b \pmod{n} \iff [a] = [b] \text{ in } \mathbb{Z}/(n).$$

Observe that each of these three formulations involves the word "mod" in a slightly different way. In the first, "mod n" is used as a unary operator which inputs an integer a and outputs its remainder upon division by n. In the second, mod is used as an equivalence relation through which we can decide if two integers are "the same" enough for our purposes. Mirroring the distinction between divisibility and division, note that $37 \equiv 17 \pmod{20}$ is a true statement, whereas the expression $-15 \bmod 4$ equals 1. Finally, in $\mathbb{Z}/(n)$, the word "mod" appears in the name of the ring in which we are working and encourages us to think of these congruence classes as a new type of number operating with its own rules of arithmetic. Moving forward with our arithmetic goals, fluency in switching back and forth between these perspectives, both conceptually and notationally, will be of tremendous benefit.

Despite its rather lackluster appearance, Lemma 4.2.4 provides the single most important property of modular arithmetic: in any calculation involving mod-n addition or multiplication, we can replace any summand or factor with anything congruent to it mod n.

▶ **Example 4.2.5** Compute $39 \cdot 99 + 95 \cdot 64 \bmod 19$.

Solution First, observe that a naive solution technique certainly exists. We could compute, in $\mathbb{Z}$, the value of $39 \cdot 99 + 95 \cdot 64$, and then use long division to find the remainder when this result is divided by 19. Much more expedient is to employ Lemma 4.1.5, which shows us that the answer to this question remains unchanged if we replace any of 39, 99, 95, or 64 with any other congruent-mod-19 value. In particular, mental arithmetic provides for us

$$39 \equiv 1 \bmod 19, \quad 99 \equiv 4 \bmod 19, \quad 95 \equiv 0 \bmod 19, \quad \text{and} \quad 64 \equiv 7 \bmod 19,$$

and so making these substitutions, we find

$$39 \cdot 99 + 95 \cdot 61 \equiv 1 \cdot 4 + 0 \cdot 7 \equiv 4 \pmod{19}.$$

We conclude that $39 \cdot 99 + 95 \cdot 64 \bmod 19 = 4$. □

The general name of this process, where we perform arithmetic operations as usual except that we are free to replace components with their values mod n (or other congruent values), is referred to as *modular arithmetic*. The goal of modular arithmetic is typically to *reduce* an integer mod n, i.e., to replace it with its remainder modulo n. The example above could have instead asked us to reduce $39 \cdot 99 + 95 \cdot 64$ modulo 19, or in the language of equivalence classes, to find the unique integer r, $0 \le r < 19$, such that $[39 \cdot 99 + 95 \cdot 64] = [r]$ in $\mathbb{Z}/(19)$. The latter, being excessively cumbersome notationally, explains why we will typically work in $\mathbb{Z}/(n)$ using the language of modular congruence as opposed to the language of equivalence classes.

4.3 Reduce First and ask Questions Later

The single most important step in mastering modular arithmetic is observing that when a problem depends only on the congruence classes of the inputs (e.g., when we are asked for a mod-n remainder), we should as quickly as possible reduce the allowable inputs modulo n. Remember this philosophy as you see fit, either as the section title, or as the phrase *reduce then compute*[4] :

▶ **Example 4.3.1** The bread and butter calculations of modular arithmetic mirror those of Example 4.2.5, involving only addition and multiplication of integers. A streamlined version of this might look like

$$38 \cdot 17 + 40 \equiv 3 \cdot 3 + 5 \equiv 9 + 5 \equiv 2 + 5 \equiv 0 \pmod{7},$$

which shows that $7 \mid 38 \times 17 + 40$.

▶ **Example 4.3.2** Compute 17^3 mod 5.

Solution Since exponentiation is simply repeated multiplication, we find

$$17^3 \equiv 17 \cdot 17 \cdot 17 \equiv 2 \cdot 2 \cdot 2 \equiv 2^3 \equiv 3 \pmod{5},$$

so 17^3 mod $5 = 3$.   □

This last example highlights how the technique generalizes cleanly from reducing expressions involving only addition and multiplication to any polynomial expression with integer coefficients (as evaluating such a polynomial amounts to repeated applications of addition and multiplication).

Lemma 4.3.3

If $f \in \mathbb{Z}[x]$ and $a \in \mathbb{Z}$, then

$$f(a \bmod n) \equiv f(a) \pmod{n}$$

◀

As always, there are several equivalent ways of writing this relationship, including $[f(x)] = [f(x) \bmod n]$ or $f(x) \bmod n = f(x \bmod n) \bmod n$. The case $f(x) = x^k$ generalizes the content of Example 4.3.2, showing that we can reduce the base of an exponential before computing, furthering the scope of the "reduce then compute" mantra. For example, this shows that $10^n \equiv 1 \bmod 9$ for any n, and this one simple trick[5] is responsible for the famous mod-9 divisibility phenomenon.

---

[4] *Mod then multiply? arithmetical fluency from modular congruency*? Invent your own!

[5] Which, contrary to popular belief, mathematicians absolutely *do* want you to know!

▶ **Example 4.3.4** To compute 71,423 mod 9, we note

$$
\begin{aligned}
71{,}423 &= 7 \cdot 10^4 + 1 \cdot 10^3 + 4 \cdot 10^2 + 2 \cdot 10^1 + 3 \\
&\equiv 7 \cdot 1^4 + 1 \cdot 1^3 + 4 \cdot 1^2 + 2 \cdot 1^1 + 3 \\
&\equiv 7 + 1 + 4 + 2 + 3 \\
&\equiv 17 \\
&\equiv 8 \quad (\text{mod } 9).
\end{aligned}
$$

In words, any natural number is congruent mod 9 to the sum of its digits. The proof is so concise that it fits within the statement of the theorem.

Theorem 4.3.5 (Divisibility-by-9 Test)
If $n = a_k a_{k-1} \cdots a_2 a_1 a_0$ is the base-10 representation of n, then mod 9 we have

$$
\begin{aligned}
n &= a_k 10^k + a_{k-1} 10^{k-1} + \cdots + a_2 10^2 + a_1 10 + a_0 \\
&\equiv a_k 1^k + a_{k-1} 1^{k-1} + \cdots + a_2 1^2 + a_1 1 + a_0 \\
&\equiv a_k + a_{k-1} + \cdots + a_2 + a_1 + a_0 \quad (\text{mod } 9).
\end{aligned}
$$

In particular, $9 \mid n$ if and only if 9 divides the sum of its digits!

▶ **Remark 4.3.6** Since $10^k \equiv 1 \pmod 3$, the analogous test works for divisibility by 3 as well. In fact, it's not hard to see that such a divisibility-by-n test exists *if and only if* $10 \equiv 1 \pmod n$. In turn, this holds if and only if $n \mid 10 - 1$, i.e., only when $n = 3$ or $n = 9$ (fine, fine, and $n = 1$). One more clean divisibility test is worth mentioning (Exercise 4.2), coming from the observation that $10^k \equiv (-1)^k \bmod 11$.

A potent use of modular arithmetic is to take an expression and reduce it modulo various bases, i.e., to think of the modulus n as a variable and see how the reductions change as n varies. As a first example of this, recall from Chapter 2 that the Diophantine equation $x^2 + y^2 = 5$ has several integer solutions, whereas $x^2 + y^2 = 3$ does not. While we could deduce the latter statement by exhaustive search, this technique would certainly be too burdensome for an equation like $x^2 + y^2 = 10000003$. On the other hand, it is completely resolved by a mod-4 calculation.

Lemma 4.3.7

The Diophantine equation $a^2 + b^2 = n$ has no solutions if $n \equiv 3 \pmod 4$. ◀

Proof We consider the possible values of $a^2 + b^2$ mod 4, which in turn are constrained by the possible value of *any* square mod 4. We compute

$$[0]^2 = [0] \quad [1]^2 = [1] \quad [2]^2 = [0] \quad [3]^2 = [1],$$

which shows, since any integer is in one of those four congruence classes, that any integer squared is congruent to either 0 or 1 mod 4. Thus, since a^2 mod 4 and b^2 mod 4 are both either 0 or 1, it could not be that $a^2 + b^2 \equiv 3$ mod 4. □

With practice, impossibility arguments for Diophantine equations become fluent dismissals: "The equation $x^2 + y^2 = 10000003$ has no solutions since it reduces mod 4 to $x^2 + y^2 \equiv 3$, which has no solutions since the squares mod 4 are 0 and 1." It is easy to generate a slew of Diophantine equations with no solutions using this technique:

- The equation $2x^2 + 4y^2 = n$ has no solutions in $\mathbb{Z}$ if n is odd.
- The equation $x^2 + 2y^2 = n$ has no solutions in $\mathbb{Z}$ if $n \equiv 5 \pmod{8}$.
- The equation $x^3 + 7y^5 = n$ has no solutions in $\mathbb{Z}$ if $n \equiv 4 \pmod{7}$.

We can verify these claims by reducing the equations mod 2, 8, and 7, respectively, and considering the possible values of each term with respect to those moduli. For example, when $n \equiv 4 \pmod{7}$, the equation $x^3 + 7y^5 = n$ reduces modulo 7 to $x^3 + 0y^5 \equiv 4 \pmod{7}$, and we can check by hand that no element $[x]$ of $\mathbb{Z}/(7)$ satisfies $[x]^3 = [4]$.

The culmination of this line of thinking is given in the following.

Definition 4.3.8

Given a polynomial $f \in \mathbb{Z}[x]$, we obtain the **mod-n reduction** of that polynomial $\overline{f} \in \mathbb{Z}/(n)[x]$ by reducing all of its coefficients modulo n. ◄

By convention, when we reduce a polynomial mod n we often suppress the formal equivalence class notation for its coefficients. For example, taking $f(x) = x^3 + 12x + 9$, since 12 mod 5 = 2 and 9 mod 5 = 4, the reduction of f mod 5 is written $\overline{f}(x) = x^3 + 2x + 4$ (rather than $x^3 + [2]x + [4]$). Note that we can perform this process analogously for polynomials in any number of variables, so we can generally make sense of "reducing a Diophantine equation modulo n."

▶ **Example 4.3.9** If E is the elliptic curve $y^2 = x^3 + 16x - 22$, reducing E mod 7 gives us the reduced curve $y^2 = x^3 + 2x + 6$. In looking for the set of solutions $E(\mathbb{Z}/(7))$, we start by examining the possible values of $y^2 \in \mathbb{Z}/(7)$,

$$[0]^2 = [0], \quad [1]^2 = [6]^2 = [1], \quad [2]^2 = [5]^2 = [4], \text{ and } [3]^2 = [4]^2 = [2].$$

Thus, we need to find values of $x \in \mathbb{Z}/(7)$ such that $x^3 + [2]x + [6] \in \{[0], [1], [2], [4]\}$. Since $\mathbb{Z}/(7)$ is pretty small, we can just work our way through the possibilities. For $x = [0], [1], \ldots, [6]$, we get $x^3 + [2]x + [6]$ is equal to $[6], [2], [4], [4], [1], [1]$, and $[3]$, respectively. Matching these outputs with values of y^2 we find there are no solutions with $x = [0]$ or $[6]$, and all other values of x yield two solution points $(x, y) \in E(\mathbb{Z}/(7)) = \{([1], [3]), ([1], [4]), ([2], [2]),$ $([2], [5]), ([3], [2]), ([3], [5]), ([4], [1]), ([4], [6]), ([5], [1]), ([5], [6])\}$ for a total of 10 points (out of 49 possible in $\mathbb{Z}/(7) \times \mathbb{Z}/(7)$).

By Lemma 4.3.3, the existence of a solution to a Diophantine equation implies the existence of a solution to each of its mod-n reductions, and so we end up with an elegant and powerful test for Diophantine equations:

Theorem 4.3.10 (Mod-n Root Test)
If a Diophantine equation has a solution in $\mathbb{Z}$, then it has a solution in $\mathbb{Z}/(n)$ for each $n \in \mathbb{N}$. The contrapositive is even more striking: if there exists even *one* $n \in \mathbb{N}$ for which the reduced equation has no solution in $\mathbb{Z}/(n)$, then the equation has no integer solutions.

The converse, however, eludes our grasp. It is *not* true that having a solution to a reduced equation mod n guarantees us the existence of a solution to the original equation in $\mathbb{Z}$. It is not even true that having a solution to an equation modulo *every* natural number n guarantees us such a solution. We will have to work diligently to resolve this state of affairs over the upcoming chapters, with some milestone results in this direction occurring in Chapter 8. In the meantime, the power of "reducing equations mod n" suggests that it is in our interest to better understand the solving of equations (or the impossibility thereof) in the ring $\mathbb{Z}/(n)$. As such, the remainder of the chapter largely focuses on expanding our understanding of the operations of arithmetic in the setting of $\mathbb{Z}/(n)$ beyond those of addition and multiplication, namely, to include those of division, exponentiation, and factorials.

Exploration E

Units Mod n ◄

The Python worksheet "Units mod *n*" provides an outline for this Exploration and is available from the companion website. Pending theoretical developments that will answer questions like those below for all *n*, for small values of *n* we can proceed by brute-force calculation.

Recall that an element $[a] \in \mathbb{Z}/(n)$ is a unit if and only if $[a] \mid [1]$, i.e., if we can solve the equation
$$ax \equiv 1 \bmod n.$$

E.1 Find the units of $\mathbb{Z}/(5)$, $\mathbb{Z}/(6)$, $\mathbb{Z}/(7)$, $\mathbb{Z}/(8)$, $\mathbb{Z}/(9)$, and $\mathbb{Z}/(10)$.

E.2 Make a conjecture about which elements of $\mathbb{Z}/(n)$ are units. For which values of *n* would your conjecture imply that $\mathbb{Z}/(n)$ a field? Prove that your conjecture is correct if you replace "field" with "integral domain."

E.3 For $2 \leq n \leq 8$, divide the elements of $\mathbb{Z}/(n)$ into zero, units, primes, and composites. Come up with some conjectures.

E.4 For $3 \leq n \leq 8$, what is the sum of all the elements in $\mathbb{Z}/(n)$? Document and explain any patterns you notice.

E.5 For $3 \leq n \leq 8$, what is the product of all the non-zero elements of $\mathbb{Z}/(n)$? What patterns do you notice?

4.4 Division, Exponentiation, and Factorials in $\mathbb{Z}/(n)$

The ubiquity of division, exponentiation, and exclamation marks largely excuses us from having to justify the study of their modular analogs, but the behavior of these operations in the world of modular arithmetic is bizarre enough for us to cover them in some detail. This section is, therefore, an exploration of various ways that things work differently in the modular world, and the adjustments we need to make to acclimate to this setting. We begin with the notion of modular division, a concept necessary to solve linear congruences like the one below.

▶ **Example 4.4.1** Solve $3x \equiv 5 \pmod{11}$.

As always, we see that the setting of the problem is crucial. If this were an equality of real numbers for us to solve, we would solve for x by dividing both sides by 3 to get $x = \frac{5}{3}$. At the other extreme, if we consider the same equation $3x = 5$ in the ring of integers, we see there is no solution, as there is no integer x that when multiplied by 3 gives 5. Thus we can't *divide by* 3 to solve for x in $\mathbb{Z}$. This raises the question of division in an arbitrary ring, a question which we've fortunately already prepared ourselves to answer. Instead of "dividing by 3," we multiply both sides by 3^{-1}, the multiplicative inverse of 3. In general, we interpret division as just multiplication by the multiplicative inverse of an element when such an inverse exists, and not defined otherwise.

▶ **Remark 4.4.2** Note that if the ring is non-commutative, we must take care to multiply on the same side: $ax = b$ implies $a^{-1}ax = a^{-1}b$, not $a^{-1}ax = ba^{-1}$. The non-commutative case will not occupy much of our attention, but serves as explanation for the decision to consistently write "$a^{-1}b$" instead of "$\frac{b}{a}$" in most ring settings, as the fraction notation gives us the heebie-jeebies in a general ring.

Elements of rings with multiplicative inverses, and hence the elements we can divide by, are precisely the units of the ring. If $[3]$ is a unit in $\mathbb{Z}/(11)$ we can mirror the technique used for real numbers to solve $3x \equiv 5 \pmod{11}$. As it turns out, $[3]$ is indeed a unit in this setting since $3 \cdot 4 \equiv 1 \pmod{11}$, and thus $[4]$ is the multiplicative inverse of $[3]$. We can now proceed.

Solution (to Example 4.4.1) Multiplying both sides of $3x \equiv 5 \pmod{11}$ by 4, as $[3]^{-1} = [4]$ in $\mathbb{Z}/(11)$, solves the congruence much as we solved it in $\mathbb{R}$:

$$3x \equiv 5$$
$$4 \cdot 3x \equiv 4 \cdot 5$$
$$1x \equiv 4 \cdot 5$$
$$x \equiv 20 \equiv 9 \pmod{11}.$$

And indeed, we verify that $3 \cdot 9 \equiv 5 \pmod{11}$. Note that we can phrase our answer either in the language of $\mathbb{Z}/(11)$—that $x = [9]$ is the unique element of $\mathbb{Z}/(11)$

satisfying $[3]x = [5]$—or in the language of $\mathbb{Z}$, that the integers x such that $3x$ mod $11 = 5$ are those satisfying x mod $11 = 9$. $\square$

Which elements are units in the modular world? We've already noted two extremes on this front: units in $\mathbb{Z}$ and $\mathbb{Z}[i]$ are few and far between, whereas units comprise almost the entirety of $\mathbb{Q}$ and $\mathbb{R}$ (see the discussion after Definition 3.3.2). Where does the ring $\mathbb{Z}/(n)$ fall along this spectrum? The experimentation begun on Exploration E leads to a natural conjecture, which we'll now prove with a little help from the Fundamental Theorem of GCDs (Theorem 3.2.25).

Theorem 4.4.3
The element $[a] \in \mathbb{Z}/(n)$ has a multiplicative inverse if and only if $\gcd(a, n) = 1$.

Proof By Bézout's Identity, if $\gcd(a, n) = 1$, there exist x and y so that $ax + ny = 1$, and reducing this equation mod n gives $ax \equiv 1 \pmod{n}$. That is, $[x] = [a]^{-1}$ provides a multiplicative inverse. Conversely, if $ax \equiv 1 \pmod{n}$, then by definition of modular congruence $n \mid 1 - ax$, and so $1 - ax = ny$ for some $y \in \mathbb{Z}$, giving $ax + ny = 1$. Thus by the Fundamental Theorem of GCDs, $\gcd(a, n) = 1$. $\square$

The proof of Theorem 4.4.3, in addition to proving the existence of a multiplicative inverse, also provides a recipe for finding one: finding $[a]^{-1}$ amounts to solving the linear Diophantine equation $ax + ny = 1$ and then taking $[a]^{-1} = [x]$ in $\mathbb{Z}/(n)$. Happily, the Extended Euclidean Algorithm already provides us a very efficient means for doing just this.

▶ **Example 4.4.4** Find $[7]^{-1}$ in $\mathbb{Z}/(93)$.

In Example 3.2.19, an application of the Extended Euclidean Algorithm provided that $\gcd(93, 7) = 1$ with the explicit linear combination $1 = 40 \cdot 7 + (-3) \cdot 93$. Thus $40 \cdot 7 \equiv 1 \pmod{93}$, and so $[7]^{-1} = [40]$.

On the spectrum from $\mathbb{Z}$ and $\mathbb{Z}[i]$ (having almost no units) to $\mathbb{Q}$ and $\mathbb{R}$ (comprised almost exclusively of units), Theorem 4.4.3 places a generic $\mathbb{Z}/(n)$ somewhere in between, and it is intriguing to see how this placement varies with n. In $\mathbb{Z}/(8)$, for example, precisely half of the elements ($[1]$, $[3]$, $[5]$, and $[7]$) have multiplicative inverses, whereas in $\mathbb{Z}/(11)$, *every* non-zero element has a multiplicative inverse, since all of the integers in the set $\{1, 2, \ldots, 10\}$ are relatively prime to 11. This extreme case, that every non-zero element is a unit, is captured in the following corollary.

Corollary 4.4.5

For a natural number $n > 1$, the following statements are equivalent:

  (i) n is prime.
 (ii) $\mathbb{Z}/(n)$ is a field.
(iii) $\mathbb{Z}/(n)$ is an integral domain.

◀

Proof The implication (i) $\Rightarrow$ (ii) is a direct consequence of Theorem 4.4.3: if n is prime, any non-zero $[a] \in \mathbb{Z}/(n)$ is represented by an integer a, $1 \le a < n$, which is necessarily relatively prime to n, and so is a unit. The implication (ii) $\Rightarrow$ (iii) follows for any ring (Lemma 3.1.11), and for (iii) $\Rightarrow$ (i), proceeding via the contrapositive, if n were not prime, then $n = ab$ for some $a, b < n$, and so $[a][b] = 0 \in \mathbb{Z}/(n)$. $\square$

▶ **Remark 4.4.6** Also in Lemma 3.1.11 is the observation that not every integral domain is a field (e.g., $R = \mathbb{Z}$). However, while we do not prove it here, it *is* true that for rings with finitely many elements, the two notions are equivalent, and the above Corollary is a special case of this result.

Next, a quick appetizer question before the main course:

▶ **Question 4.4.7** What is the sum of all of the elements of $\mathbb{Z}/(n)$?

Experimentation (like that done in Problem E.4) quickly gives a result of $[0]$ when n is odd and $[n/2]$ if n is even. This is readily explained by the fact that each element in $\mathbb{Z}/(n)$ pairs with its additive inverse to contribute $[0]$ to the sum, *except* for the term $[n/2]$ in $\mathbb{Z}/(n)$ when n is even. This element is its own additive inverse, and thus it is left unpaired, leaving it to contribute $[n/2]$ to the sum. While the additive version of this problem reveals no deep secrets, the multiplicative analog is much more intriguing.

▶ **Question 4.4.8** What is the product of all of the non-zero elements in $\mathbb{Z}/(n)$?

Let us first restrict our attention to the case where $n = p$ is a prime (the composite case, Exercise 4.49, being somewhat less interesting). In this case, we can employ the same trick we did with additive inverses: since each element of $\mathbb{Z}/(p)$ has a multiplicative inverse, we can multiply them in pairs, with the one caveat that, just as in addition, we have to know in advance which elements are their own multiplicative inverse. A quick lemma dispatches this issue entirely.

Lemma 4.4.9

Let p be prime. Then if $a \in \mathbb{Z}$ satisfies $[a] = [a]^{-1} \in \mathbb{Z}/(p)$, then $a \equiv \pm 1$ (mod p). ◄

Proof Given $a \in \mathbb{Z}$, suppose $[a] = [a]^{-1}$, or in other words, $[a]^2 = [1]$. Casting this equality in terms of modular congruence and rearranging, we have $0 \equiv a^2 - 1 \equiv (a+1)(a-1)$ (mod p). Thus $p \mid (a+1)(a-1)$, and so by the Prime Divisor Property, either $p \mid a+1$ or $p \mid a-1$; i.e., $a \equiv \pm 1$ mod p.   □

▶ **Remark 4.4.10** Alternative proof: ± 1 are clearly solutions to $x^2 - 1 \equiv 0$ (mod p). Further, since $\mathbb{Z}/(p)$ is an integral domain, the polynomial $x^2 - 1 \in \mathbb{Z}/(p)[x]$ has at most two roots by Lemma 3.1.13, so $[1]$ and $[-1]$ are the *only* two solutions in $\mathbb{Z}/(p)$.

As with many results of this ilk, the previous lemma fails horribly (or wonderfully, depending on your perspective) when n is not prime. In $\mathbb{Z}/(8)$, for example, all of 1, 3, 5, and 7 satisfy $x^2 \equiv 1$ (mod 8), and thus are each their own inverse. This also finally provides a good counter-example to Lemma 3.1.13 in the case the ring is not an integral domain (as $\mathbb{Z}/(8)$ is not) since the degree-two polynomial $x^2 - 1 \in \mathbb{Z}/(8)[x]$ has four roots.

The ability to pair off elements when n is prime allows us to answer Question 4.4.8, which we can interpret as a statement about factorials mod p.

Theorem 4.4.11 (Wilson's Theorem)
Let p be a prime number. Then

$$(p-1)! \equiv -1 \quad (\text{mod } p).$$

Proof By Lemma 4.4.9 and Corollary 4.4.5, every element in the product

$$(p-1)! \equiv 1 \cdot 2 \cdot 3 \cdots (p-1) \text{ mod } p$$

has a multiplicative inverse that also appears elsewhere in the product except for 1 and $(p-1)$, which are their own inverses. Thus, the factors cancel in pairs with their inverses and we are left with $1 \cdot (p-1) \equiv -1$ (mod p).   □

It is occasionally worth pausing to observe how far we've come. For example, it is now a trivial calculation for us that the number 100!—an unfathomably large number—is just one less than a multiple of 101 by Wilson's Theorem. Let's take stock of where we stand with respect to doing modular arithmetic with exponents.

▶ **Example 4.4.12** Reduce 11^{62} mod 5.

We can multiply either before or after reducing mod 5, so could *either*:

(a) Compute the decimal expansion of 11^{62} first, and then do long division to compute its remainder when divided by 5 or
(b) Remember to reduce first and ask questions later, in which case we note $11^{62} \equiv 1^{62} \equiv 1 \pmod 5$.

Other solution trajectories might exist for such problems, and the method one chooses is often, efficiency aside, a matter of personal taste. Troubling, however, is that option (a) above exists for any exponent computation, whereas option (b) depended heavily on a happy coincidence between the base of the expression and the modulus, namely, that the base was one more than a multiple of the modulus. So the existence of option (b) provides little relief in the case that we are faced with a similar expression in which no such coincidence occurs.

▶ **Example 4.4.13** Reduce 11^{62} mod 37.

Let's discuss some solution options, sorted from most exhausting to most inspired[6] and consider how long they would take to work through by hand.

- Compute the integer 11^{62} and then divide by 37, finding the remainder.
  Estimated time to complete: Several hours.
- We could prevent the appearance of astronomical numbers by iteratively multiplying together 62 factors of 11 and reducing mod 37 after each one:

n	11^n mod 37
1	11
2	$11^2 = 121 \equiv 10$
3	$11^3 = 11^2 \cdot 11 \equiv 10 \cdot 11 \equiv 36$
4	$11^4 = 11^3 \cdot 11 \equiv 36 \cdot 11 \equiv 26$
$\vdots$	$\vdots$
62	$11^{62} = 11^{61} \cdot 11 \equiv$???

Estimated time to complete: The better[7] part of an hour.
- Even better, we realize we don't need *all* of the 11^n for $1 \leq n \leq 62$: To minimize the number of such products needed, we could just keep squaring, computing in turn 11^2, 11^4, 11^8, etc., then we could just observe that

$$11^{62} = 11^{32} \cdot 11^{16} \cdot 11^8 \cdot 11^4 \cdot 11^2$$

---

[6] From clueless to clueful? inept to outept?

[7] Or really, worse

and compute this product mod 37. This is known as the *binary exponentiation* method ("binary" since we have used the binary expansion $62 = 2^5 + 2^4 + 2^3 + 2^2 + 2^1$), and in general it requires relatively few multiplications.

Estimated time to complete: Probably less time than it took to understand the method in the first place.

- Finally, even better, we note by hook or by crook that $11^6 \equiv 1 \pmod{37}$, and then reason that

$$11^{62} \equiv (11^6)^{10} \cdot 11^2 \equiv 1^{10} \cdot 10 \equiv 10 \pmod{37}.$$

Estimated time to complete: Literally already done, so about negative five seconds at this point.

We're left with the problem of trying to jump directly to that last, most inspired, approach. Where did that magical number 6 come from? In fact, we weren't far off in one of our more naive attempts: in our table above we had calculated that $11^3 \equiv 36$ mod 37, but had neglected at the time to notice that $36 \equiv (-1) \pmod{37}$, and thus that $11^3 \equiv -1 \pmod{37}$. Squaring both sides of this gives the magic number, $11^6 \equiv (-1)^2 \equiv 1$ mod 37. Indeed, had we gone through just a couple more rows in that table, we would have seen this number appear without any "magic" needed at all. The goal now becomes clear: we need a mechanism for predicting in advance, with as little computation as possible, when and how these magic exponents will arise. It is to this topic we turn next.

Exploration F

Units, Exponents, and Orders ◄

For $[a] \in \mathbb{Z}/(n)$, the least positive exponent k such that $a^k \equiv 1 \bmod n$ is called the *order of a mod n* (if such a k exists).

F.1 Find the order of each element of $\mathbb{Z}/(7)$.

F.2 Explain why if $[a] \in \mathbb{Z}/(n)$ has an order, then $[a]$ must be a unit. Find the order of each unit of $\mathbb{Z}/(n)$ for each n, $4 \leq n \leq 10$.

F.3 Use the previous problem for the following computations:

$$2^{76} \bmod 7 \qquad 5^{40} \bmod 9 \qquad 3^{215} \bmod 8.$$

Finally, find the last digit of 7^{169}.

F.4 Find the order of $[2] \in \mathbb{Z}/(17)$ and compute $2^{165} \bmod 17$.

F.5 Let $\varphi(n)$ denote the number of units in $\mathbb{Z}/(n)$. Compute $\varphi(n)$ for $4 \leq n \leq 10$. What is $\varphi(p)$ for a prime p? How could we compute $\varphi(p^2)$? $\varphi(p^3)$? $\varphi(pq)$ for distinct primes p and q?

Given the importance of the equation $a^k \equiv 1 \bmod n$, it is worth considering the analogous equation for the additive identity: an element $[a] \in \mathbb{Z}/(n)$ is *nilpotent* if $a^k \equiv 0 \bmod n$ for some $k \in \mathbb{Z}$. For example, $[9]$ is nilpotent in $\mathbb{Z}/(27)$ since $9^2 \equiv 0 \bmod 27$.

F.6 Find all nilpotent elements of $\mathbb{Z}/(8)$, of $\mathbb{Z}/(7)$, and of $\mathbb{Z}/(12)$. For which n do there exist nilpotent elements in $\mathbb{Z}/(n)$? How do you find them? Could a unit be nilpotent?

4.5 Group Theory and the Ring of Integers Modulo n

As we saw in the last section, finding a positive integer k such that $x^k \equiv 1 \pmod{n}$ allows us to perform modular exponentiation very efficiently. But does such a number k always exist, and when it does, how do we find it? The key to this question lies, once again, in the underlying algebraic structure. Sadly, the set of units of a ring do not themselves form a ring, for while it is true that the product of two units is a unit, it is not true that the sum of two units is a unit (e.g., in $\mathbb{Z}$, one can verify that $1 + 1 = 2$). Instead, we introduce another algebraic structure, this one requiring only one working operation.

Definition 4.5.1

A **group** is a set G with a single binary operation $\cdot$ (or $+$, $*$, $\times$, etc.) that satisfies the following properties:

(1) Closure: for all $g, h \in G$, $g \cdot h \in G$.
(2) Associativity: $g \cdot (h \cdot k) = (g \cdot h) \cdot k$ for all $g, h, k \in G$.
(3) Identity: there is an identity element $e \in G$ satisfying $e \cdot g = g \cdot e = g$ for all $g \in G$.
(4) Inverses: for each $g \in G$, there is an inverse element g^{-1} such that $g \cdot g^{-1} = g^{-1} \cdot g = e$.

If, in addition, the operation is commutative (i.e., $g \cdot h = h \cdot g$ for all $g, h \in G$), then we call the group **abelian**[8] . The number of elements in the set G is called the **order of the group** and is denoted $|G|$. ◀

As with rings, it will take some time to get used to working in groups, though our experience with rings will surely speed things up a little. For example, a ring could be thought of as an abelian group under the operation $+$ that just happens to also have an associative multiplication that distributes over addition. Similarly, a field is precisely a commutative ring in which the non-zero elements form an abelian group under multiplication (Exercise 4.52). The principal difference between groups and rings lies in the number of operations, and a notational subtlety arises depending on whether we think of that one operation as an additive one or a multiplicative one. If additive, we would use the notation $+$ for the operation and adopt the symbols $e = 0$ for the identity and $-a$ for the inverse of a. If multiplicative, we use $\times$ or $\cdot$ for the operation and the symbols $e = 1$ for the identity and a^{-1} for the inverse of a. While we encounter additive groups periodically, our principal groups of interest (units in a ring) are all multiplicative, so we will adopt the multiplicative conventions for dealing with groups generally and encourage readers to think through the additively written analogs on your own.

---

[8] This unusual word stems from the name of 19th-century Norwegian mathematician Niels Abel, a pioneer of group theory. It has been remarked that it is among the highest of mathematical honors to have one's name terminologically enshrined as a lower-case adjective.

> **Theorem 4.5.2**
> Let R be a ring with unity. Then the set of units forms a multiplicative group called the **group of units of** R, denoted $R^\times$.

We spent much of Chapter 3 reasoning through the units in various rings, so we have a handy supply of groups at our fingertips.

▶ **Example 4.5.3** We have the following groups of units:

- $\mathbb{R}^\times = \mathbb{R} - \{0\}$
- $\mathbb{Q}^\times = \mathbb{Q} - \{0\}$
- $\mathbb{C}^\times = \mathbb{C} - \{0\}$,

- $\mathbb{Z}^\times = \{\pm 1\}$.
- $\mathbb{Z}[i]^\times = \{\pm 1, \pm i\}$.
- $\mathbb{R}[x]^\times = \{\text{non-zero constants}\}$.

But before we get too far ahead of ourselves, let's prove Theorem 4.5.2 that these sets of units form multiplicative groups.

Proof Let R be a ring with unity, with multiplicative identity 1, and let $R^\times$ be the set of units of R. We show that R is a group (using the multiplication of R as its operation). First, $1 \in R^\times$ and is clearly the identity of $R^\times$. Multiplication in $R^\times$ is associative because R is a ring (and one of the axioms of a ring is associativity of multiplication). The inverse of a unit of R is also a unit of R, and so $R^\times$ contains a multiplicative inverse for each element. All that remains to show is closure, that the product of two units in R is again a unit in R. Given units u_1, u_2 of R, we claim that $u_1 u_2$ is again a unit, with inverse $u_2^{-1} u_1^{-1}$. Indeed we compute, using associativity, that

$$(u_1 u_2)(u_2^{-1} u_1^{-1}) = u_1 (u_2 u_2^{-1}) u_1^{-1} = u_1 1 u_1^{-1} = u_1 u_1^{-1} = 1,$$

and likewise $(u_2^{-1} u_1^{-1})(u_1 u_2) = 1$. Thus, $u_1 u_2$ is a unit and an element of $R^\times$, and we conclude that $R^\times$ is a group.   □

Most important for our purposes, with the specific goal of mastering modular exponentiation in mind, will be the group $\mathbb{Z}/(n)^\times$.

▶ **Example 4.5.4** By Theorem 4.4.3, $\mathbb{Z}/(n)^\times$ consists of those $[a] \in \mathbb{Z}/(n)$ with $\gcd(a, n) = 1$. For example:

$\mathbb{Z}/(6)^\times = \{[1], [5]\}$ $\mathbb{Z}/(7)^\times = \{[1], [2], [3], [4], [5], [6]\}$
$\mathbb{Z}/(10)^\times = \{[1], [3], [7], [9]\}$ $\mathbb{Z}/(11)^\times = \{[1], [2], [3], [4], [5], [6], [7], [8], [9], [10]\}$.

To see how our modular arithmetic questions are housed naturally in the language of group theory, we note that we can make sense of exponentiation in any (multiplicatively written) group G by defining

$$g^n = \underbrace{g \cdot g \cdots g}_{n \text{ times}}$$

for positive n, $g^0 = 1$, and for negative n,

$$g^n = \underbrace{g^{-1} \cdot g^{-1} \cdots g^{-1}}_{|n| \text{ times}}.$$

One readily checks that most standard rules of exponentiation hold, e.g.,

$$g^m \cdot g^n = g^{m+n} \qquad \text{and} \qquad (g^m)^n = g^{mn}$$

for any $m, n \in \mathbb{Z}$. This allows for the following lemma generalizing the key step in our cleverest approach to modular exponentiation.

Lemma 4.5.5

Let G be a group and let $g \in G$. If $g^b = 1$ for some positive integer b, then for any integer k we have

$$g^k = g^{k \bmod b}. \qquad \blacktriangleleft$$

Proof By the Division Algorithm we can write $k = bq + r$ with $0 \le r < b$; thus $r = k \bmod b$. Then

$$g^k = g^{bq+r} = (g^b)^q \cdot g^r = 1^q \cdot g^r = g^{k \bmod b},$$

as desired. $\qquad \square$

While our focus is on the group of units $\mathbb{Z}/(n)^\times$, it is worth emphasizing that this lemma highlights the importance of modular arithmetic in doing algebraic operations in *any* group.

▶ **Example 4.5.6** Since $i^4 = 1$ and $(-1)^2 = 1$, Lemma 4.5.5 provides our well-known rules

$$i^k = i^{k \bmod 4} \qquad \text{and} \qquad (-1)^k = (-1)^{k \bmod 2}$$

in the group $\mathbb{Z}[i]^\times$. The lemma can be viewed as a vast generalization of these patterns.

Note that Lemma 4.5.5 refers to *any* positive integer b such that $g^b = 1$. The *minimal* such exponent, the "magic number" of the previous section, is called the **order** of g.

Definition 4.5.7

Let G be a group. The **order of an element** $g \in G$, denoted $|g|$, is the smallest positive integer k such that $g^k = 1$. If no such element exists, we say g has infinite order and write $|g| = \infty$. ◄

The use of "order" for both the number of elements in a group and a parameter defined for each element of the group is less questionable[9] than it seems at first glance, as we'll see in Section 4.6.

Note that if there exists any natural number exponent b such that $g^b = 1$ then any multiple of b also has this property (and thus there must exist infinitely many such exponents). The converse of this is true as well, in that all exponents such that $g^b = 1$ are multiples of the smallest one, which must exist by the Well-Ordering Principle.

Lemma 4.5.8

Given a group G and an element $g \in G$, suppose $g^b = 1$ for some positive integer b. Then b is a multiple of $|g|$. ◄

Proof It is clear that $|g| \leq b$ since $|g|$ is by definition the *least* positive exponent such that $g^{|g|} = 1$. For divisibility, we observe that $g^{b \bmod |g|} = 1$ by Lemma 4.5.5, which contradicts that $|g|$ was the smallest positive such exponent unless $b \bmod |g| = 0$. A slightly more general version of the lemma is left to Exercise 4.41.   □

This brings up an interesting analogy between the rings $\mathbb{Z}/(n)$ and $\mathbb{C}$. Our intuitive response to the goal of solving $a^b = 1$, formed from many years of exposure to the real numbers, would have us do something like "take log base a of both sides." In the real setting, we would find $b = \log_a(1) = 0$, and that is indeed the unique value of b for which that equation holds. Typically, one of the major upheavals in your understanding of the mathematical universe is that this argument becomes more intricate in the world of the complex numbers. For example, solving the equation $e^x = 1$ (with $e \approx 2.718\ldots$ representing the base of the natural logarithm) gives not only $x = 0$, but an *infinitude* of complex solutions: $x = 2k\pi i$ for all $k \in \mathbb{Z}$. In this sense, exponentiation in $\mathbb{Z}/(n)$ is much more like exponentiation in $\mathbb{C}$ than exponentiation in $\mathbb{R}$—in fact, one perspective from this discussion is that it is the *real* numbers that are the "odd ring out" when it comes to exponentiation being injective.

Creeping back toward modular exponentiation, we have argued in Exploration F that elements of $\mathbb{Z}/(n)$ that have a finite order must be units (since $x^k = 1$ implies $x \cdot x^{k-1} = 1$, providing the inverse $x^{-1} = x^{k-1}$). The converse, that every unit has finite order, is a property of every finite group (and so, in particular, of each $\mathbb{Z}/(n)^\times$).

---

[9] Not *not* questionable, just *less* questionable.

Lemma 4.5.9

In a finite group G, every element has finite order. ◄

Proof Let $g \in G$ and consider the sequence

$$1, \, g, \, g^2, \, g^3, \, \ldots$$

Since G is both closed under the group operation and finite, this infinite list of elements of G can't all be different, so there must be some $g^m = g^n$ with $m > n$. Multiplying both sides by g^{-n} gives $g^{m-n} = 1$, so g has order at most $m - n$. □

With sufficient group-theoretic background now at our disposal, we can return to the exclusive context of modular arithmetic, i.e., working in the family of groups $\mathbb{Z}/(n)^{\times}$. We have quite a few outstanding questions about such groups, and perhaps none more obviously lacking an answer than that of figuring out how big these groups are. While a complete computation for $|\mathbb{Z}/(n)^{\times}|$ is still pending, we note that we have already come across notation for this exact quantity in Exploration F, with some special cases already resolved.

Definition 4.5.10

Euler's totient[10] **function** is defined for integers $n > 1$ by $\varphi(n) = |\mathbb{Z}/(n)^{\times}|$, that is, the number of units of $\mathbb{Z}/(n)$. We also take $\varphi(1) = 1$. ◄

Of course, by Theorem 4.4.3, this is just the number of integers $0 \le a < n$ with $\gcd(a, n) = 1$. The totient function is also often referred to as the Euler φ-function (Euler phi function).

▶ **Example 4.5.11** From Example 4.5.4, we see that $\varphi(6) = 2, \varphi(7) = 6, \varphi(10) = 4$, and $\varphi(11) = 10$. In general, for a prime p, by Theorem 4.4.3 we have $\varphi(p) = p - 1$.

More generally, we compute the following.

Lemma 4.5.12

Consider distinct primes p and q, and an integer $k \ge 1$. Then

(i) $\varphi(p^k) = p^k - p^{k-1}$;
(ii) $\varphi(pq) = (p - 1)(q - 1)$.

◄

---

[10] The word "totient" is an archaic one, but is a relative of "totting" and "totalling," as in "After totting up all the units mod n, we found there were $\varphi(n)$ of them."

Table 4.1 Unit Fractions and Orders of 10 in $\mathbb{Z}/(n)^{\times}$

n	3	7	9	11	13	17
$\frac{1}{n}$	$0.\overline{3}$	$0.\overline{142857}$	$0.\overline{1}$	$0.\overline{09}$	$0.\overline{076923}$	$0.\overline{0588235294117647}$
$\|[10]\|$	1	6	1	2	6	16

Proof Since $\gcd(a, p^k) = 1$ if and only if $\gcd(a, p) = 1$, computing $\varphi(p^k)$ involves counting the number of integers from 0 to $p^k - 1$ that are not multiples of p. Since there are precisely p^{k-1} numbers in that range that *are* multiples of p, the answer is $p^k - p^{k-1}$. Similarly, to compute $\varphi(pq)$, we begin with the list of integers from 0 to $pq - 1$ from which we must remove the multiples of p and of q. We first remove the p multiples of q in that range and then remove the remaining $(q - 1)$ multiples of p (only $q - 1$ since we have already removed the multiple $0p = 0$), leaving

$$\varphi(pq) = pq - p - (q - 1) = pq - p - q + 1 = (p - 1)(q - 1)$$

elements in the list.   □

We will continue to work on evaluating $\varphi(n)$, but the second major outstanding question about the groups $\mathbb{Z}/(n)^{\times}$ is the orders of their elements. Quite generally, given a and n we can ask for the order of a mod n. While we typically fix n and consider the orders of the various elements $[a] \in \mathbb{Z}/(n)^{\times}$, there is a noteworthy aside that emerges from fixing $a = 10$ and considering the order of $[10]$ in $\mathbb{Z}/(n)^{\times}$ for various n relatively prime to 10. To get started, consider Table 4.1, showing the decimal expansions of the first few unit fractions $\frac{1}{n}$ (for $\gcd(10, n) = 1$) and the order of $[10]$ in $\mathbb{Z}/(n)^{\times}$.

One (of many!) striking patterns about decimal expansions can be explained in the language of orders and group theory. Let's use the phrase **period length** of a rational number for the number of digits in the repeating block of its decimal expansion (called the **repetend**). The following theorem explains this emergent phenomenon.

Theorem 4.5.13
Let $n \in \mathbb{N}$ with $\gcd(n, 10) = 1$. Then the period length of the decimal expansion of $\frac{1}{n}$ is precisely the order of $[10]$ in $\mathbb{Z}/(n)^{\times}$.

Proof We leave it to the reader to argue that if $\gcd(n,10)=1$, then there is no pre-periodic part of the decimal expansion of $1/n$. So write $\frac{1}{n} = 0.\overline{a_1...a_k}$, where k is the period length, so that $10^k(1/n) = a_1...a_k.\overline{a_1...a_k}$ and thus $(10^k - 1)(1/n) = a_1...a_k \in \mathbb{Z}$. This shows that n divides $10^k - 1$, so $10^k \equiv 1 \pmod{n}$. As k is the

Table 4.2 Tables of modular orders

$\mathbb{Z}/(6)^\times$	$\varphi(6) = 2$		
element	[1]	[5]	
order	1	2	

$\mathbb{Z}/(7)^\times$	$\varphi(7) = 6$					
element	[1]	[2]	[3]	[4]	[5]	[6]
order	1	3	6	3	6	2

$\mathbb{Z}/(8)^\times$	$\varphi(8) = 4$			
element	[1]	[3]	[5]	[7]
order	1	2	2	2

$\mathbb{Z}/(9)^\times$	$\varphi(9) = 6$					
element	[1]	[2]	[4]	[5]	[7]	[8]
order	1	6	3	6	3	2

$\mathbb{Z}/(10)^\times$	$\varphi(10) = 4$			
element	[1]	[3]	[7]	[9]
order	1	4	4	2

$\mathbb{Z}/(11)^\times$	$\varphi(11) = 10$									
element	[1]	[2]	[3]	[4]	[5]	[6]	[7]	[8]	[9]	[10]
order	1	10	5	5	5	10	10	10	5	2

least power of 10 for which $(10^k - 1)/n$ would be an integer, it is precisely the order of 10 mod n. □

To finish off this story, we consider the case that $\gcd(10, n) \neq 1$. Here, we write $n = 2^{v_2(n)} 5^{v_5(n)} n'$ where n' is relatively prime to both 2 and 5. Then the period length of n' determines the period length of n, and the factors of 2 and 5 affect only the number of pre-periodic digits. See Exercise 4.64.

The reverse scenario, fixing n and trying to find a practical method for finding orders of units in $\mathbb{Z}/(n)^\times$, is our principal quest moving forward. To summarize, we know that there exists a positive integer k such that $a^k \equiv 1 \pmod{n}$ if and only if $[a]$ is a unit in $\mathbb{Z}/(n)$, which happens if and only if a and n are relatively prime, and Euler's totient function $\varphi(n)$ lets us know how many such elements there are. Lagrange's Theorem, the upcoming group-theoretic result, reveals a miraculous relationship between the *orders* of units in $\mathbb{Z}/(n)$ and the *number* of units in $\mathbb{Z}/(n)$. Table 4.2 displays several groups of units, showing the size of the group, and the orders of each of its elements.

Equivalence Relations and Equivalence Classes ◀

The upcoming proof of Lagrange's Theorem, tying together orders of elements and the order of the group they live in, hinges on partitioning sets into equivalence classes. This is the process we employed in Section 4.2 to construct $\mathbb{Z}/(n)$, using the equivalence relation of congruence mod n to partition $\mathbb{Z}$. Here we get some practice with other examples.

G.1 Consider the group $G = (\mathbb{Z}/11)^{\times}$ under multiplication, and let H be the subset $\{[1], [3], [4], [5], [9]\}$ of G.

1. Check that H is a group under multiplication modulo 11.
2. Check that the relation defined for $g_1, g_2 \in G$ by

$$g_1 \equiv g_2 \iff g_2^{-1} g_1 \in H$$

   defines an equivalence relation on G.
3. Find the distinct equivalence classes for this relation. How many elements does each have? How would you define multiplication of these equivalence classes?

G.2 Consider the group $G = \mathbb{R}[x]$ under addition, and let H be the subset of G consisting of all multiples of $x^2 + 1$. Define a relation by $p(x) \equiv q(x)$ if $x^2 + 1 | (p(x) - q(x))$.

1. Quick checks: H is itself a group, and $\equiv$ is an equivalence relation on G.
2. Show that every element of $\mathbb{R}[x]$ is congruent to a constant or linear polynomial, i.e., every equivalence class can be expressed in the form $[a + bx]$ for some $a, b \in \mathbb{R}$.
3. Let C be the collection of equivalence classes. Check that the definition

$$[a + bx] + [c + dx] = [(a + c) + (b + d)x]$$

   equips C with the structure of an additive group.
4. In fact, in this case C can further be made into a ring. How should we complete the following formula with a linear polynomial to give a reasonable notion of multiplication in C?

$$[a + bx] \cdot [c + dx] = [???].$$

5. Use your definition to compute $[2 + 3x] \cdot [1 - 2x]$. Find a solution in C to the equation $y^2 = -1$. Perhaps at this point the ring C starts to feel familiar. Discuss.

4.6 Lagrange's Theorem and the Euler Totient Function

The order of a group is related to the order of its elements by Lagrange's Theorem. Exploration G set up the principal tool for us to prove this result by applying our knowledge of equivalence relations in the setting of an abstract group. As in Exploration G, we start by considering equivalence relations defined by membership in a special subset of the group.

Definition 4.6.1

Given a group G, a subset $H \subseteq G$ forms a **subgroup of** G if and only if the set H forms a group under the same operation as G. ◄

It is typically quite routine to decide if a subset is a subgroup. For example, associativity in H comes for free from associativity in G, and if H is closed under inversion and the group operation, it must also contain the identity (since $h \cdot h^{-1} = e$). It is even briefer for finite groups, since every element has a finite order: If h has order k, then its inverse is h^{k-1}, so if H is closed under the group operation, then it is closed under inversion, and hence a subgroup.

▶ **Example 4.6.2** Inside $G = \mathbb{Z}/(11)^\times$, the subset $H = \{[1], [3], [5]\}$ is not a subgroup since it is not closed under multiplication (or inverses), but since $H_1 = \{[1], [3], [4], [5], [9]\}$ and $H_2 = \{[1]\}$ and $H_3 = \{[1], [10]\}$ are closed under multiplication, they are subgroups.

A large swath of subgroups of any group can be produced simply by taking a single element of the group and considering the set of all of its powers.

Theorem 4.6.3

Given a group G and an element $g \in G$, the set of elements $\langle g \rangle = \{g^k : k \in \mathbb{Z}\}$ forms a subgroup of G called the **cyclic subgroup generated by** g.

Proof Associativity in $\langle g \rangle$ is inherited from the larger group G, we have the identity $g^0 = 1 \in \langle g \rangle$, and we also have inverses since $(g^k)^{-1} = g^{-k}$. Finally, closure under multiplication follows from $g^m \cdot g^n = g^{m+n}$.   □

▶ **Example 4.6.4** Inside $\mathbb{Z}/(7)^\times$, we have the cyclic subgroups

$$\langle[1]\rangle = \{[1]\} \qquad\qquad \langle[4]\rangle = \{[4], [2], [1]\}$$

$$\langle[2]\rangle = \{[2], [4], [1]\} \qquad\qquad \langle[5]\rangle = \{[5], [4], [6], [2], [3], [1]\}$$

$$\langle[3]\rangle = \{[3], [2], [6], [4], [5], [1]\} \quad \langle[6]\rangle = \{[6], [1]\}.$$

Though the ordering of the elements in a set is irrelevant, they are ordered above in ascending powers of the generator.

The powering-up process stops when we reach 1 (since at this point the pattern of powers will begin to loop back to the beginning). Here we see a close relationship between the two uses of the word *order*: the order $|G|$ of a group and the order $|g|$ of one of its elements. (Note the unfortunate similarity in notation between the order of the element g and the absolute value function. We will need to remain alert for context to discern which interpretation to use.) If the group in question is $\langle g \rangle$, the cyclic subgroup is generated by the element g, then the two uses are synonymous (compare the previous example to the order data for $\mathbb{Z}/(7)^\times$ in Table 4.2).

Corollary 4.6.5

Suppose $g \in G$ has finite order. Then the order of the cyclic subgroup generated by g is equal to the order of g in G, i.e., $|\langle g \rangle| = |g|$. ◀

Proof Exercise 4.40. □

Definition 4.6.6

Let G be a group and H a subgroup of G. For $a, b \in G$ we say a is **(left) congruent to b mod** H if $a^{-1}b \in H$ and write

$$a \equiv b \bmod H.$$

◀

Theorem 4.6.7

Given a group G and a subgroup H, congruence mod H is an equivalence relation on the set G.

Proof To check reflexivity, we have for all $a \in G$ that $a \equiv a \bmod H$ since $a^{-1}a = e$, and $e \in H$ since H is a subgroup of G. For symmetry, suppose $a \equiv b \bmod H$. Then $a^{-1}b \in H$, and since H is closed under inverses, $(a^{-1}b)^{-1} = b^{-1}a \in H$, so $b \equiv a \bmod H$. Finally, for transitivity, suppose that $a \equiv b \bmod H$ and $b \equiv c \bmod H$. Then $a^{-1}b \in H$ and $b^{-1}c \in H$, so their product $a^{-1}bb^{-1}c = a^{-1}c$ is also in H, giving $a \equiv c \bmod H$.   □

For a given element $a \in G$, the equivalence class of a under congruence mod H is just the set of elements in G congruent to a, $\{g \in G : a^{-1}g \in H\}$. However, if $a^{-1}g = h$ for some $h \in H$, then $g = ah$. Thus, the equivalence class of a is just the set $\{ah : h \in H\}$. These sets have a special name and role in group theory.

Definition 4.6.8

Given a group G with element a and subgroup H, the set $aH = \{ah : h \in H\}$ is called the **left coset of H containing** a. ◄

Right cosets[11] can be similarly defined via $Ha = \{ha : h \in H\}$, and while in general $aH \neq Ha$, in abelian groups (including all the $\mathbb{Z}/(n)^\times$), there is no distinction between left and right cosets, and we arbitrarily choose to continue working with left cosets.

▶ **Example 4.6.9** Consider $G = \mathbb{Z}/(14)^\times = \{[1], [3], [5], [9], [11], [13]\}$. We verify that the subset $H = \{[1], [9], [11]\}$ is in fact a subgroup of G, and construct the cosets:

$$[1]H = \{[1], [9], [11]\} \quad [3]H = \{[3], [13], [5]\} \quad [5]H = \{[5], [3], [13]\}$$
$$[9]H = \{[9], [11], [1]\} \quad [11]H = \{[11], [1], [9]\} \quad [13]H = \{[13], [5], [3]\}$$

We see that there are in fact only two distinct left cosets, $[1]H = \{[1], [9], [11]\}$ and $[3]H = \{[3], [13], [5]\}$ and that they partition all of G. Furthermore, we can read off congruence from the partitions: $3 \equiv 5 \bmod H$ since they appear in the same coset, or, equivalently, since $[3]^{-1} \cdot [5] = [25] = [11] \in H$.

▶ **Example 4.6.10** Consider $G = \mathbb{R}^\times = \mathbb{R} - \{0\}$ (under multiplication), and H the subgroup of positive real numbers. The two distinct left cosets are H and $-1H$, the set of negative real numbers.

---

[11] Corresponding to the equivalence classes of *right* congruence mod H, the equivalence relation defined as $a \equiv b \bmod H$ if and only if $ba^{-1} \in H$

Up to this point we have defined subgroups and cosets in the setting of *multiplicative* groups. We could just as easily have used the context of an *additive* group, as Exercise 4.18 and the example below show.

▶ **Example 4.6.11** Consider the group formed by $\mathbb{Z}$ under addition, and the the cyclic subgroup $H = \langle 6 \rangle$. As $\mathbb{Z}$ is an additive group, the cyclic subgroup generated by 6 contains all *multiples* of 6 (corresponding to repeated addition of 6 or its additive inverse -6) rather than all *powers* of 6 (corresponding to repeated multiplication). For this reason, as a subgroup of $\mathbb{Z}$, $\langle 6 \rangle = 6\mathbb{Z} = \{6x \mid x \in \mathbb{Z}\}$. In additive notation, "a and b are congruent mod H" means $(-a) + b \in H$, and we write the coset of H containing a as $a + H$. In this case, $(-a) + b \in \langle 6 \rangle$ means that $6 | b - a$, so congruence mod $\langle 6 \rangle$ is just standard congruence mod 6. Since we already know how this partitions the integers, we find that the cosets are

$$0 + \langle 6 \rangle = [0] \quad 2 + \langle 6 \rangle = [2] \quad 4 + \langle 6 \rangle = [4]$$
$$1 + \langle 6 \rangle = [1] \quad 3 + \langle 6 \rangle = [3] \quad 5 + \langle 6 \rangle = [5]$$

Of course, we can choose any element of a coset to represent it. For example, $3 + \langle 6 \rangle = 9 + \langle 6 \rangle = 603 + \langle 6 \rangle$.

▶ **Example 4.6.12** The set H of all multiples of $x^2 + 1$ in $\mathbb{R}[x]$ forms a subgroup of $\mathbb{R}[x]$ (see Exploration G and/or Exercise 4.36). By polynomial long division, for every polynomial $p(x) \in \mathbb{R}[x]$ there exist polynomials $q(x)$ and $r(x)$ in $\mathbb{R}[x]$ such that $p(x) = (x^2 + 1)q(x) + r(x)$ with $\deg r(x) < 2$. Thus, $p(x) - r(x) = (x^2 + 1)q(x) \in H$, and under congruence mod H, $p(x) \equiv r(x)$ mod H. This means that every polynomial in $\mathbb{R}[x]$ is congruent to a polynomial of the form $r(x) = ax + b$, and so the cosets of H can all be represented in the form $[ax + b]$ for $a, b \in \mathbb{R}$.

▶ **Remark 4.6.13** While the emphasis here is on groups, the last couple of examples hint at how we might analogously define congruence in a ring modulo a certain type of subring. For both groups and rings, under suitable conditions the set of congruence classes under these relations can themselves be given the structures of groups and rings (as when we constructed $\mathbb{Z}/(n)$ using congruence mod n). In the congruence mod multiples of $(x^2 + 1)$ example above, one can check that if $f(x) \equiv a + bx$ and $g(x) \equiv c + dx$, then $f(x)g(x) \equiv (ac - bd) + (ad + bc)x$, an operation strikingly similar to multiplication in $\mathbb{C}$ (the field constructed expressly to allow $x^2 + 1$ to have a solution!). See Exercise 4.68 for more on these "quotient rings."

Lemma 4.6.14

Let G be a finite group and H a subgroup of G. Then for all $a \in G$, we have

$$|aH| = |H|.$$

That is, all left cosets of H have the same size as H. ◄

If you take all the elements of H and multiply each by a, then intuitively you have the same number of elements as when you started—unless of course you don't. What if some of the elements in this new set aH are duplicates? If you take each element of $\mathbb{Z}/(4)$ and multiply by two, for example, you only end up with two distinct elements. What we need is a more precise version of the argument showing that multiplication by a is a bijection.

Proof We show that the function $f \colon H \to aH$ defined by $f(h) = ah$ for all $h \in H$ is a bijection. Since surjectivity follows for free from the definition of aH, we need only show injectivity. Suppose $f(h_1) = f(h_2)$ for some h_1, h_2 in H. Thus $ah_1 = ah_2$. Multiplying both sides by a^{-1} on the left gives

$$a^{-1}(ah_1) = a^{-1}(ah_2)$$

$$(a^{-1}a)h_1 = (a^{-1}a)h_2$$

$$h_1 = h_2,$$

so f is injective, and hence a bijection, so $|H| = |aH|$.   □

(Note that the $\mathbb{Z}/(4)$ example doesn't contradict this lemma since $\mathbb{Z}/(4)$ is not a group under multiplication. If we add [2] to each element of $\mathbb{Z}/(4)$ you do find $|\mathbb{Z}/(4)| = |[2] + \mathbb{Z}/(4)|$.)

We now have all the parts we need to conclude one of the most powerful theorems of finite group theory.

Theorem 4.6.15 (Lagrange's Theorem)

If H is a subgroup of a finite group G, then the number of elements in H divides the number of elements in G, i.e., $|H| \big| |G|$.

We will try to minimize the frequency of the aesthetic unpleasantness of using $|\cdot|$ for orders juxtaposed with $|$ for divides, but there's a cautionary tale to be told here about the necessity of good handwriting[12].

---

[12] For example, if your 1 looks too much like a ı, then writing that the cyclic subgroup generated by 11 is a subgroup of that generated by 1 ends up looking something like $|\langle \text{ıı} \rangle| \, | \, |\langle \text{ı} \rangle|$

Proof As the left cosets of H are the equivalence classes under congruence modulo H, by Lemma 4.1.5 they partition the set:

$$G = a_1 H \cup a_2 H \cup \cdots \cup a_k H,$$

where $a_1 H, \ldots, a_k H$ represent the distinct, and therefore disjoint, cosets. It follows from the previous lemma that

$$|G| = |a_1 H| + |a_2 H| + \cdots + |a_k H| = |H| + |H| + \cdots + |H| = k|H|.$$

Thus $|H|$ divides $|G|$. $\qquad\square$

In the proof above, k represents the number of distinct cosets of H. This value, $k = \frac{|G|}{|H|}$, is referred to as the **index** of H in G and denoted $[G : H]$.

Where does this leave us in our quest to simplify exponentiation in $\mathbb{Z}/(n)^\times$? We know for all $[a] \in \mathbb{Z}/(n)^\times$, $\langle [a] \rangle$ is a subgroup of $\mathbb{Z}/(n)^\times$ and that $|[a]| = |\langle [a] \rangle|$. Alongside Lagrange's Theorem, these observations provide the following incredibly useful theorem. Recall Euler's totient function, $\varphi(n) = |\mathbb{Z}/(n)^\times|$, which gives the number of positive integers less than or equal to n that are relatively prime to n.

Theorem 4.6.16 (Euler's Theorem)
For all $n \in \mathbb{N}$ and $a \in \mathbb{Z}$ with $\gcd(a, n) = 1$, we have

$$a^{\varphi(n)} \equiv 1 \bmod n.$$

Proof The hypotheses imply that $[a] \in \mathbb{Z}/(n)^\times$. In this group we have $|[a]| = |\langle [a] \rangle|$ by Corollary 4.6.5, and so by Lagrange's Theorem, $|[a]|$ divides $|\mathbb{Z}/(n)^\times| = \varphi(n)$. Write $\varphi(n) = |[a]|k$ for some $k \in \mathbb{Z}$, and now

$$[a]^{\varphi(n)} = ([a]^{|[a]|})^k = [1]^k = [1],$$

as desired. $\qquad\square$

While we have not solved *all* of the mysteries of Euler's φ-function, there are some special cases where the result makes for an even more pristine version of Euler's Theorem; for example, the following corollary typically attributed to Fermat.

Corollary 4.6.17 (Fermat's Little Theorem)

For p prime, if $\gcd(a, p) = 1$, then $a^{p-1} \equiv 1 \pmod{p}$. For all $a \in \mathbb{Z}$, $a^p \equiv a \bmod p$. ◄

Proof The first claim is a special case of Euler's Theorem with $\varphi(p) = p - 1$. Multiplying both sides by a gives $a^p \equiv a \bmod p$, and this identity holds for $a \equiv 0 \bmod p$ as well. ☐

Corollary 4.6.18

If $[a] \in \mathbb{Z}/(n)^\times$, then $|[a]| \mid \varphi(n)$. ◄

Proof This gem comes directly from the proof of Euler's Theorem. ☐

At long last, we have the key to modular exponentiation calculations. Combining Euler's Theorem with Lemma 4.5.5, we deduce the remarkable corollary.

Corollary 4.6.19

For all $a \in \mathbb{Z}$, if $\gcd(a, n) = 1$, then

$$a^k \equiv a^{k \bmod \varphi(n)} \bmod n.$$

◄

▶ **Example 4.6.20** To compute $15^{80} \bmod 41$, we compute $\varphi(41) = 40$ since 41 is prime, and so
$$15^{80} \equiv 15^{80 \bmod 40} = 15^0 = 1 \pmod{41}.$$

▶ **Example 4.6.21** To compute $71^{83} \bmod 34$, we compute $\varphi(34) = 16$ by Lemma 4.5.12(ii) and note $71 \bmod 34 = 3$, so
$$71^{83} \equiv 3^{83 \bmod 16} = 3^3 = 27 \pmod{34}.$$

▶ **Example 4.6.22** To compute $133^{59} \bmod 27$, we compute $\varphi(27) = 18$ by Lemma 4.5.12(i) and note $133 \bmod 27 = 25$, so
$$133^{59} \equiv 25^{59 \bmod 18} \equiv (-2)^5 = -32 \equiv 22 \pmod{27}.$$

▶ **Remark 4.6.23** One normally has to be careful putting footnotes on mathematical expressions, as they can be confused with exponents, but the following footnote shows that sometimes things work out fine regardless, e.g., that modulo 13 we have $7^{49} \equiv 7$.[13]

Finally, we note that the ability to reason through orders of elements has benefits not only for explicit computation, but also in clarifying algebraic properties of the modular worlds. For example, in solving Diophantine equations by working mod n, we saw that not every element of $\mathbb{Z}/(n)$ is a square (for example, only $[0], [1] \in \mathbb{Z}/(4)$ were squares). In fact, for every n, there must be at least one element of $\mathbb{Z}/(n)$ that is not a square since in the list $[0]^2, [1]^2, [2]^2, \ldots [n-1]^2$ there are many repeats (as $[a]^2 = [-a]^2$), and so we could not get all n possible outputs. The following result stands in stark contrast.

Corollary 4.6.24

If p is a prime with $p \bmod 3 = 2$, then every element of the ring $\mathbb{Z}/(p)$ is a cube.

◀

Proof Define $\varphi : \mathbb{Z}/(p) \to \mathbb{Z}/(p)$ by $\varphi(x) = x^3$. We claim that φ is a bijection, for which it suffices, since $\mathbb{Z}/(p)$ is finite, to show it is an injection. Suppose $a, b \in \mathbb{Z}/(p)$ satisfy $a^3 = b^3$. Since $\mathbb{Z}/(p)$ is a field (and hence has no zero divisors), if $a^3 = b^3 = [0]$, then $a = b = [0]$. More interestingly, if $a^3 = b^3 \neq [0]$, then $a, b \neq [0]$ and so each is a unit mod p, and we can write

$$(ab^{-1})^3 = [1].$$

Now if $a \neq b$, then ab^{-1} has order 3, and so Lagrange's Theorem implies that $3 \mid |\mathbb{Z}/(p)| = p - 1$. But this contradicts that $p \equiv 2 \bmod 3$. So $a = b$, proving injectivity, hence bijectivity, hence surjectivity. □

A second structural consequence is a striking analog of the Binomial Theorem in the mod-p world. Recall that for real numbers x, y and a natural number n, we have

$$(x + y)^n = x^n + \binom{n}{1}x^{n-1}y^1 + \binom{n}{2}x^{n-2}y^2 + \cdots + \binom{n}{n-1}x^1y^{n-1} + y^n,$$

where the coefficients $\binom{n}{k}$ are defined by Pascal's Triangle and the combinatorial formula $\binom{n}{k} = \frac{n!}{k!(n-k)!}$. For contrast, mod p we have the following remarkable result, which we lovingly refer to as the nØØb's Binomial Theorem, or perhaps the nØØbinomial Theorem, in honor of that point in all of our mathematical trajectories where we lost points on an exam for trying to get away with identities like $(x+y)^3 = $

---

[13] Since $a^{13} \equiv a \bmod 13$ for any integer a, by Fermat's Little Theorem.

$x^3 + y^3$. Little did we know that we needed only work mod 3 for our claims to be vindicated[14].

Theorem 4.6.25 (The n ∅∅ b's Binomial Theorem)
For any $x, y \in \mathbb{Z}$ and p a prime, we have

$$(x + y)^p \equiv x^p + y^p \pmod{p}.$$

Proof Fermat's Little Theorem tells us that $a^p \equiv a \pmod{p}$ for *every* a in $\mathbb{Z}$, so applying this to each of $x + y$, x, and y, we learn

$$(x + y)^p \equiv x + y \equiv x^p + y^p \pmod{p}.$$

▶ **Remark 4.6.26** A second common proof of this theorem follows from the usual Binomial Theorem. When p is prime, each of the binomial coefficients

$$\binom{p}{k} = \frac{p!}{k!(p-k)!},$$

for $1 \leq k \leq p-1$, is a multiple of p since p divides the numerator of this fraction but not the denominator. When we reduce the Binomial Theorem mod p, the only terms that remain are thus x^p and y^p.

---

[14] We suggest revisiting your calculus instructor and demanding those exam points back for being unclear as to which ring you were supposed to do your calculus in.

Mixed Modulus Musings ◄

Our focus so far has been for fixed n to consider the world $\mathbb{Z}/(n)$ and investigate the reductions of various integers a modulo n. Below we consider a somewhat reversed perspective, that is, for fixed a, the relationship between the quantities $a \bmod n$ for various n.

H.1 Suppose $a \equiv 1 \pmod 5$. What can you say about $a \bmod 10$? $a \bmod 25$? $a \bmod 3$?

H.2 Suppose $a \equiv 7 \pmod{30}$. What can you say about $a \bmod 2$? $a \bmod 3$? $a \bmod 5$? $a \bmod 7$?

Let's adopt a new temporary notation for keeping track of two moduli at once. We write

$$k \equiv (a, b) \quad (\bmod\ (m, n))$$

to mean $k \equiv a \pmod m$ and $k \equiv b \pmod n$. For example,

$$41 \equiv (2, 1) \quad (\bmod\ (3, 5))$$

since $41 \equiv 2 \pmod 3$ and $41 \equiv 1 \pmod 5$.

A series of mixed-modulus problems using the above notation:

H.3 Find $n \bmod (3, 4)$ for all $[n] \in \mathbb{Z}/(12)$. Use your results to prove the following claim: that if $n \equiv 1 \pmod 3$ and $n \equiv 1 \pmod 4$, then $n \equiv 1 \pmod{12}$ (and likewise if we replace each 1 with -1).

H.4 A slightly less brute-force approach to the same problem. Suppose $n = 3k + 1$ and $n = 4\ell + 1$, and see if you can deduce a "mod 12 condition" by writing n in the form $n = 12 \_\_ + \_\_$.

H.5 Prove that if $a \equiv \pm 1 \pmod m$ and $a \equiv \pm 1 \pmod n$ with $\gcd(m, n) = 1$, then $a \equiv \pm 1 \pmod{mn}$.

H.6 There is promise that this type of argument will work much more generally. If $a = 11k + 4$ and $a = 5\ell + 3$, deduce the value of $a \bmod 55$.

Table 4.3 Mixed modulus data

x mod 12	0	1	2	3	4	5	6	7	8	9	10	11
x mod $(3, 4)$	(0,0)	(1,1)	(2,2)	(0,3)	(1,0)	(2,1)	(0,2)	(1,3)	(2,0)	(0,1)	(1,2)	(2,3)

4.7 Sunzi's Remainder Theorem and $\varphi(n)$

Having nigh on mastered modular arithmetic for a fixed modulus, there is significant merit in studying what happens when *varying* from one modulus to another: given $a \in \mathbb{Z}$, what are the relationships between a mod m and a mod n as m, n vary? Our first case is when one of the two divides the other, say $m \mid n$ with $n = mk$. Now, for example, if $x \equiv 5$ mod n, then it must be that $x \equiv 5$ mod m as well as

$$x = qn + 5 \Rightarrow x = (qk)m + 5.$$

The following lemma generalizes this observation.

Lemma 4.7.1

If $m \mid n$ and $x \equiv a$ mod n, then $x \equiv a$ mod m. ◄

More interestingly, if m and n are relatively prime, it seems there is little relationship between x mod m and x mod n. One interpretation of the results in this section is a provable version of this statement—the values of x mod m and x mod n are completely independent when $\gcd(m, n) = 1$. As in Exploration H, we introduce an ad hoc notational device for simultaneously keeping track of both x mod m and x mod n, writing

$$x \equiv (a, b) \quad (\text{mod } (m, n))$$

to mean that $x \equiv a$ (mod m) and $x \equiv b$ (mod n). Summarizing one of the results from Exploration H, we found that knowing the value of x mod 12 was equivalent to knowing *both* x mod 3 and x mod 4 (see Table 4.3).

More explicitly, we note that every element of $\mathbb{Z}/(12)$ corresponds to a distinct ordered pair (a, b) with $a \in \mathbb{Z}/(3)$ and $b \in \mathbb{Z}/(4)$. Further, *every* possible such pair appears, i.e., every pair corresponds to precisely one element of $\mathbb{Z}/(12)$. These statements imply that we have a bijection between the ring $\mathbb{Z}/(12)$ and this collection of ordered pairs. This link is strengthened further by the realization that this collection of ordered pairs can also be viewed as a ring in a natural way.

Theorem 4.7.2

Let A and B be rings, with operations $+$ and $\cdot$. The **Cartesian product** $A \times B = \{(a, b) : a \in A, b \in B\}$ is also a ring with the component-wise defined operations

$$(a, b) + (c, d) = (a + c, b + d)$$
$$(a, b) \cdot (c, d) = (a \cdot c, b \cdot d),$$

where the operations within first and second entries are the respective operations of the rings A and B.

Proof Almost all properties follow from A and B themselves being rings. For example, the additive identity of $A \times B$ is $(0_A, 0_B)$ where 0_A is the additive identity of A and 0_B is the additive identity of B. We leave the details to Exercise 4.38.   □

Our immediate goal is to determine when this function $[x] \rightarrow ([x \bmod m], [x \bmod n])$ from $\mathbb{Z}/(mn)$ to $\mathbb{Z}/(m) \times \mathbb{Z}/(n)$ is a bijection. For m and n much larger than 3 or 4, the process of *listing* all the possible values becomes cumbersome, so we will need a more systematic approach. We begin by re-examining the connection between $\mathbb{Z}/(12)$ and $\mathbb{Z}/(3) \times \mathbb{Z}/(4)$. Suppose $x \equiv 1 \pmod 3$ and $x \equiv 1 \pmod 4$. How might we determine $x \bmod 12$? Given our assumed congruences modulo 3 and 4 we know that $x = 3k + 1$ and $x = 4\ell + 1$ for some $k, \ell \in \mathbb{Z}$. Here's a sneaky way of deducing $x \bmod 12$ from these two pieces of information:

$$x = 4x - 3x = 4(3k + 1) - 3(4\ell + 1) = 12k + 4 - 12\ell - 3 = 12(k - \ell) + 1,$$

so $x \bmod 12 = 1$. This example was chosen to be particularly simple (and to allow our sneaky idea), but the general process is only one degree of difficulty higher.

▶ **Example 4.7.3** If $x \equiv 4 \pmod{11}$ and $x \equiv 3 \pmod 5$, find $x \bmod 55$.

Solution Write $x = 11k + 4$ and $x = 5\ell + 3$. Then $5x = 55k + 20$ and $11x = 55\ell + 33$. We can deduce that

$$x = 11x - 10x = (55\ell + 33) - 2(55k + 20) = 55(\ell - 2k) - 7 \equiv 48 \pmod{55}$$

by writing x as a linear combination of $11x$ and $5x$ ($x = 11x + (-2)5x$).   □

The key to the process is slowly being revealed. Knowing $x \bmod 11$ and $x \bmod 5$ we can find representations of x in terms of multiples of 11 and 5, respectively. By multiplying each of these representations of x by the opposite modulus ($x = 11k + 4$ by 5 and $x = 5\ell + 3$ by 11), we find expressions for $5x$ and $11x$ in terms of the desired common modulus (in this case 55). Then, finding a linear combination of these expressions equalling x—or, equivalently, a linear combination of 5 and 11 equal to 1—we were able to write x in terms of the common modulus. By Bézout's Identity, this can be done if and only if 5 and 11 are relatively prime (which they are!). When m and n are not relatively prime, we should thus expect our approach to fail, as the following example illustrates.

▶ **Example 4.7.4** Suppose $x \equiv 1 \pmod 4$ and $x \equiv 4 \pmod 6$. Find $x \bmod 24$.

Solution ...or lack thereof! Not only does our procedure fail, but it is moreover impossible to find any $x \in \mathbb{Z}$ that satisfies $x \equiv 1 \pmod 4$ and $x \equiv 4 \pmod 6$, as the first congruence would force x to be odd, and the second would force x to be even.   □

Note that regardless of whether m and n are relatively prime, there *is* a bijection from $\mathbb{Z}/(mn)$ to $\mathbb{Z}/(m) \times \mathbb{Z}(n)$—they are both, after all, finite sets with the same number of elements—but there may be no bijection whose definition is particularly natural, and none that preserve the algebraic structure. For example, $\mathbb{Z}/(16)$ and $\mathbb{Z}/(4) \times \mathbb{Z}/(4)$ both have 16 elements, but the two rings are structurally quite different, as evidenced, for example, by the observation that for any $r \in \mathbb{Z}/(4) \times \mathbb{Z}/(4)$ we have that $4r = 0$, whereas this is not true for an arbitrary element of $\mathbb{Z}/(16)$.

When $\gcd(m, n) = 1$, however, there is a *natural* bijection, stemming from Table 4.3. Take, for example, the elements [7] and [11] in $\mathbb{Z}/(12)$, and note that 7 and 11 are paired in the table with $(1, 3)$ and $(2, 3)$, respectively. Now two pleasant things happen: first, consider the sum $[7] + [11] = [6]$. From the table, we see 6 corresponds to $(0, 2)$, a convenient result since this is precisely the sum

$$(1, 3) + (2, 3) \equiv (0, 2) \pmod{(3, 4)}.$$

Similarly for their product $[7] \cdot [11] = [77] = [5] \in \mathbb{Z}/(12)$, we find 5 corresponds to $(2, 1)$ and the corresponding multiplicative miracle also occurs as given below:

$$(1, 3) \cdot (2, 3) \equiv (2, 1) \pmod{(3, 4)}.$$

Sunzi's Remainder Theorem is the generalization of this result to any pair of relatively prime moduli. If one were to build analogous tables, with rows for $\mathbb{Z}/(mn)$ and $\mathbb{Z}/(m) \times \mathbb{Z}/(n)$, then this preservation of structure always occurs—adding or multiplying the elements in the top row mod mn, or performing the same operation on the corresponding elements in the bottom row mod (m, n) gives corresponding results.

Theorem 4.7.5 (Sunzi's Remainder Theorem)
Suppose $m, n \in \mathbb{N}$ are relatively prime. Then the function

$$\psi \colon \mathbb{Z}/(mn) \longrightarrow \mathbb{Z}/(m) \times \mathbb{Z}/(n)$$

defined by $\psi([x]) = ([x \bmod m], [x \bmod n])$ is a well-defined bijection that preserves the ring operations: for $[x], [y] \in \mathbb{Z}/(mn)$, we have

$$\psi([x] + [y]) = \psi([x]) + \psi([y])$$

and

$$\psi([x] \cdot [y]) = \psi([x]) \cdot \psi([y]).$$

Proof Since both m and n divide mn, ψ is well defined by Lemma 4.7.1. We then verify

$$
\begin{aligned}
\psi([x] + [y]) &= ([x + y \bmod m], [x + y \bmod n]) \\
&= ([x \bmod m] + [y \bmod m], [x \bmod n] + [y \bmod n]) \\
&= ([x \bmod m], [x \bmod n]) + ([y \bmod m], [y \bmod n]) \\
&= \psi([x]) + \psi([y])
\end{aligned}
$$

by the definition of the ring operations in Cartesian products (Theorem 4.7.2) and using Lemma 4.2.4. Verifying the result for multiplication follows similarly.

Next, to establish that ψ is a bijection, it's clear that $\mathbb{Z}/(mn)$ and $\mathbb{Z}/(m) \times \mathbb{Z}/(n)$ have the same number of elements (namely, mn), and so to show that our function is a bijection it suffices to show it is surjective. To this end, suppose $([a], [b]) \in \mathbb{Z}/(m) \times \mathbb{Z}/(n)$. Since $\gcd(m, n) = 1$, we can write

$$sm + tn = 1$$

for some $s, t \in \mathbb{Z}$. Let $x = bsm + atn$. Then $x \equiv 0 + atn \equiv a(1) \equiv a \pmod{m}$ and $x \equiv bsm + 0 \equiv b(1) + 0 \equiv b \pmod{n}$. Thus, by definition $\psi([x]) = ([a], [b])$, as desired. Since $([a], [b])$ was arbitrary, we conclude ψ is surjective. $\qquad\square$

▶ **Remark 4.7.6** This theorem frequently goes by the name of *The Chinese Remainder Theorem*, but it has been remarked that this is somewhat like referencing the Pythagorean Theorem as *The Greek Triangle Theorem* or Fermat's Little Theorem as *The French Power Postulate*. In the interest of giving credit where credit is due, we include it here with the name of the 3rd-century Chinese mathematician, Sunzi, to whom the result is often attributed.

▶ **Remark 4.7.7** An algebraic perspective on Sunzi's Remainder Theorem is that when $\gcd(m, n) = 1$, the rings $\mathbb{Z}/(mn)$ and $\mathbb{Z}/(m) \times \mathbb{Z}/(n)$ are essentially "the same." Namely, if every time someone mentioned $\mathbb{Z}/(12)$ you secretly thought of the corresponding elements of $\mathbb{Z}/(3) \times \mathbb{Z}/(4)$, no one would ever be the wiser. While we will not introduce the phrase as a term of art, the algebraic way of saying this is that the rings $\mathbb{Z}/(mn)$ and $\mathbb{Z}/(m) \times \mathbb{Z}/(n)$ are *isomorphic* (from the Greek: *iso* (same) and *morph* (form)), and the bijection that demonstrates this—in this case, the one described in Sunzi's Remainder Theorem—is called an *isomorphism* from one ring to the other.

While the theorem is phrased abstractly as a relationship between rings, the proof reveals how to interpret it as a computational tool, and the principal step is (yet again!) Bézout's Identity, finding $s, t \in \mathbb{Z}$ such that $sm + tn = 1$. To find an integer solution x to the system

$$x \equiv a \bmod m$$
$$x \equiv b \bmod n,$$

we take

$$x = bsm + atn$$

where s and t are integers such that $[s] = [m]^{-1} \in \mathbb{Z}/(n)$ and $[t] = [n]^{-1} \in \mathbb{Z}/(m)$. In this form, the result generalizes by induction to systems with even more relatively prime moduli. Suppose we seek an integer solution x to the system of congruences

$$x \equiv a_1 \pmod{n_1}$$
$$x \equiv a_2 \pmod{n_2}$$
$$x \equiv a_3 \pmod{n_3}$$
$$\vdots \equiv \vdots$$

with each $\gcd(n_i, n_j) = 1$, $1 \leq i \lessgtr j \leq k$. Then the more general version of Sunzi's Remainder Theorem says that such a solution exists and can be found as follows: Let $N = n_1 \cdots n_k$, and set

$$x = a_1 s_1 \tfrac{N}{n_1} + a_2 s_2 \tfrac{N}{n_2} + \cdots + a_n s_n \tfrac{N}{n_k},$$

where $[s_i] = \left[\tfrac{N}{n_i}\right]^{-1}$ in $\mathbb{Z}/(n_i)$. As in the proof above, we can verify the solution by simply reducing this expression for x modulo n_i: almost every term is a multiple of n_i, with only the one exception $\tfrac{N}{n_i}$. Hence

$$x \equiv a_i s_i \tfrac{N}{n_i} \equiv a_i \pmod{n_i},$$

by choice of s_i.

▶ **Example 4.7.8** Solve the system

$$x \equiv 3 \pmod 7$$
$$x \equiv 2 \pmod 9$$
$$x \equiv 4 \pmod{11}.$$

Solution We first embed the problem in the notation above. We have $n_1 = 7, n_2 = 9$, and $n_3 = 11$. Let $N = 7 \cdot 9 \cdot 11 = 693$. We compute the requisite inverses: s_1 is any integer such that

$$[s_1] = \left[\frac{693}{7}\right]^{-1} = [99]^{-1} = [1]^{-1} = [1] \quad \text{in } \mathbb{Z}/(7),$$

so we can take $s_1 = 1$. A similar process gives $s_2 = 2$ and $s_3 = 7$. This gives

$$x = 3(1)(99) + 2(2)(77) + 4(7)(63) = 2369 \equiv 290 \pmod{693}.$$

The bijection in (the general form of) Sunzi's Remainder Theorem implies [290] is the unique congruence class in $\mathbb{Z}/(693)$ that satisfies these equations, so the set of integer solutions to this system is $\{x \in \mathbb{Z} : x \bmod 693 = 290\}$. □

Returning to the motivation for Sunzi's Remainder Theorem, we consider the question of inverses in $\mathbb{Z}/(m) \times \mathbb{Z}/(n)$. Since multiplication is defined component-wise, elements $([a], [b])$ and $([c], [d])$ in $\mathbb{Z}/(m) \times \mathbb{Z}/(n)$ are multiplicative inverses if

$$(a, b) \times (c, d) \equiv (1, 1) \pmod{(m, n)},$$

which just means that $ac \equiv 1 \pmod{m}$ and $bd \equiv 1 \pmod{n}$. That is, $([a], [b])^{-1} = ([a]^{-1}, [b]^{-1})$, assuming a is invertible mod m and b is invertible mod n. This gives us an easy count of the number of units of the ring $\mathbb{Z}/(m) \times \mathbb{Z}/(n)$.

Corollary 4.7.9

Suppose $\gcd(m, n) = 1$. Then an element $[x] \in \mathbb{Z}/(mn)$ is a unit if and only if $x \bmod m$ and $x \bmod n$ are units mod m and n, respectively. That is, the bijection of Sunzi's Remainder Theorem also provides a bijection from $\mathbb{Z}/(mn)^\times$ to $\mathbb{Z}/(m)^\times \times \mathbb{Z}/(n)^\times$. Consequently, we have

$$\varphi(mn) = \varphi(m)\varphi(n). \qquad \blacktriangleleft$$

▶ **Remark 4.7.10** If $f : \mathbb{Z} \to \mathbb{Z}$ is a function such that $f(mn) = f(m)f(n)$ whenever $\gcd(m, n) = 1$, we say that f is a **multiplicative function**. We have argued previously that the "number of divisors" function is multiplicative, and Corollary 4.7.9 provides φ as a second example. We will explore other multiplicative functions in the exercises.

Again, an induction argument extends the result to products of several relatively prime moduli, and in conjunction with Lemma 4.5.12(i) thereby gives a complete formula for evaluating $\varphi(n)$ for all $n \in \mathbb{N}$.

Corollary 4.7.11

If $n = \prod p_i^{a_i}$, then

$$\varphi(n) = \prod \varphi(p_i^{a_i}) = \prod p_i^{a_i - 1}(p_i - 1) = \prod (p_i^{a_i} - p_i^{a_i - 1}). \qquad \blacktriangleleft$$

▶ **Example 4.7.12**

$$\varphi(1512) = \varphi(2^3 \cdot 3^3 \cdot 7) = \varphi(2^3)\varphi(3^3)\varphi(7) = (2^3 - 2^2)(3^3 - 3^2)(7 - 1) = 432.$$

At long last, combined with Euler's Theorem, we now have a complete picture of how to expedite modular exponentiation.

▶ **Example 4.7.13** Compute 11^{100} mod 90.

Solution By the corollary, we have $\varphi(90) = \varphi(2)\varphi(3^2)\varphi(5) = 1 \cdot 6 \cdot 4 = 24$, and so we can reduce exponents in a mod 90 calculation mod 24:

$$11^{100} \equiv 11^{100 \bmod 24} \equiv 11^4 \equiv 121^2 \equiv 31^2 \equiv 61 \bmod 90.$$

□

A very significant interpretation of Sunzi's Remainder Theorem is that to deduce a mod n it suffices to compute a mod $p^{v_p(n)}$ for each of the finitely many primes dividing n. As an aside, this has practical application in distributed computing: to determine the integer solution a to a suitable problem, one might employ many parallel processors, each computing a mod k for one small k, and then piece that information together to find a itself via Sunzi's Remainder Theorem. A more theme-fitting application of this perspective is that Sunzi's Remainder Theorem offers a way of raising arbitrary elements of $\mathbb{Z}/(n)$ to powers, not just units.

▶ **Example 4.7.14** Compute 14^{71} mod 120.

Solution Since $120 = 2^3 \cdot 3 \cdot 5$, it suffices to find the values of 14^{71} modulo 8, 3, and 5, and put them together using Sunzi's Theorem. Since 14 is not a unit modulo 120, we cannot simply apply Corollary 4.6.19, but working modulo prime powers offers a second method using nilpotency. Since $v_2(14) = 1$, we have $v_2(14^{71}) = 71 > 3$, and so $2^3 \mid 14^{71}$. The other two moduli are relatively straightforward and we collect

$$14^{71} \equiv 0 \bmod 8$$
$$14^{71} \equiv (-1)^{71} \equiv 2 \bmod 3$$
$$14^{71} \equiv (-1)^{71} \equiv 4 \bmod 5.$$

Applying the formula coming from Sunzi's Remainder Theorem provides the unique element $[104] \in \mathbb{Z}/(120)$ satisfying these three congruence conditions, and we conclude 14^{71} mod $120 = 104$.   □

Table 4.4 Powers of 3 mod 7

k	1	2	3	4	5	6
3^k mod 7	3	2	6	4	5	1

Table 4.5 Powers of 2 mod 7

k	1	2	3	4	5	6
2^k mod 7	2	4	1	2	4	1

4.8 Phis, Polynomials, and Primitive Roots

In some sense, the natural culmination of the study of modular exponentiation would be the study of modular logarithms. While we will not embrace the language of logarithms outside of this introductory thought experiment, we can at least ask the question: What should we mean by the value of $\log_{[3]}([5])$ in $\mathbb{Z}/(7)$? The answer, following the obvious analog with standard logarithms, should be an integer k such that $[3]^k = [5] \in \mathbb{Z}/(7)$, or equivalently such that $3^k \equiv 5 \bmod 7$. Since there are many such k as soon as there is at least one, we will follow our conventions for orders and find the least positive such power. Table 4.4 shows that $\log_{[3]}([5]) = 5$. The table shows further that *every* element of $\mathbb{Z}/(7)^\times$ can be log-base-three'd, as every element of $\mathbb{Z}/(7)^\times$ has a representative appearing somewhere in the bottom row as a power of 3. And again, as with standard logarithms, the base-[3] logarithm is the inverse to base-[3] exponentiation.

On the other hand, the idea of $\log_{[2]}$ is a significantly weaker notion in this group. Table 4.5 shows that *no* power of 2 is equal to 5, so we would have to begrudgingly agree that like $\ln(-3)$ in $\mathbb{R}$, $\log_{[2]}([5])$ doesn't exist in $\mathbb{Z}/(7)$.

What special property did 3 possess that 2 did not? It is precisely its *order* that matters: since [3] had order 6 in $\mathbb{Z}/(7)$, its first 6 powers necessarily ran through all 6 elements of $\mathbb{Z}/(7)^\times$, whereas since [2] had order 3, the first 6 powers of [2] started repeating before they had a chance to obtain all the different values. We encode this special property in a definition.

Definition 4.8.1

An element $g \in \mathbb{Z}/(p)^\times$ is a **primitive root mod** p if has order $p-1$, i.e., if every element of $\mathbb{Z}/(p)^\times$ is a power of g. We also say that an integer g is a primitive root mod p if its reduction mod p is. ◄

Our example above showed that since [3] has order 6 in $\mathbb{Z}/(7)$, the number 3 is a primitive root mod 7, and making the analogous tables (or just order computations) shows that 5 is also a primitive root mod 7. On the other hand, 1, 2, 4, and 6 are not primitive roots mod 7 as they have orders strictly less than 6. One computational benefit of finding a primitive root, which we do not pursue in detail here, is that it admits efficient modular arithmetic algorithms. In contexts with very large numbers, it may be computationally expensive (relatively speaking) to repeatedly compute the product of arbitrary elements $a, b \in \mathbb{Z}/(n)$. On the other hand, if we pre-compute a list of each element's base-g logarithm, then multiplication is very fast: We know $a = g^i$ and $b = g^j$ for some i, j, and so $ab = g^{i+j}$ can be read off the list. That is, we can make use of the important properties of standard logarithm functions, e.g.,

$$\log_g(ab) = \log_g(a) + \log_g(b),$$

but here interpreted modulo $p - 1$.

We turn to the principal theoretical result of this section that there is a primitive root mod p for each prime p. The argument is very simple in spirit, but requires a reasonable amount of checking to prove completely. We illustrate it with an example before generalizing.

▶ **Example 4.8.2** Show that there must be a primitive root mod 11.

Solution We show there must be an element of $\mathbb{Z}/(11)^\times$ of order 10. By Euler's Theorem, each of that group's 10 elements has order 1, 2, 5, or 10. The elements of order 5 are precisely the roots of the mod 11 polynomial $x^5 - 1 \in \mathbb{Z}/(11)[x]$, of which there are at most 5 (Lemma 3.1.13). Likewise, there is at most 1 element of order 1 and 2 elements of order 2. Since at most 8 of the 10 elements have orders 1, 2, or 5, there must be at least one element remaining of order 10 (in fact, at least 2). □

To write down a proof generalizing the example requires only a counting argument showing that the total number of elements of order *other* than $p - 1$ is always less than $p - 1$, leaving at least one element of $\mathbb{Z}/(p)^\times$ to have order exactly $p - 1$. The principal counting mechanism is the following combinatorial identity.

Theorem 4.8.3
For a natural number n, we have

$$\sum_{d \mid n} \varphi(d) = n,$$

where the sum runs over all divisors d of n.

Proof We partition the integers from 1 to n into sets S_d defined, for each $d \mid n$, by $k \in S_d$ if and only if $\gcd(k, n) = d$. Then since $\gcd(k, n) = d \iff \gcd(k/d, n/d) = 1$, we see that S_d has the same number of elements as integers up to $\frac{n}{d}$ that are relatively prime to $\frac{n}{d}$, that is, $|S_d| = \varphi(n/d)$. Since the sets S_d partition $\{1, \ldots, n\}$, the union of the S_d contains exactly n elements, so

$$n = \sum_{d|n} \varphi\left(\frac{n}{d}\right) = \sum_{d|n} \varphi(d),$$

the last equality following since $\frac{n}{d}$ runs through all divisors of n as d does.   □

▶ **Example 4.8.4** We consider the partition described in the proof for the case $n = 20$. Our sets are

$$
\begin{aligned}
S_1 &= \{1, 3, 7, 9, 11, 13, 17, 19\}, &|S_1| &= 8 = \varphi(20/1) = \varphi(20) \\
S_2 &= \{2, 6, 14, 18\}, &|S_2| &= 4 = \varphi(20/2) = \varphi(10) \\
S_4 &= \{4, 8, 12, 16\}, &|S_4| &= 4 = \varphi(20/4) = \varphi(5) \\
S_5 &= \{5, 15\}, &|S_5| &= 2 = \varphi(20/5) = \varphi(4) \\
S_{10} &= \{10\}, &|S_{10}| &= 1 = \varphi(20/10) = \varphi(2) \\
S_{20} &= \{20\}, &|S_{20}| &= 1 = \varphi(20/20) = \varphi(1)
\end{aligned}
$$

Note the total of the cardinalities is 20 as the sets S_d partition $\{1, \ldots, 20\}$.

This remarkable combinatorial identity allows us to deduce that there are not enough non-primitive elements of $\mathbb{Z}/(p)^\times$ to account for the whole group. The proof below, slightly modernized, is essentially due to Gauss.

Theorem 4.8.5 (Primitive Root Theorem)
Let p be prime. For a divisor d of $p - 1$, the number of elements of order d in $\mathbb{Z}/(p)^\times$ is $\varphi(d)$. In particular, there are $\varphi(p-1)$ primitive roots (so at least one!) modulo each prime p.

Proof Let d be a divisor of $p - 1$, let T_d be the set of elements of $\mathbb{Z}/(p)^\times$ of order d. Since every element of $\mathbb{Z}/(p)^\times$ has order dividing $p - 1$ (Corollary 4.6.18), these sets partition $\mathbb{Z}/(p)^\times$. The set T_d is a subset of the elements $x \in \mathbb{Z}/(p)$ satisfying $x^d = [1]$, and by Lemma 3.1.13, there are at most d such elements since $\mathbb{Z}/(p)$ is an integral domain. If a is any element of T_d, then since a has order d the set $\{a, a^2, \ldots, a^d\}$ consists of exactly d distinct elements of $\mathbb{Z}/(p)^\times$ that satisfy the equation $x^d = [1]$, so must represent all such solutions. This implies that T_d must consist entirely of powers a^k of a, and in particular the ones with order exactly d— those with $\gcd(k, d) = 1$ (Exercise 4.44). Since there are $\varphi(d)$ of these, we conclude that $|T_d|$ is either 0 (if no such a existed) or $\varphi(d)$ (if there is at least one such a). To deduce the statement of the theorem, we have left to show that $|T_d|$ is never zero, for which we employ Theorem 4.8.3: since the T_d partition $\mathbb{Z}/(p)^\times$ and $|T_d| \leq \varphi(d)$

for each $d \mid n$, the chain

$$p - 1 = \sum_{d \mid p-1} |T_d| \leq \sum_{d \mid p-1} \varphi(d) = p - 1$$

implies that each summand on the left of the middle inequality must in fact be equal to its corresponding summand on the right, and so $|T_d| = \varphi(d)$ for each $d \mid n$.   □

Righteous closing

While this text does not focus heavily on applications of number theory outside of its own hallowed halls, the content of this chapter provides the foundation for one of the most important advances of the 20th century. If you are curious to see how the algebraic structures covered in this chapter have revolutionized cryptography, computing, and commerce, you may want to dive into Sections 9.3.1 and 9.3.2 now. The applications explored in these sections are provocative examples of the ways mathematics plays a critical role in modern society and will likely continue to do so for some time to come. Calculus and continuous mathematics may rule the world of classical physics and engineering, but in the digital age, number theory reigns supreme!

4.9 Exercises

Calculation and Short Answer

Exercise 4.1 Discuss the possibilities for the value p mod 6 for p a prime number. Use this to classify all "triple primes," instances of three consecutive odd numbers that are all prime.

Exercise 4.2 Devise divisibility-by-11 and reduction-mod-11 tests analogous to those we have for the modulus 9 (see Remark 4.3.6).

Exercise 4.3 Use your test from the previous problem to verify that each of the following is divisible by 11.

$$7777 \quad 178 \quad 1234554321 \quad 1234512345.$$

Come up with another interesting family of examples of multiples of 11.

Exercise 4.4 For an integer n, what are the possible values of n^2 mod 10? Explain how this allows you to instantly deduce that the number 71,415,258,713 is not a square (in $\mathbb{Z}$).

Exercise 4.5 Find requirements on n mod 3 for there to exist integers x and y such that $n = x^2 + 3y^2$.

Exercise 4.6 Reason mod 4 to show that the sum of two odd squares cannot be a square.

Exercise 4.7 Compute the following reductions (efficiently!):

(a) $1024 \cdot 3728 + 1111^2 \mod 10$;   (c) $3^{42} \mod 7$;
(b) $2^{10000} \mod 3$;   (d) $\sum_{n=1}^{100} n! \mod 12$.

Exercise 4.8 Compute $\sum_{n=1}^{100} n! \mod 100$.

Exercise 4.9 Oh, the perils of rendering fabulous homework! Your glitter gel pen ink smudged and you can't read a digit in the following equation:

$$512 \cdot 19?3125 = 1000000000.$$

You don't have a calculator on hand so you can't simply divide 1000000000 by 512. Fix the smudged digit. (Hint: Work mod 9.)

Exercise 4.10 Make a table of inverses for $\mathbb{Z}/(11)$. For each $[a] \in \mathbb{Z}/(11)^\times$, solve the equation $ax \equiv 5 \mod 11$.

Exercise 4.11 Use the Extended Euclidean Algorithm to find the inverse of $[52]$ in $\mathbb{Z}/(77)$. Use this to solve the equation

$$52x \equiv 3 \pmod{77}.$$

Exercise 4.12 Use Sunzi's Remainder Theorem to solve each system of congruences.

(a) $x \equiv 3 \pmod 5$, $x \equiv 7 \pmod 9$;
(b) $x \mod 4 = 3$, $x \mod 7 = 2$, and $x \mod 11 = 4$;
(c) $[x] = [1]$ in $\mathbb{Z}/(3)$, $[x] = [2]$ in $\mathbb{Z}/(5)$, $[x] = [4]$ in $\mathbb{Z}/(7)$, and $[x] = [3]$ in $\mathbb{Z}/(11)$.

Exercise 4.13 Suppose a department of 17 professors is dividing up the grading of a bunch of exam problems (say somewhere between 2 and 6, 000 problems). When they attempt to divide the number of problems evenly among them, they find there are three problems left over. In the subsequent bickering over who would grade the extra problems, one of the professors resigned (we are a delicate bunch). Again they tried to divide the problems evenly, only to find that now ten problems were left unassigned, and again a professor resigned amid the tumult. Happily, the remaining professors found that the problems could now be evenly divided.

How many problems were there?

Exercise 4.14 Compute $\varphi(n)$ for each $n \in \{30, 144, 143, 108\}$.

Exercise 4.15 Find all solutions n to the following equations:

1. $\varphi(2n) = \varphi(n)$;

2. $\varphi(2n) = 2\varphi(n)$;
3. $\varphi(3n) = \varphi(n)$;
4. $\varphi(3n) = 2\varphi(n)$;
5. $\varphi(3n) = 3\varphi(n)$.

Exercise 4.16 Compute the following:

(a) 3^{27} mod 7;   (c) 12^{682} mod 55;
(b) 8^{426} mod 11;   (d) 18^{735417} mod 187.

Exercise 4.17 Write down the analog of Lemma 4.5.5 for a group written in additive notation.

Exercise 4.18 Write down the definition of $\langle g \rangle$, the cyclic subgroup generated by g, for a group written in additive notation. Translate the proof of Theorem 4.6.3 to the additive setting.

Exercise 4.19 Determine with justification whether each subset is a subgroup of the indicated *additive* group.

(a) $\{[1], [3], [5], [7]\}$ in $\mathbb{Z}/(8)$;
(b) $\{[0], [2], [4], [6]\}$ in $\mathbb{Z}/(8)$;
(c) $\{[0], [2], [4], [6]\}$ in $\mathbb{Z}/(9)$.

Exercise 4.20 Determine with justification whether each subset is a subgroup of the indicated *multiplicative* group.

(a) $\{[1], [2], [4]\}$ in $\mathbb{Z}/(7)^\times$;
(b) $\{[1], [2], [4]\}$ in $\mathbb{Z}/(5)^\times$;
(c) $\{[1], [4], [9], [16]\}$ in $\mathbb{Z}/(17)^\times$.

Exercise 4.21 The sets H_1 and H_2 given below are subgroups of $\mathbb{Z}/(31)^\times$. For each, write down the partition of $\mathbb{Z}/(31)^\times$ consisting of the cosets of those subgroups.

$$H_1 = \{[1], [2], [4], [8], [16]\} \qquad H_2 = \{[1], [5], [6], [25], [26], [30]\}.$$

Exercise 4.22 Let us say that $a \in \mathbb{Z}/(n)^\times$ is a square if there exists $b \in \mathbb{Z}/(n)^\times$ with $b^2 = a$. Find the squares of $\mathbb{Z}/(17)^\times$ and show that the set of squares in $\mathbb{Z}/(n)^\times$ always forms a subgroup.

Exercise 4.23 Find the order of each element of $\mathbb{Z}/(13)^\times$.

Exercise 4.24 Find all of the cyclic subgroups of $\mathbb{Z}/(24)^\times$. Find a subgroup of $\mathbb{Z}/(24)^\times$ that is not cyclic (i.e., is not the cyclic subgroup generated by any element of $\mathbb{Z}/(24)^\times$).

In the following exercises, we make use of the idea of reducing a polynomial mod n. Recall that for $f \in \mathbb{Z}[x]$, its *reduction mod n* is the polynomial $\overline{f} \in \mathbb{Z}/(n)[x]$ obtained by reducing all the coefficients of f modulo n.

Exercise 4.25 Let $f(x) \in \mathbb{Z}[x]$ and $a \in \mathbb{Z}$. Write down a formal interpretation of the loose claim "Plugging a into f and then reducing the value mod n gives the same answer as if I plugged a mod n into $\overline{f}$."

Exercise 4.26 Consider, for p an odd prime, the polynomial $f(x) = x^p - x$. This polynomial has three integer roots. How many roots does its reduction $\overline{f}$ have mod p? Deduce from this the factorization of $\overline{f}$ in $\mathbb{Z}/(p)[x]$.

Exercise 4.27 Find counter-examples to each of the following claims about degrees of reductions:

(a) $\deg(f) = \deg(\overline{f})$.
(b) If $\deg(f) = \deg(\overline{f})$ and $\deg(g) = \deg(\overline{g})$, then $\deg(fg) = \deg(\overline{fg})$.
(c) If $\overline{f}$ and $\overline{g}$ have the same number of roots, then $\deg(\overline{f}) = \deg(\overline{g})$.

Exercise 4.28 Find all roots of the polynomial $x^2 + x + 1$ in $\mathbb{Z}/(2)$ and then in $\mathbb{Z}/(3)$.

Exercise 4.29 Find all points in $\mathbb{Z}/(11) \times \mathbb{Z}/(11)$ that lie on the elliptic curve $y^2 = x^3 - 8x$.

Formal Proofs

Exercise 4.30 Verify from first principles that if $a \equiv b \pmod{n}$ and $c \equiv d \pmod{n}$, then $ac \equiv bd \pmod{n}$.

Exercise 4.31 Prove that for all primes $p > 3$, we have $24 \mid p^2 - 1$.

Exercise 4.32 Prove that for all $a \in \mathbb{Z}$, if $7 \nmid a$ then 7 divides $a^3 + 1$ or 7 divides $a^3 - 1$.

Exercise 4.33 Prove that for all $a \in \mathbb{Z}$, if p is prime then $p \mid a^p + (p-1)!a$.

Exercise 4.34 Prove that for all $a \in \mathbb{Z}$, we have $a^7 \equiv a$ mod 42.

Exercise 4.35 Prove that for all $n \in \mathbb{N}$, the sequence of mod-n Fibonacci numbers, F_k mod n, is periodic.

Exercise 4.36 Consider the set H of polynomial multiples of $x^2 + 1$ in $\mathbb{R}[x]$. Prove that H is a subgroup of $\mathbb{R}[x]$ under addition.

Exercise 4.37 The congruence relation defined in Definition 4.6.6 is often called *left congruence*. Given a group G and a subgroup H of G, we define the *right congruence relation* by setting $a \sim b$ if and only if $ba^{-1} \in H$, for all $a, b \in G$. Prove that this relation is an equivalence relation on the set of elements in G.

Exercise 4.38 Prove Theorem 4.7.2 that the Cartesian product of two rings is itself a ring using the component-wise operations.

Exercise 4.39 Prove that a finite group with a prime number of elements must be cyclic (i.e., is the cyclic subgroup generated by one of its elements).

For Exercises 4.40 through 4.43, recall that $|\langle g \rangle|$ denotes the order (number of elements in) the group $\langle g \rangle$, while $|g|$ is the order (least positive power k giving $g^k = 1$) of the element g.

Exercise 4.40 Prove Corollary 4.6.5: Given $g \in G$ of finite order, prove that $|\langle g \rangle| = |g|$.

Exercise 4.41 Suppose $g \in G$ and $a, b \in \mathbb{Z}$. Show that if $g^a = 1$ and $g^b = 1$, then $g^{\gcd(a,b)} = 1$. Conclude that

$$g^n = 1 \iff |g| \mid n.$$

Exercise 4.42 Prove that if $a, b \in \mathbb{Z}/(n)^\times$ and $\gcd(|a|, |b|) = 1$, then $|ab| = |a||b|$. Find a counter-example in the case $\gcd(|a|, |b|) \neq 1$. (Here, $|a|$ denotes the order of the element a in $\mathbb{Z}/(n)^\times$ and not an absolute value, and similarly for $|b|$ and $|ab|$.)

Exercise 4.43 Given an element a of an arbitrary group G, let $|a| = d$. Prove that $|a^k| = d$ if and only if $\gcd(k, d) = 1$.

Exercise 4.44 Generalizing the previous problem, show that if the order of a is d, then the order of a^k is $\dfrac{d}{\gcd(d, k)}$.

Exercise 4.45 Prove that for all $d, n \in \mathbb{N}$, if $d \mid n$ then $\varphi(d) \mid \varphi(n)$. Is the converse true?

Exercise 4.46 Show that if $a \in \mathbb{Z}/(n)$ is nilpotent (see Problem F.6), then a is not a unit.

Exercise 4.47 Prove that in every Pythagorean triple, at least one side length is a multiple of 5.

Exercise 4.48 Prove that there are no solutions to the Diophantine equation $x^3 + 117y^3 = 5$.

Exercise 4.49 Prove the inverse of Wilson's Theorem: if n is not prime, then

$$(n - 1)! \not\equiv -1 \pmod{n}.$$

Find a formula for $(n - 1)! \bmod n$.

Exercise 4.50 Use the previous problem to deduce the following *exact* formula for the prime-counting function:

$$\pi(n) = \sum_{j=2}^{n} \left\lfloor \cos^2 \left(\pi \cdot \frac{(j-1)! + 1}{j} \right) \right\rfloor.$$

Note the two different uses of the letter π in this problem—the one on the left is the obviously important one, but the one on the right does have some applications to circles or something.

Exercise 4.51 Define an equivalence relation on $\mathbb{Z}[i]$ by

$$\alpha \equiv \beta \iff (2+i) \mid (\alpha - \beta).$$

1. Verify that $4 + 3i \equiv 9 - 2i$, but $4 + 3i \not\equiv 12 + 10i$.
2. Show that every element of $\mathbb{Z}[i]$ is congruent to one of $0, 1, 2, i$, or $1 + i$.

Exercise 4.52 Prove that a field is a commutative ring with unity in which the non-zero elements form an abelian group under multiplication.

Remark 4.7.10 provided a couple of examples of *multiplicative functions*. The following exercise provides another.

Exercise 4.53 For a natural number n, let $\sigma(n)$ be the sum of all the positive divisors of n.

(i) Show that if $\gcd(m, n) = 1$, every divisor of mn can be written uniquely in the form $m'n'$ with $m' \mid m$ and $n' \mid n$.
(ii) Show that σ is multiplicative.
(iii) For a prime power p^k, find and prove a formula for $\sigma(p^k)$.
(iv) Use the previous part to find a formula for $\sigma(n)$ given the prime factorization of n.
(v) Compute $\sigma(28)$ and $\sigma(496)$ and comment on anything interesting.

Exercise 4.54 For $n \in \mathbb{N}$, the *Möbius μ function* $\mu(n)$ is defined as follows:

- If $v_p(n) > 1$ for some prime p, then $\mu(n) = 0$.
- Otherwise, n is a product of k distinct primes, and we set $\mu(n) = (-1)^k$.

Prove that μ is multiplicative.

Exercise 4.55 Prove that the product of two multiplicative functions is multiplicative. Conclude that for all $k \in \mathbb{N}$, the function

$$\sigma_k(n) = \sum_{d \mid n} d^k$$

is multiplicative.

Computation and Experimentation

Exercise 4.56 Write a program to compute the Euler totient function, $\varphi(n)$. The Python worksheet "Euler Totient Function" provides an outline.

Exercise 4.57 Lagrange's Polynomial Congruence Theorem says that if $f \in \mathbb{Z}[x]$ is a polynomial of degree n and p is prime, then $f(x) \equiv 0 \pmod{p}$ has at most n solutions mod p. Use a computer algebra system to perform the calculations below. The Python worksheet "Lagrange's Polynomial Congruence" provides an outline.

1. Check that $x = 52$, $x = 82$, and $x = 107$ make up three solutions to the degree-2 equation $x^2 - 2x - 4 \equiv 0 \pmod{11}$. Why doesn't this violate Lagrange's Polynomial Congruence Theorem?
2. Demonstrate that the "at most" clause is needed by giving a degree-3 polynomial $f(x)$ and a prime p such that $f(x) \equiv 0 \pmod{p}$ has fewer than 3 solutions.
3. Demonstrate that p being prime is crucial in the statement of the theorem by finding a degree-2 polynomial with more than 2 roots in $\mathbb{Z}/(21)$.

In short, the rings $\mathbb{Z}/(p)$ for a prime p are very congenial places to work: everything has an inverse, and polynomial equations have an eminently reasonable number of solutions.

Exercise 4.58 Let us return to the elliptic curve $y^2 = x^3 - 8x$, but this time over $\mathbb{Z}/(101)$. Write a program that finds all points $(x, y) \in \mathbb{Z}/(101) \times \mathbb{Z}/(101)$ that lie on this elliptic curve. The Python worksheet "Elliptic Curves mod n" provides an outline.

Exercise 4.59 Write a program that finds primitive roots mod a prime p. The Python worksheet "Primitive Roots" provides an outline.

Exercise 4.60 Write a program that implements the Extended Euclidean Algorithm (writing the gcd of a and b as a linear combination of a and b). If you wrote a program earlier to implement the Euclidean Algorithm, you may wish to build on that. The Python worksheet "Extended Euclidean Algorithm" provides an outline.

General Number Theory Awareness

Exercise 4.61 Explore how one could potentially use Fermat's Little Theorem as a primality test. Look up the definitions for *pseudoprimes* and *Carmichael numbers*. Explain how the existence of such numbers ruins your otherwise reasonable-sounding primality test.

Exercise 4.62 Look into the life of Carl Friedrich Gauss. When was his talent first recognized? What summation formula is he said to have discovered in class as a child? (Give another proof of the answer to Question 4.4.7 using this formula). Do spend some time making yourself feel better about your accomplishments to date by recalling that Gauss had no Internet access, and thus much more free time.

Exercise 4.63 The door to the math club's secret candy stash can be equipped with a password. To prevent the rampant corruption and candy laundering in a typical math club leadership, no *one* member can be entrusted with the password, but we would like it to be that any three members can collectively deduce the password and extract confectionery bliss. Research how Sunzi's Remainder Theorem (usually under the name Chinese Remainder Theorem) can be used to share a secret password in such a way.

Exercise 4.64 Explore more about decimal expansions: how does the period length of $\frac{1}{n}$ relate to the Euler φ function? Given the prime factorization of n, how do you determine the period length and the number of pre-periodic digits of $\frac{1}{n}$? What are some extremal values of the order of 10 mod n?

Exercise 4.65 Look into what happens when the moduli in Sunzi's Remainder Theorem are not relatively prime. Can anything be salvaged? What are some applications of the results?

Exercise 4.66 The section covers divisibility tests by 3, 9, and 11. Divisibility tests for 2, 4, and 8 (and other powers of 2) are relatively easy, and for 5 is even easier. Explore: What else is out there? Why do we not need a separate divisibility test for 6 or 22 or 36? Is there a divisibility-by-7 test analogous to the ones for 9 and 11?

Exercise 4.67 State and prove Lucas's Theorem. Discuss combinatorial applications and relations to theorems of this section.

Exercise 4.68 Follow-up on the thought experiment begun in Remark 4.6.13. What is the relationship between the congruence classes of $\mathbb{R}[x]$ modulo the multiples of $x^2 + 1$ and the field of complex numbers?

Exercise 4.69 A somewhat whimsical application of modular arithmetic is to be able to deduce the day of the week of any date, past or present. Look into how these algorithms employ modular arithmetic (and/or get good at implementing it mentally). Work out your day of birth without referencing a calendar.

Exercise 4.70 A second whimsical application is to the magic of card tricks. Research the "21 card trick" and explain it using the language of modular arithmetic. Generalize the trick to more cards and/or columns.

Exercise 4.71 A more significant application is to *checksums*, an application of modular arithmetic to the branch of mathematics known as coding theory. Look up how modular arithmetic is used in codes like UPCs, ISBNs, etc.

Gaussian Number Theory: $\mathbb{Z}[i]$ of the Storm

5

...wherein we prepare ourselves for the upcoming hurricane.

5.1 The Calm Before

We have in our midst a veritable maelstrom of pedagogical threads: what is a number? How do we solve Diophantine equations? How do the Fundamental Theorem of Arithmetic and surrounding notions generalize to more exotic number systems? What role does modular arithmetic have to play? The ring $\mathbb{Z}[i]$, recurring in every chapter thus far, turns out to be the eye of this storm, providing us shelter to collect our thoughts on all of these fronts before we venture out into the awe-inspiring wild.

With the usual caveat that any given Diophantine equation is usually just a convenient placeholder for the study of something deeper, let us choose to focus for the time being on the Diophantine equations

$$x^2 + y^2 = n$$

for various integers n, asking which numbers n can be written as the sum of two integer squares (and how to do so). We argued in the last chapter that the equation has no solutions if $n \equiv 3 \bmod 4$, as reducing the equation mod 4 returns $x^2 + y^2 \equiv 3$, impossible since x^2 and y^2 can each only be either 0 or 1 mod 4. But if $n \equiv 1 \bmod 4$, this technique gives us nothing (save the observation that one of x, y must be even, the other odd). This is an important and inherent limitation to the whole idea of using modular arithmetic to solve Diophantine equations—the technique can only provide impossibility statements, not guarantee the existence of solutions. For example, it's easy to see there are no solutions to the Diophantine equation

$$24x^2 + 36y^2 = 12,$$

but literally *any* integers x and y form a solution to this equation when reduced modulo 4.

© Springer Nature Switzerland AG 2022
C. McLeman et al., *Explorations in Number Theory*, Undergraduate Texts in Mathematics,
https://doi.org/10.1007/978-3-030-98931-6_5

As it turns out (spoiler alert!), this phenomenon does *not* occur for the Diophantine equation $x^2 + y^2 = p$ when p is prime: if we can solve the equation modulo 4, then we can solve it in the integers. Most excitingly, to attain this result we eschew passing to the *smaller* ring $\mathbb{Z}/(4)$ and instead work in the *bigger* ring $\mathbb{Z}[i]$. Moreover, the basic idea is quite simple: since $a^2 + b^2$ can be factored in $\mathbb{Z}[i]$ as the expression $(a+bi)(a-bi)$, being able to write a number as a sum of two squares is closely related to whether we can factor it in $\mathbb{Z}[i]$. For example, since $13 = (3+2i)(3-2i)$, we obtain the representation $13 = 3^2 + 2^2$ as a sum of two squares. We are left, therefore, with a need to understand in $\mathbb{Z}[i]$ that same infrastructure of $\mathbb{Z}$ which we have spent the several previous chapters mastering: what are the primes in $\mathbb{Z}[i]$? How do we factor Gaussian integers into primes? What could expressions like $(13 + 2i) \bmod (5 + i)$ possibly mean?

Finally, to return to the opening metaphor of this section, we mention that developing these ideas for $\mathbb{Z}[i]$ will be the forerunner to developing them for a much larger class of rings in the subsequent chapters. We encourage the reader to pay special attention to the *process* of abstraction. Beginning with a fixed idea in $\mathbb{Z}$ (e.g., the notion of a prime), what stays the same as we try to conceptualize that idea in the context of $\mathbb{Z}[i]$? What changes? How do we accommodate these changes? In particular, if we wish to mirror the development of the Fundamental Theorem of Arithmetic in $\mathbb{Z}[i]$, we should pay careful attention to its development in $\mathbb{Z}$.

5.2 Gaussian Divisibility

The process of replicating *The Path* (Figure 5.1) for $\mathbb{Z}[i]$ amounts to documenting what aspects of arithmetic in $\mathbb{Z}$ continue to hold in $\mathbb{Z}[i]$, and which instead need updating. The first major theme is that much of the basic arithmetic of $\mathbb{Z}$ holds in *any* commutative ring (see, e.g., Explorations A and G, and Section 3.3), and so while we will be interested specifically in $\mathbb{Z}[i]$ for this chapter, we may as well adopt the more general approach for later use. Recall Definition 3.1.14.

Definition 5.2.1

Given a commutative ring R and elements $\alpha, \beta \in R$, we say that α **divides** β, and write $\alpha \mid \beta$, if there exists $\gamma \in R$ such that $\beta = \alpha\gamma$. ◄

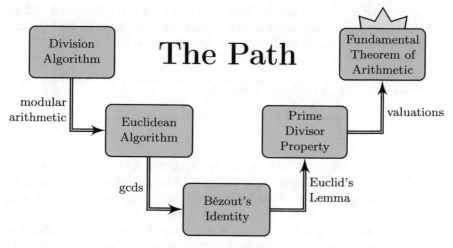

Fig. 5.1 The Path to Unique Factorization in $\mathbb{Z}$

▶ **Example 5.2.2** Since $1 + 13i = (3 + 5i)(2 + i)$, we have $3 + 5i \mid 1 + 13i$ in $\mathbb{Z}[i]$. Note that $3 \nmid 1$ and $5 \nmid 13$, so Gaussian divisibility is not as direct as checking the divisibility of real and imaginary parts.

As was the case in $\mathbb{Z}$, to check for divisibility in $\mathbb{Z}[i]$ we can simply perform the division $\frac{\beta}{\alpha}$, as long as $\alpha \neq 0$, and see if the quotient is a Gaussian integer.

▶ **Example 5.2.3** Since $7 + 4i = (1 + 2i)(3 - 2i)$, we have $3 - 2i \mid 7 + 4i$. Does $3 + 2i \mid 7 + 4i$? No, since we compute

$$\frac{7 + 4i}{3 + 2i} = \frac{(7 + 4i)(3 - 2i)}{(3 + 2i)(3 - 2i)} = \frac{29 - 2i}{13} = \frac{29}{13} - \frac{2}{13}i \notin \mathbb{Z}[i].$$

▶ **Example 5.2.4** Recall that the units of $\mathbb{Z}[i]$ are $\mathbb{Z}[i]^\times = \{\pm 1, \pm i\}$. Since $\pm 1, \pm i$ are units, they divide *everything* in $\mathbb{Z}[i]$. As in $\mathbb{Z}$, the Gaussian integer 0 divides only itself, and all Gaussian integers divide 0.

▶ **Remark 5.2.5** A reasonable concern is that we now in principle have two different definitions of the symbol $3 \mid 6$, depending on whether we parse it as "Does 3 divide 6 in $\mathbb{Z}$?" or "Does 3 divide 6 in $\mathbb{Z}[i]$?" Fortunately, the two questions are equivalent (see Exercises 5.17 and 5.18), so we need not be concerned… for now.

It is instructive to revisit proofs of the basic arithmetical results from $\mathbb{Z}$ and see how they carry over nearly verbatim to a general ring. For example, the following result precisely mirrors the analogous one (Lemma 3.2.4) in $\mathbb{Z}$, and we reuse the name.

Lemma 5.2.6 (Linear Combination Lemma)

Let R be a commutative ring, and take $\alpha, \beta, \mu, \nu, \delta \in R$. If $\delta \mid \alpha$ and $\delta \mid \beta$, then $\delta \mid (\alpha\mu \pm \beta\nu)$. ◄

We emphasize that proofs for these generalized results are often not merely *analogous* to those for the corresponding results in $\mathbb{Z}$, but essentially *verbatim* the same. We will thus often omit derivations for basic divisibility results like this one (e.g., that if $\alpha \mid \beta$ and $\beta \mid \gamma$, then $\alpha \mid \gamma$). For results that do not carry directly forward to $\mathbb{Z}[i]$, we have a second major theme: the norm function from $\mathbb{Z}[i]$ to $\mathbb{Z}$ allows us to bootstrap theorems and concepts from $\mathbb{Z}$ to construct analogs in $\mathbb{Z}[i]$. For this, we recall that the norm is multiplicative $(N(z_1)N(z_2) = N(z_1 z_2))$, that the only Gaussian integer of norm 0 is $z = 0$, and that the only Gaussian integers of norm 1 are the units of $\mathbb{Z}[i]$. Here is a first result along these lines.

Lemma 5.2.7

If $\alpha \mid \beta$ in $\mathbb{Z}[i]$, then $N(\alpha) \mid N(\beta)$ in $\mathbb{Z}$. ◄

Proof Write $\beta = \alpha\gamma$, take the norm of both sides, and use multiplicativity to get $N(\beta) = N(\alpha)N(\gamma)$, so $N(\alpha) \mid N(\beta)$. □

The lemma's contrapositive provides a quick way to verify non-divisibility.

▶ **Example 5.2.8** Does $(5 + 2i) \mid (7 + 5i)$? No, as $N(5 + 2i) = 29$, which does not divide $N(7 + 5i) = 74$.

▶ **Example 5.2.9** Note, however, that the converse to the lemma is false. For example, we have already seen that $3 + 2i \nmid 4i + 7$, but we do have divisibility of their respective norms: $13 \mid 65$.

Of fundamental import in $\mathbb{Z}$ was thinking of the prime numbers as the building blocks of the rest of the integers via prime factorization. As we begin this discussion in the Gaussian realm, note that the notion of prime in $\mathbb{Z}$ and $\mathbb{Z}[i]$ are *not* synonymous. The integer 5 is prime in $\mathbb{Z}$, but the factorization $5 = (2 + i)(2 - i)$ will show, once the terms have been properly defined, that 5 is not prime in $\mathbb{Z}[i]$. On the other hand, we will see that 7 is prime in both $\mathbb{Z}$ and $\mathbb{Z}[i]$. In cases when the extra clarity is called for, we will distinguish these two statements by saying that 5 is a *rational prime* but not a *Gaussian prime*, whereas 7 is both a rational prime and a Gaussian prime. The adjective "rational" here references that 5 is also an element of $\mathbb{Q}$.

Definition 5.2.10

Let $\pi \in \mathbb{Z}[i]$ be a Gaussian integer that is neither zero nor a unit. We say π is **prime** (and hence a *Gaussian prime*) if its only divisors are units and associates of π; otherwise, π is **composite**. ◄

Equivalently, π is prime if and only if whenever $\pi = \alpha\beta$ for some $\alpha, \beta \in \mathbb{Z}[i]$, either α or β must be a unit. Note that by norm multiplicativity, associates always have the same norm. Analogous to the classification of integers (see the list before Definition 3.2.2), we have a classification of Gaussian integers into exactly one of the following four types:

- **Zero**: Again, only 0 is 0.
- **Units**: The units are precisely $\{\pm 1, \pm i\}$.
- **Prime numbers**: $1 + 2i, 7, 3 + 2i$, etc.[1] .
- **Composite numbers**: By definition everything not yet addressed.

How do we decide if a Gaussian integer is prime? For $\mathbb{Z}$ we had the sieve of Eratosthenes as an efficient algorithm for mechanically checking primality, but the generalization to $\mathbb{Z}[i]$ is less clear. Again, the norm map helps save the day, providing a test for primeness in $\mathbb{Z}[i]$ via primeness in $\mathbb{Z}$.

Theorem 5.2.11

Let $\pi \in \mathbb{Z}[i]$. If $N(\pi)$ is prime in $\mathbb{Z}$, then π is prime in $\mathbb{Z}[i]$.

Proof Let π be an arbitrary Gaussian integer such that $N(\pi)$ is a rational prime p, and suppose $\alpha \mid \pi$. Then $\pi = \alpha\beta$ for some $\beta \in \mathbb{Z}[i]$, and taking norms of both sides gives $N(\alpha)N(\beta) = N(\pi) = p$. Now since p is prime in $\mathbb{Z}$, one of $N(\alpha)$ or $N(\beta)$ must equal p, and the other must equal 1. Thus α is either a unit or associate of π, and since α was an arbitrary divisor of π, we conclude that π is a Gaussian prime. ☐

---

[1] The contents of this list and, in particular, this "etc." are at least as mysterious as the analogous list for $\mathbb{Z}$, and will be where we turn our focus now.

This justifies the inclusion of two of the examples of primes in our list above: $1 + 2i$ and $3 + 2i$ have prime norms in $\mathbb{Z}$, so the theorem tells us they are both themselves prime in $\mathbb{Z}[i]$. The converse of Theorem 5.2.11 is highly non-true[2] as, among many other examples, 7 is a Gaussian prime but $N(7) = 49$ is not a rational prime. So to decide whether a rational prime p "stays prime" in $\mathbb{Z}[i]$, we will need to do something more nuanced than simply taking norms. Fortunately this ties in precisely to the Diophantine equations that started the chapter.

Theorem 5.2.12
The rational prime $p \in \mathbb{Z}$ is a Gaussian prime if and only if $p \neq a^2 + b^2$ for all $a, b \in \mathbb{Z}$.

Proof Suppose $p \in \mathbb{Z}$ is not the sum of two squares, and suppose that α is a divisor of p in $\mathbb{Z}[i]$. Then taking norms, we see $N(\alpha) \mid N(p) = p^2$, so $N(\alpha) = 1, p,$ or p^2. We show the middle case is impossible: if $N(\alpha) = p$ then writing $\alpha = a + bi$ shows that $a^2 + b^2 = p$, contradicting our assumption that p could not be written as a sum of squares. Thus, it must be the case that $N(\alpha) = 1$ or p^2. This shows that an arbitrary divisor α of p is either a unit or an associate of p in $\mathbb{Z}[i]$, and thus p is by definition a Gaussian prime. Finally, for the converse, if $p = a^2 + b^2 = (a + bi)(a - bi)$, then $a + bi$ is a divisor of p with norm p and thus is neither a unit nor an associate of p, so p is not prime. $\square$

Combining this result with Lemma 4.3.7 gives the following.

Corollary 5.2.13

Given a rational prime $p \in \mathbb{Z}$, if $p \equiv 3 \bmod 4$, then p is a Gaussian prime. ◄

We now have an ample source of Gaussian primes—for example, in addition to the Gaussian prime 7 given above, there is also 11. This gives us a starting point to think about what primes look like in the Gaussian integers by describing precisely which rational primes stay prime when viewed as Gaussian integers, but this is only a partial answer. To complete the story we will need to start making inroads on *The Path* described in 5.1. Before doing so, it is very helpful to develop a more geometric visualization of Gaussian integers and their arithmetic properties.

---

[2] Technically, something is either true or false, without levels of non-trueness. But boy howdy is this not true.

Exploration I

The Gaussian Lattice ◀

The standard visualization of the Gaussian lattice is the points of the complex plane with integer coordinates, with the point (a, b) representing the number $a + bi$.

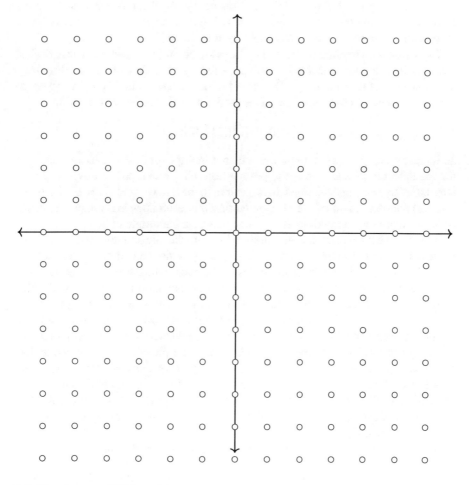

I.1 Plot the set of $\mathbb{Z}[i]$-multiples of $1 + i$, enough to get a feel for the geometry of these multiples.

I.2 Repeat, on the same grid but in a different color, for the multiples of $2 + i$. Are there any common multiples of $1 + i$ and $2 + i$? Any rational integer common multiples?

Fig. 5.2 Multiples of 3 in $\mathbb{Z}$

5.3 Gaussian Modular Arithmetic

As will happen googolplexes[3] of times in the upcoming chapters, we make progress on a Gaussian version of a concept by mirroring the integer version of it. We begin by revisiting our construction of modular arithmetic in $\mathbb{Z}$.

Pay particular attention to the dashed lines added on to the picture. These dashed lines partition the number line $\mathbb{Z}$ into a repeating geometric pattern, dividing it up into translates of the interval $[0, 3)$. The set of integer points in this region serves as a natural system of coset representatives mod 3 as well, leading to the group

$$\mathbb{Z}/(3) = \{[0], [1], [2]\},$$

an integer n being in $[x]$ if n is x more than a multiple of 3. Particularly important for the upcoming discussion is the observation that these representatives were not inevitable. By shifting the dashed lines one unit to the left, we could have alternatively selected representatives of $-1, 0, 1$, reflecting that each integer is either a multiple of 3, one more than a multiple of 3, or one less than a multiple of 3.

Moving to $\mathbb{Z}[i]$, we replace our number line with the integer lattice in the complex plane, as shown in Figure 5.3. The multiples of 3 are now much more plentiful: we have all of the previous integer multiples of 3 (indeed, the entirety of Figure 5.2 is contained in the horizontal axis of Figure 5.3), but also multiples like $3i, 6i, 9i$, etc., on the vertical axis, and further a full rectangular lattice of them including multiples like $3(1 + i) = 3 + 3i$. Also as shown in Figure 5.2, we include dashed lines to divide the complex plane into repeating geometric patterns. Each square contains precisely one black dot, representing a Gaussian multiple of 3, and eight white dots representing possible positions relative to such a multiple. For example, to the right of the black dot in each square is a Gaussian integer that is one more than a multiple of 3, above each black dot is a Gaussian integer that is i more than a multiple of 3, above-right of each dot is a Gaussian that is $1 + i$ more than a multiple of 3, and so on for each of the dots in the square.

We conclude that while the standard mod-3 equivalence relation

$$m \equiv n \iff 3 \mid (m - n)$$

gives rise to the three elements of $\mathbb{Z}/(3) = \{[0], [1], [2]\}$, the Gaussian analog of the identical equivalence relation would have nine equivalence classes:

$$\mathbb{Z}[i]/(3) = \{[0], [1], [2], [i], [i + 1], [i + 2], [2i], [2i + 1], [2i + 2]\},$$

---

[3] More or less. Probably less.

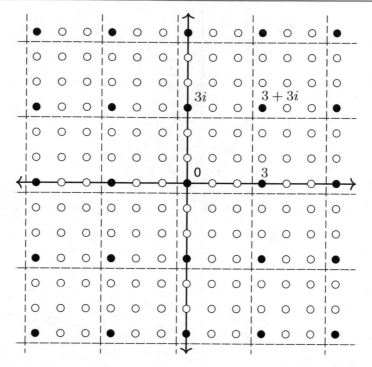

Fig. 5.3 Multiples of 3 in $\mathbb{Z}[i]$

one per dot in the square. As before, these equivalence classes form a partition of $\mathbb{Z}[i]$. We formalize this construction below.

Definition 5.3.1

Given $\beta \neq 0 \in \mathbb{Z}[i]$, define the equivalence relation **congruence mod** β on $\mathbb{Z}[i]$ as follows: for $\mu, \nu \in \mathbb{Z}[i]$, we have

$$\mu \equiv \nu \bmod \beta \iff \beta \mid (\mu - \nu).$$

A **system of coset representatives** modulo β is a choice of one element from each equivalence class. Given such a system, for $\mu \in \mathbb{Z}[i]$ the symbol $\mu \bmod \beta$ denotes the unique element of this system to which μ is congruent. ◄

Directly mirroring the arguments of Section 4.2 (see upcoming Exploration J), we can show that addition and multiplication of mod-β equivalence classes is well defined and hence these equivalence classes can be made into a ring completely analogous to $\mathbb{Z}/(n)$.

Definition 5.3.2

For $\beta \in \mathbb{Z}[i]$, we define the ring $\mathbb{Z}[i]/(\beta)$ to be the set of equivalence classes of the mod-β congruence relation on $\mathbb{Z}[i]$. ◄

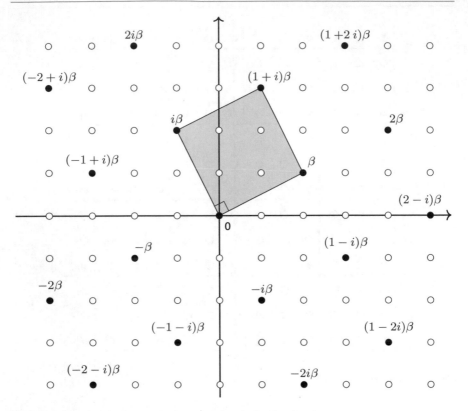

Fig. 5.4 The square lattice and fundamental domain for $\beta = 2 + i$

Geometrically, $\mathbb{Z}/(\beta)$ can be visualized as the collection of points in what we call a **fundamental domain** for β, a region of the complex plane whose translates tile the plane and contain precisely one representative from each mod-β equivalence class (that is, each element of $\mathbb{Z}/(\beta)$). Any of the squares in Figure 5.3 could serve as a fundamental domain modulo 3 in $\mathbb{Z}[i]$. Or, like we did for the multiples of 3 in $\mathbb{Z}$, we could shift all of the dashed lines one unit down and left, and then choosing the square centered at the origin to be the fundamental domain, giving the system of mod-3 coset representatives $\{0, \pm 1, \pm i, \pm 1 \pm i\}$.

Unlike $\mathbb{Z}/(n)$, though, where we always have the natural set $\{0, 1, \ldots, n - 1\}$ of coset representatives, it is less clear how to systematically enumerate such a list modulo an arbitrary $\beta \in \mathbb{Z}[i]$. Let's explore a more complicated example.

In Figure 5.4 we plot the multiples of $2 + i$, choose the shaded square as a fundamental domain, see that $\mathbb{Z}[i]/(2+i)$ consists of five elements (the four elements interior to the fundamental domain and any one corner), and identify a system of coset representatives modulo $2 + i$:

$$\mathbb{Z}[i]/(2 + i) = \{[0], [i], [1 + i], [2i], [1 + 2i]\}.$$

To reiterate, by translating the shaded square in Figure 5.4, we can cover the entire plane in such a way that every Gaussian integer is in the same position within a translated square as one of $\{0, i, 1+i, 2i, 1+2i\}$, our choice of coset representatives. We can find from Figure 5.4 that, for example,

$$2 + 2i \equiv i \bmod 2 + i, \quad -1 - i \equiv 1 \bmod 2 + i, \quad \text{and} \quad 2 - 4i \equiv 0 \bmod 2 + i.$$

Or, using our selected system of coset representatives, write $(2+2i) \bmod (2 + i) = i$, etc.

Establishing this framework represents a major landmark on our journey down *The Path*, as all of the number-theoretic benefits of modular arithmetic are now at our disposal for understanding the structure of the Gaussian integers.

5.4   Gaussian Division Algorithm: The Geometry of Numbers

Mathematicians use the phrase *the geometry of numbers* to describe the general use of a picture like that below, a lattice sitting inside of the complex plane (or a higher dimensional analog), to better understand its algebraic structure. For example, a closer look at the fundamental domain lends some insight as to the number of equivalence classes modulo a Gaussian integer.

To begin, we have been somewhat glib using a picture as a proxy for a formal argument, so let's pause to establish some facts that we took for granted in the last section—namely, that the multiples of β form a lattice within $\mathbb{Z}[i]$, and that the fundamental domain generated by any β (not just $\beta = 2 + i$) is indeed always a square. The justification hinges on the geometric interpretation of complex number arithmetic begun in Section 1.3. Namely, if $\beta \in \mathbb{Z}[i]$, then the real integer multiples 2β, 3β, etc. appear as equally spaced lattice points on the line through the origin and β. Further, taking $\beta = a + bi$, since $i\beta = i(a + bi) = -b + ia$ is the rotation of β by 90 degrees counter-clockwise around the origin, the imaginary-integer multiples of β comprise a second line of lattice points perpendicular to the first. Finally, *any* Gaussian multiple of β is a sum of integer multiples of β and $i\beta$ since $(c + id)\beta = c\beta + di\beta$, and so the collection of all multiples of β is a square lattice (much like $\mathbb{Z}[i]$ itself). Indeed, the multiples $\beta\mathbb{Z}[i]$ of β in $\mathbb{Z}[i]$ can be viewed as a sublattice of $\mathbb{Z}[i]$ obtained by stretching and rotating the full lattice.

Now to extract something from this picture: again writing $\beta = a + bi$, we see that the distance from β to the origin is $\sqrt{a^2 + b^2}$, the square root of $N(\beta) = a^2 + b^2$ (Figure 5.5). We conclude that the fundamental domain has area $N(\beta)$. As it turns out, this is precisely the number of equivalence classes in the fundamental domain— we have $N(3) = |\mathbb{Z}[i]/(3)| = 9$ and $N(2 + i) = |\mathbb{Z}[i]/(2 + i)| = 5$. This is more or less clear when β is a rational integer since the fundamental domain is a square with sides parallel to the axes, and likewise when β is prime, as then the only lattice points on the boundary of the fundamental domain are the vertices of the square. The general claim can be argued in several ways, and we refer the interested reader to Pick's Theorem, an elementary counting tool in geometry, for a complete answer (Exercise 5.29).

Finally, the real punchline of this section: not only can we write down an explicit list of representatives for the mod-β congruence classes, but we're offered an upper bound on how far away an arbitrary Gaussian integer can be from a multiple of any

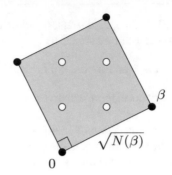

Fig. 5.5 A fundamental domain

given β. Namely, since the fundamental domain is a square (of side length $\sqrt{N(\beta)}$), an upper bound for how far any $\alpha \in \mathbb{Z}[i]$ could be from a lattice point is half the length of the diagonal of the fundamental domain, $\frac{1}{2}\sqrt{2N(\beta)}$, as shown in Figure 5.6.

We conclude that the difference, $\alpha - \chi\beta$, between α and the closest multiple $\chi\beta$ of β must be a Gaussian integer ρ such that

$$\sqrt{N(\rho)} \leq \frac{1}{2}\sqrt{2N(\beta)} < \sqrt{N(\beta)} \leq N(\beta).$$

We arrive at our desired result:

Theorem 5.4.1 (Gaussian Division Algorithm)
If $\alpha, \beta \in \mathbb{Z}[i]$ with $\beta \neq 0$, then there exist Gaussian integers χ and ρ with $0 \leq N(\rho) < N(\beta)$ and[4]

$$\alpha = \chi\beta + \rho.$$

Equivalently, for all $\beta \neq 0$, there exists a system of coset representatives mod β where each representative has norm less than $N(\beta)$.

Before we prove this, let's pause to compare this theorem to its analog in $\mathbb{Z}$, Theorem 3.2.9. The statement there provided us with (a) a definitive list of possible remainders (namely, from 0 to one less than the dividend) and (b) a unique value of the quotient and remainder. We have discussed in Section 5.3 the situation of remainders—it is easy to choose a set of remainders, but there is no clear, universally accepted choice of them. After making such a choice, we can evaluate expressions

---

[4] The ideal Gaussian analog to the notation $a = qb + r$ would require the sadly non-existent Greek analog of the letter q. There is an *ancient* Greek letter Ϙ ("koppa") serving that role, but since it would be a pain to typeset the symbol Ϙ in LATEX, we use χ instead.

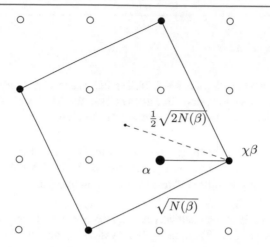

Fig. 5.6 Off the grid

of the type α mod β. We do not, alas, have uniqueness (Exercise 5.4). As we will see in the next section, neither of these complications impedes progress. All that is crucial is the guarantee of the theorem that upon applying the Division Algorithm the norm of the remainder is less than the norm of the divisor. This guarantees that a Euclidean Algorithm analog for the Gaussian integers will necessarily terminate.

Proof (of Theorem 5.4.1) By analogy with the proof in $\mathbb{Z}$, we define $S = \{\alpha - \chi\beta : \chi \in \mathbb{Z}[i]\}$, and let $\mathcal{N} = \{N(z) : z \in S\}$. Note that $\mathcal{N}$ is a non-empty set of non-negative integers, so has a least element r by the Well-Ordering Principle. Thus we can write $r = N(\rho)$, where $\rho \in S$, and so there exists some χ such that $\rho = \alpha - \chi\beta$, which implies $\alpha = \chi\beta + \rho$. Now, if $N(\rho) \geq N(\beta)$, then the distance between α and $\chi\beta$ is at least the side length of the square fundamental domain for β. This implies that some vertex $\chi'\beta$ in that lattice is nearer to α than is $\chi\beta$, which contradicts the minimality of r. Thus $N(\rho) < N(\beta)$. □

The careful formulation of the result also produces a second more computational approach to finding quotients and remainders. We note that for $\chi, \rho \in \mathbb{Z}[i]$, we have $\alpha = \chi\beta + \rho$ with $N(\rho) < N(\beta)$ if and only if

$$\frac{\alpha}{\beta} = \chi + \frac{\rho}{\beta} \qquad \text{with} \qquad N\left(\frac{\rho}{\beta}\right) < 1.$$

Thus, as long as we know that any $\frac{\alpha}{\beta} \in \mathbb{Q}[i]$ is within 1 unit of some $\chi \in \mathbb{Z}[i]$, there will be an appropriate χ and ρ in $\mathbb{Z}[i]$ to satisfy the conclusion of the Division Algorithm. To achieve this in practice we *divide and round*, that is, we compute the quotient $\frac{\alpha}{\beta}$ in $\mathbb{Q}[i]$ and obtain χ by rounding the real and imaginary parts to a nearest integer.

▶ **Example 5.4.2** To enact the Division Algorithm for $\alpha = 7+4i$ and $\beta = 3+2i$, we recall from Example 5.2.3 that $\frac{7+4i}{3+2i} = \frac{29}{13} - \frac{2}{13}i$ and round the real and imaginary parts to the nearest integers to obtain χ. We get $\chi = 2 - 0i = 2$, and then $\rho = \alpha - \beta\chi = 1$.

Exploration J

Gaussian Miscellany ◄

J.1 Earlier we gave the set $\{0, i, 1 + i, 2i, 1 + 2i\}$ as a system of representatives for $\mathbb{Z}[i]/(2+i)$. While a valid system, this does not satisfy the requirements of Theorem 5.4.1. Find one that does. How small can you force $N(\rho)$ to be?

J.2 How many equivalence classes are there modulo $1 + i$? Find a set of representatives modulo $1 + i$ satisfying the requirements of the Division Algorithm. Finally, propose a "divisibility-by-$(1 + i)$" test for Gaussian integers.

J.3 Choose a system of representatives modulo $3 + i$ that satisfy the conditions of the Division Algorithm. Then, using that system, reduce the following Gaussian integers mod $3 + i$:

$$24 - 2i \qquad 42 + 17i \qquad 300 + 101i.$$

J.4 Propose a definition of the gcd for two Gaussian integers and what it should mean for two Gaussian integers to be relatively prime.

J.5 Consider the claim that for each $\beta \neq 0$ in $\mathbb{Z}[i]$, the set $\mathbb{Z}[i]/(\beta)$ is a ring. Which ring axioms are obvious? Which steps of the proof are completely analogous to those for $\mathbb{Z}/(n)$? Is there anything that requires extra care?

5.5   A Gausso-Euclidean Algorithm

We're well on our way to a Euclidean algorithm for Gaussian integers, which has been the cornerstone of most of our mathematical discoveries, including the next step on *The Path*, Bézout's Identity. A caution is in order here when defining gcds: if γ divides both α and β, then so do $i\gamma$, $-\gamma$, and $-i\gamma$, all of equal norm. As we have already confronted this phenomenon before, labeling these four elements as *associates* of one another, we have a straightforward solution to the problem: we acknowledge that gcds are non-unique, and use the phrase "unique up to units" or "unique up to associates" to encapsulate this ambiguity.

Definition 5.5.1

Given $\alpha, \beta \in \mathbb{Z}[i]$ (not both zero), we define a **greatest common divisor** of α and β to be any of the common divisors of α and β of greatest norm. As a slight abuse of notation, we use $\gcd(\alpha, \beta)$ to refer to any one of these choices, all of which are associates of one another (Exercise 5.22). We say α and β are **relatively prime** if their only common divisors are units, or equivalently, when $N(\gcd(\alpha, \beta)) = 1$ (or just $\gcd(\alpha, \beta) = 1$). ◄

One of the principal benefits of abstraction in mathematics is that the more encompassing a definition becomes, the more becomes proved with each new proof idea. For example, Lemma 5.2.6 shows that divisibility is preserved upon taking linear combinations in *any* ring, and so in particular this holds in $\mathbb{Z}[i]$. The proofs of the following theorems are nearly verbatim copies of their integer counterparts, and we largely leave the fleshing out of details to the reader[5].

For example, a direct application of Lemma 5.2.6 is to simplify gcd computations by subtracting off multiples. Section 5.3 primed us for this by showing that we can interpret the expression $\alpha \bmod \beta$ as a Gaussian integer ρ of norm less than $N(\beta)$ that is obtainable by subtracting from α a multiple of β.

Lemma 5.5.2 (The Reduction Lemma)

For $\alpha, \beta \in \mathbb{Z}[i]$, let ρ be any remainder guaranteed by the Division Algorithm (Theorem 5.4.1). Then

$$\gcd(\alpha, \beta) = \gcd(\beta, \rho).$$ ◄

Theorem 5.5.3 (The Euclidean Algorithm for Gaussian Integers)

For all $\alpha, \beta \in \mathbb{Z}[i]$, $\beta \neq 0$, set $\rho_{-1} = \alpha$ and $\rho_0 = \beta$. For each $j \geq 0$, recursively define $\rho_{j+1} = \rho_{j-1} \bmod \rho_j$, a remainder as given by the Gaussian

---

[5] Yeah, yeah, we know, we know. When you inevitably write your own textbook, you can do this too. It's quite liberating.

Division Algorithm	Reduction Lemma	$N(\rho)$
$11 + 3i = -i(1 + 8i) + (3 + 4i)$	$\gcd(11 + 3i, 1 + 8i) = \gcd(1 + 8i, 3 + 4i)$	25
$1 + 8i = 1(3 + 4i) + (-2 + 4i)$	$\gcd(1 + 8i, 3 + 4i) = \gcd(3 + 4i, -2 + 4i)$	20
$3 + 4i = -i(-2 + 4i) + (-1 + 2i)$	$\gcd(3 + 4i, -2 + 4i) = \gcd(-2 + 4i, -1 + 2i)$	5
$-2 + 4i = 2(-1 + 2i) + 0$	$\gcd(-2 + 4i, -1 + 2i) = \gcd(-1 + 2i, 0)$	0

Fig. 5.7 The Gausso-Euclidean Algorithm

Division Algorithm. Then for some $n \geq 1$ the process terminates with $\rho_n = 0$ and ρ_{n-1} equal to a greatest common divisor of α and β.

The following example illustrates the process, almost completely analogous to our calculations of gcd's in the rational integers.

▶ **Example 5.5.4** Find $\gcd(11 + 3i, 1 + 8i)$.

Solution As with $\mathbb{Z}$, we systematically reduce the larger Gaussian integer (as measured by norms) modulo the other, as shown in Figure 5.7. We conclude that $\gcd(11 + 3i, 1 + 8i) = -1 + 2i$. In general, the gcd will be the last non-zero remainder in this process. Further, retracing our steps and solving for ρ_j at the jth step, we can find a linear combination of α and β giving this gcd.

$$
\begin{aligned}
-1 + 2i &= (3 + 4i) + i(-2 + 4i) \\
&= (3 + 4i) + i((1 + 8i) - 1(3 + 4i)) \\
&= (1 - i)(3 + 4i) + i(1 + 8i) \\
&= (1 - i)((11 + 3i) + i(1 + 8i)) + i(1 + 8i) \\
&= (1 - i)(11 + 3i) + (1 + 2i)(1 + 8i).
\end{aligned}
$$

Thus $(1 - i)(11 + 3i) + (1 + 2i)(1 + 8i) = -1 + 2i$.   □

As before, the process of performing the Euclidean algorithm while keeping track of the steps used to get there, thereby writing the greatest common divisor as a linear combination of the two initial Gaussian integers, is called the *extended Euclidean Algorithm* and leads to Bézout's Identity in $\mathbb{Z}[i]$.

Theorem 5.5.5 (Bézout's Identity)
Let $\alpha, \beta \in \mathbb{Z}[i]$, not both zero, and let γ be a gcd of α and β. Then there exist $\mu, \nu \in \mathbb{Z}[i]$ such that

$$\alpha\mu + \beta\nu = \gamma.$$

Proof Let $S = \{\gamma = \mu\alpha + \nu\beta \neq 0 : \mu, \nu \in \mathbb{Z}[i]\}$. The set $N(S) = \{N(\gamma) : \gamma \in S\}$ is a non-empty subset of $\mathbb{N}$, and by the Well-Ordering Principle there exists a smallest element $d \in N(S)$. Let δ be any element of S with $N(\delta) = d$. By definition of S, we have $\delta = \alpha\sigma + \beta\tau$ for some $\sigma, \tau \in \mathbb{Z}[i]$. By the Division Algorithm, there exist Gaussian integers χ and ρ, with $N(\rho) < N(\delta)$, such that $\alpha = \chi\delta + \rho$, and so

$$\rho = \alpha - \chi\delta = \alpha - \chi(\alpha\sigma + \beta\tau) = \alpha(1 - \chi\sigma) + \beta(-\chi\tau).$$

This shows that $\rho \in S$ if $N(\rho) > 0$, which would contradict the fact that δ had minimal norm (since $N(\rho) < N(\delta)$). It follows that $N(\rho) = 0$, so $\rho = 0$, and thus $\delta \mid \alpha$. An analogous argument proves that $\delta \mid \beta$, and so $\delta \mid \gamma$ as well (recalling that γ is the given gcd of α and β we seek to write as a linear combination of α and β). Write $\gamma = \phi\delta$. We get

$$\gamma = \phi\delta = \phi(\alpha\sigma + \beta\tau) = \alpha(\phi\sigma) + \beta(\phi\tau),$$

as desired.   □

Corollary 5.5.6

If $a, b \in \mathbb{Z}$ have $\gcd(a, b) = 1$ in $\mathbb{Z}$, then we have $\gcd(a, b) = 1$ in $\mathbb{Z}[i]$ as well.
◀

Proof If $am + bn = 1$ in $\mathbb{Z}$, then $am + bn = 1$ in $\mathbb{Z}[i]$, too!   □

A caveat is in order here. In general, if d is a linear combination of a and b, that does *not* mean that d is the gcd of a and b: all we know is that $\gcd(a, b) \mid d$. However, 1 is special: if 1 is a linear combination of a and b, then since their gcd must then divide 1, their gcd *is* 1.

As before, Bézout's Identity leads to Euclid's lemma and the Prime Divisor Property.

Lemma 5.5.7 (Euclid's Lemma)

Given Gaussian integers α, β and γ, if α and β are relatively prime and $\alpha \mid \beta\gamma$, then $\alpha \mid \gamma$. ◀

Lemma 5.5.8 (Gaussian Prime Divisor Property)

If $\pi \in \mathbb{Z}[i]$ is a Gaussian prime and $\pi \mid \beta\gamma$, then $\pi \mid \beta$ or $\pi \mid \gamma$. More generally, if $\pi \mid \beta_1\beta_2 \cdots \beta_n$, then $\pi \mid \beta_j$ for some $1 \leq j \leq n$. ◄

As was the case with Bézout's Identity and the Euclidean Algorithm, the proofs of these lemmas are directly analogous to their counterparts for the integers. At long last, we reach home.

Theorem 5.5.9 (Fundamental Theorem of Gaussian Arithmetic)
Every non-zero Gaussian integer can be written in the form

$$\alpha = \varepsilon\pi_1 \cdots \pi_k,$$

where ε is a unit, $k \geq 0$, and each π_j a Gaussian prime. Moreover, this form is unique up to reordering and associates.

Proof (sketch) Recall that the proof of the Fundamental Theorem of Arithmetic in $\mathbb{Z}$ has two parts: an existence proof and a uniqueness proof. As was the case in $\mathbb{Z}$ we can use strong induction to prove the existence of a prime factorization into Gaussian primes. This time, rather than inducting on the natural number a, we induct on $N(\alpha)$.

The crux of the uniqueness proof in $\mathbb{Z}$ was the Prime Divisor Property, and it is again in $\mathbb{Z}[i]$. Suppose we had two distinct factorizations of α as the product of Gaussian primes:

$$\varepsilon\pi_1 \cdots \pi_k = \varepsilon'\tau_1 \cdots \tau_\ell.$$

Canceling any like or associate factors from both sides, what remains is an equality between two products

$$\pi_1\pi_2 \cdots \pi_s = \tau_1\tau_2 \cdots \tau_t.$$

Let π_j be a prime appearing in the product on the left. Then $\pi_j \mid \tau_1\tau_2 \cdots \tau_t$ and so by Lemma 5.5.8, $\pi_j \mid \tau_k$ for some $1 \leq k \leq \ell$, which implies that π_j and τ_k are associates. This contradicts the statement that π_j and τ_k are primes that remain after canceling all like or associate factors appearing in both factorizations. Thus, the original factorizations must not have been distinct up to associates.   □

Much as the theorems are phrased analogously for both $\mathbb{Z}$ and $\mathbb{Z}[i]$, so do we repeat ourselves with our cautions about their interpretation. For example,

$$(2 + i)(2 - i) = 5 = (1 + 2i)(1 - 2i)$$

is not a counter-example to unique factorization in $\mathbb{Z}[i]$, since the identities $2 + i = (1 - 2i)i$ and $2 - i = (1 + 2i)(-i)$ reveal the two factorizations to be associate to one another. Calling these different factorizations would be as silly as calling $6 = (2)(3) = (-3)(-2)$ different factorizations.

5.6   Gaussian Primes and Prime Factorizations

As with $\mathbb{Z}$, unique factorization makes much of number theory vastly simpler—for example, finding a greatest common divisor from prime factorizations is even simpler than the efficient process of performing the Euclidean algorithm. Thus one immediate goal is being able to factor Gaussian integers into the product of Gaussian primes, a process which has as a prerequisite the ability to recognize primes in $\mathbb{Z}[i]$.

We began this process in Theorem 5.2.12, where we saw that the Diophantine equation $a^2 + b^2 = p$ lies at the heart of which rational primes p remain prime when viewed as elements of $\mathbb{Z}[i]$. By reducing mod 4, we saw that when p is a rational prime congruent to 3 mod 4 the Diophantine Equation

$$a^2 + b^2 = p$$

has no integer solutions. When $p \equiv 1 \bmod 4$, there is no such mod-4 obstruction, and indeed it's not hard (dare we say fun?) to write such primes as a sum of two squares:

$$5 = 2^2 + 1^2 \qquad 13 = 3^2 + 2^2 \qquad 17 = 4^2 + 1^2 \qquad 29 = 5^2 + 2^2 \qquad \cdots$$

We recall from the chapter introduction that given such a representation, we are also given a factorization of p in $\mathbb{Z}[i]$ demonstrating non-primeness, e.g., $5 = (2 + i)(2 - i)$, $13 = (3 + 2i)(3 - 2i)$, $17 = (4 + i)(4 - i)$, $29 = (5 + 2i)(5 - 2i)$, etc. Thus a demonstration that every prime $p \equiv 1 \bmod 4$ can be written as a sum of two squares will satisfactorily close this line of inquiry.

Theorem 5.6.1 (Fermat's Two-Square Lemma)
For every prime $p \equiv 1 \pmod 4$, there exist unique (up to sign and ordering) $a, b \in \mathbb{Z}$ such that $p = a^2 + b^2$.

The key lemma for the upcoming proof of Theorem 5.6.1 is the following.

Theorem 5.6.2 (Lagrange's Lemma)
For every prime $p \equiv 1 \pmod 4$, there exists an $m \in \mathbb{Z}$ such that $p \mid m^2 + 1$.

m	1	2	3	4	5	6	7	8	9	10	11	$\cdots$
m^2+1	2	5	$2 \cdot 5$	17	$2 \cdot 13$	37	$2 \cdot 5^2$	$5 \cdot 13$	$2 \cdot 41$	101	$2 \cdot 61$	$\cdots$

Fig. 5.8 Factorizations of Numbers of the form $m^2 + 1$

Though fairly uninspiring at first glance, this is a rather remarkable discovery. As motivation for this claim, consider this table of factorizations of the first few integers of the form $m^2 + 1$ (see Fig. 5.8).

There are no obvious patterns in these factorizations, and yet the lemma makes the bold claim that every single prime of the form $4k + 1$ eventually appears somewhere in the bottom row of the table in Fig. 5.8. Further, as we will soon see, *no* prime of the form $4k + 3$ appears in the factorization, so "dividing an $m^2 + 1$" turns out to be a multiplicative litmus test for odd primes to be 1 mod 4. The proof is a pleasingly unexpected application of Wilson's Theorem (Theorem 4.4.11), particularly remarkable given how difficult it seems to be to use modular arithmetic to positively solve equations in $\mathbb{Z}$. Furthermore, the proof is delightfully constructive (see, e.g., Exercise 5.16).

Proof (of Lagrange's Lemma) We write $p = 4k+1$ and then apply Wilson's Theorem to get

$$
\begin{aligned}
-1 &\equiv 1 \cdot 2 \cdots 4k \quad (\text{mod } p) \\
&\equiv (1 \cdot 2 \cdots 2k) \cdot ((-2k) \cdots (-2) \cdot (-1)) \quad (\text{mod } p) \\
&= (1 \cdot 2 \cdots 2k)^2 \cdot (-1)^{2k} \quad (\text{mod } p) \\
&\equiv ((2k)!)^2 \quad (\text{mod } p).
\end{aligned}
$$

Setting $m = (2k)!$, we have $m^2 \equiv -1 \ (\text{mod } p)$ and so $p \mid m^2 + 1$. $\square$

This combines with the Gaussian Prime Divisor Property (Lemma 5.5.8) to prove Fermat's 2-square Lemma.

Proof (Of Fermat's 2-square Lemma) Suppose p is a rational prime congruent to 1 (mod 4). We show that p is not a Gaussian prime. By Lagrange's Lemma, we can choose m so that $p \mid m^2 + 1$. In $\mathbb{Z}[i]$, we factor

$$
m^2 + 1 = (m + i)(m - i).
$$

Suppose p were a Gaussian prime. Then by the Prime Divisor Property in $\mathbb{Z}[i]$, since $p \mid m^2 + 1$, either $p \mid m + i$ or $p \mid m - i$. However, p does not divide either of these since $\frac{m}{p} \pm \frac{1}{p} \notin \mathbb{Z}[i]$. Thus p is not a Gaussian prime, so by Theorem 5.2.12, $p = a^2 + b^2$ for some $a, b \in \mathbb{Z}$.

To prove uniqueness, suppose $p = a^2 + b^2 = c^2 + d^2$. It follows that $(a+bi)(a-bi) = (c+di)(c-di)$ in $\mathbb{Z}[i]$. Since all four of these factors have norm p, they are all prime, and hence by unique factorization in $\mathbb{Z}[i]$, must be associates of one of another. Running through the units we find $(a, b) = (\pm c, \pm d)$ or $(a, b) = (\mp d, \pm c)$. In all cases, $\{a^2, b^2\} = \{c^2, d^2\}$. $\qquad \square$

This is a good point to pause and consider the success story of moving between worlds. We've observed that to answer the question in $\mathbb{Z}$ about representability as a sum of squares, we pass to the smaller ring $\mathbb{Z}/(4)$ to deduce that p *cannot* be written as a sum of two squares, and to $\mathbb{Z}[i]$ to deduce that it *can*, depending on its equivalence class modulo 4. One further observation strengthens the ties between the worlds: note that Lagrange's Lemma can be interpreted as providing a solution to the equation $m^2 \equiv -1 \pmod{p}$, that is, a square root of negative one mod p. The existence of such a square root—an interpretation of i mod p—is yet another decisive split between primes that are 1 versus 3 modulo 4.

Lemma 5.6.3

If p is a prime such that $p \equiv 3 \pmod{4}$, then $x^2 \equiv -1 \pmod{p}$ has no solutions.

◀

Proof By way of contradiction, suppose there exists an $x \in \mathbb{Z}$ such that $x^2 \equiv -1 \pmod{p}$ for some prime p of the form $4n + 3$. Raising both sides to the power $2n + 1$ gives

$$-1 \equiv (-1)^{2n+1} \equiv (x^2)^{2n+1} \equiv x^{4n+2} \equiv x^{p-1} \pmod{p},$$

which contradicts Fermat's Little Theorem. $\qquad \square$

It would be risky to get in the habit of writing $\sqrt{-1}$ for an element whose square is -1 in $\mathbb{Z}/(p)$, but temporarily suspending notational sanity permits a salient observation: we have deduced that

$$p \text{ factors in } \mathbb{Z}[\sqrt{-1}] \quad \Longleftrightarrow \quad \sqrt{-1} \in \mathbb{Z}/(p).$$

That is, the existence of a factor of p when we adjoin a square root of -1 is equivalent to the existence of a square root of -1 modulo p. In fact, we now have several ways of interpreting the divide between two classes of primes.

Theorem 5.6.4
The following are equivalent for all rational primes p:

  (i) $p \equiv 1$ or $2 \pmod{4}$;
 (ii) $p = a^2 + b^2$ for some unique $a, b \in \mathbb{Z}$ (up to sign and order);
(iii) $\sqrt{-1} \in \mathbb{Z}/(p)$;
 (iv) p factors in $\mathbb{Z}[i]$, i.e., p is not a Gaussian prime.

It is noteworthy that each of the statements takes place in a different ring: $\mathbb{Z}/(4)$, $\mathbb{Z}$, $\mathbb{Z}/(p)$, and $\mathbb{Z}[i]$, respectively. Before moving on from this sublime result, allow yourself a second to applaud in the comfort of your own reading environment, to give your eyes a chance to stop twinkling with delight, and to wipe away the tear falling unheeded down your cheek. Back? Wonderful, we have more great things ahead. In particular, it may seem we have spent an inordinately long time figuring out which rational primes stay prime, rather than the broader question of which Gaussian integers are prime. The following lemma shows us that we have already done the work.

Lemma 5.6.5

If $\alpha \in \mathbb{Z}[i]$ is a Gaussian prime, then α divides some rational prime p. ◄

Proof Write $\alpha = a + bi$. Then

$$\alpha \mid (a + bi)(a - bi) = a^2 + b^2.$$

By unique factorization in $\mathbb{Z}$, $a^2 + b^2$ can be factored into a product of rational primes $p_1 p_2 \cdots p_r$. By the Gaussian Prime Divisor Property, α must divide one of these rational prime factors.   □

The punchline of this is a drastic simplification of our search for Gaussian primes: we get all the primes of $\mathbb{Z}[i]$ by factoring all the primes of $\mathbb{Z}$ in $\mathbb{Z}[i]$. We already know that the rational primes that are 3 mod 4 stay prime in $\mathbb{Z}[i]$, and all that remain are the Gaussian prime factors of the other rational primes. For primes $p \equiv 1 \pmod 4$, we can write

$$p = a^2 + b^2 = (a + bi)(a - bi),$$

and both of these factors must be prime by taking norms: $N(a + bi) = a^2 + b^2 = p$ and likewise for $N(a - bi)$. As always, $p = 2$ is a little weird:

$$2 = (1 + i)(1 - i) = (1 + i)(-i)(1 + i) = -i(1 + i)^2.$$

So the rational prime 2 is actually a Gaussian *square* up to units.

Theorem 5.6.6
The primes of $\mathbb{Z}[i]$ are, up to associates, precisely:

- The prime $(1 + i)$.
- The rational primes p that are congruent to 3 mod 4.
- The primes $a + bi$ and $a - bi$, where $a^2 + b^2 = p \equiv 1 \pmod 4$ and p is a rational prime.

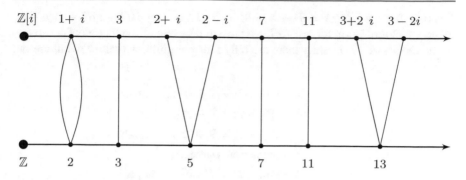

Fig. 5.9 The primes of $\mathbb{Z}$ and $\mathbb{Z}[i]$

The situation is encoded graphically in Figure 5.9. As one progresses through algebraic number theory (and especially toward algebraic geometry) it becomes handy to visualize any relationships between the primes in any two given rings. Here, the picture conveys that rational primes congruent to 1 mod 4 factor in $\mathbb{Z}[i]$ (fancy word: they *split*), while primes congruent to 3 mod 4 stay prime in $\mathbb{Z}[i]$ (fancy word: they are *inert*). Finally there is the always obstinate prime 2 as a special case in that it factors in $\mathbb{Z}[i]$, but not into two distinct primes of $\mathbb{Z}[i]$ like the split primes, but into a single prime of multiplicity 2 (fancy word: 2 is *ramified*). Note that, caring only about the algebraic structure, the primes in Figure 5.9 are not spaced to scale, and even the depiction of $\mathbb{Z}[i]$ as a linear set is questionable. Nevertheless, every prime of $\mathbb{Z}[i]$ would appear in the top row as we continued to expand the picture.

We conclude the section by noting that we now have all the tools to compute explicit prime factorizations of Gaussian integers. We use the following idea: if a Gaussian prime π divides $N(\alpha) = \alpha\overline{\alpha}$, then by the Prime Divisor Property π must divide either α or $\overline{\alpha}$ (or both). Further, whichever one π divides, $\overline{\pi}$ must divide the other, as it is easy to check that $\pi \mid \beta$ if and only if $\overline{\pi} \mid \overline{\beta}$. So to factor a Gaussian integer α, we factor $N(\alpha)$ in $\mathbb{Z}[i]$ using our knowledge of primes above, and then figure out which of those factors belong to α and which to $\overline{\alpha}$.

▶ **Example 5.6.7** Factor $7 + 5i$ in $\mathbb{Z}[i]$.

Solution We have

$$7 + 5i \mid N(7 + 5i) = 74 = 37 \cdot 2 = -i(1+i)^2(6+i)(6-i),$$

and so each Gaussian prime factor of $7 + 5i$ must come from the factorization on the right, and the product of those factors must have norm 74. Since $6 \pm i$ each have norm 37 and $i + 1$ has norm 2, it must be that one of $(6+i)(1+i)$ and $(6-i)(1+i)$ is $(7+5i)$ and the other is $(7-5i)$, up to units. We quickly find $7+5i = (1+i)(6-i)$.
$\square$

▶ **Example 5.6.8** Factor $z = -2 + 9i$ in $\mathbb{Z}[i]$.

Solution Since $z \mid N(z) = 81 + 4 = 85 = 5 \cdot 17 = (2+i)(2-i)(4+i)(4-i)$, to get a factor of norm $85 = 5 \cdot 17$, we have to take one of the factors of 5 and one of the factors of 17. Luckily there are only four possibilities (before we figure out units):

$$(2+i)(4+i) = 7 + 6i \ldots \text{ nope.}$$
$$(2-i)(4+i) = 9 - 2i \ldots \text{ nope.}$$
$$(2+i)(4-i) = 9 + 2i \ldots \text{ Ta-dah!}$$

and just for completion's sake...

$$(2-i)(4-i) = 7 - 6i \ldots \text{ nope.}$$

We can stop when we've found the associate $9 + 2i$ of $-2 + 9i$, giving

$$-2 + 9i = i(9 + 2i) = i(2+i)(4-i)$$

as the desired factorization.   □

We close with the observation that along with all the other goodies, unique factorization provides us with a well-defined notion of valuations for Gaussian integers (compare to Definition 3.4.1):

Definition 5.6.9

For non-zero $\alpha \in \mathbb{Z}[i]$ and Gaussian prime π, define the π-**adic valuation** of α, denoted $v_\pi(\alpha)$, to be the power of π that appears in the prime factorization of α. In other words, $v_\pi(\alpha)$ is the unique integer such that we can write

$$\alpha = \pi^{v_\pi(\alpha)} \alpha'$$

for some $\alpha' \in \mathbb{Z}[i]$ with $\pi \nmid \alpha'$. ◄

▶ **Example 5.6.10** From Example 5.6.7, we see that

$$v_{1+i}(74) = 2, \quad v_{6+i}(74) = v_{6-i}(74) = 1, \quad \text{and} \quad v_{3+2i}(74) = 0.$$

5.7   Applications to Diophantine Equations

Application 1: Sums of Squares

We have satisfactorily solved the Diophantine equation

$$a^2 + b^2 = n$$

in the case that n is prime. Part of this argument continues to hold when n is not prime, as when $n \equiv 3 \bmod 4$ there are still no solutions. But this is not the whole story, as the case of $n = 21 = 3 \cdot 7$ shows: We have $21 \equiv 1 \bmod 4$ but 21 is not the sum of two squares. The statement for arbitrary n is as follows.

> **Theorem 5.7.1**
>
> An integer n can be written as a sum of two integer squares if and only if $v_p(n)$ is even for every prime $p \equiv 3 \bmod 4$.

That is, primes that are 3 mod 4 form something of a toggle switch: they themselves cannot be written as a sum of two squares ($7 \neq a^2 + b^2$), but surely their squares can ($7^2 = 7^2 + 0^2$). Since being expressible as a sum of two squares is equivalent to being expressible as a norm of a Gaussian integer, we can use the multiplicativity of the norm to complete the solution.

Proof Write $n = \prod_i p_i^{e_i} \prod_j q_j^{f_j}$ where the p_i are the prime divisors of n congruent to 1 or 2 mod 4, and the q_j are 3 mod 4. Then each p_i and q_j^2 is a norm of an element of $\mathbb{Z}[i]$, and so if each f_j is even, n is the norm of the product of those elements, and thus is itself a sum of two squares. For the reverse direction, we proceed nearly identically to the prime case: suppose $n = a^2 + b^2$ and that n has at least one prime divisor $q \equiv 3 \bmod 4$ for which $v_q(n)$ is odd. If q divides both a and b, then an even power of q divides both sides of the equation $n = a^2 + b^2$, and canceling as many powers from q as possible from both sides gives a new equation $n' = (a')^2 + (b')^2$ where $q \mid n'$ (since $v_q(n)$ was odd) but $q \nmid a', b'$. Reducing this equation mod q gives $0 \equiv (a')^2 + (b')^2$, so $-1 \equiv (a'/b')^2 \pmod{q}$. But -1 being a square modulo q contradicts that $q \equiv 3 \bmod 4$. $\square$

As an interesting consequence, we can return to topics begun in Chapter 2 asking about the difference between the Euclidean circles $x^2 + y^2 = 3$ and $x^2 + y^2 = 5$ from the rational perspective. It's clear by search that the former has no integer solutions whereas the latter does. But searching is a pretty ineffective tool for finding *rational* points on this circle, so the following result is a pleasant bonus.

Corollary 5.7.2

An integer n can be written as a sum of two *rational* squares if and only if $v_p(n)$ is even for every prime $p \equiv 3 \bmod 4$. ◄

Proof We show that n is a sum of integer squares if and only if it is a sum of rational squares, after which Theorem 5.7.1 provides the result.

As a sum of integer squares is evidently a sum of rational squares, one direction is immediately covered by the theorem. It remains to show that any number that can be written as a sum of rational squares is also a sum of integer squares, so suppose $n = \left(\frac{a}{b}\right)^2 + \left(\frac{c}{d}\right)^2$. Clearing denominators gives the equality of integers

$$nb^2d^2 = (ad)^2 + (bc)^2.$$

Now since nb^2d^2 is a sum of integer squares, for any $q \equiv 3 \bmod 4$ we must have

$$v_q(nb^2d^2) = v_q(n) + 2v_q(b) + 2v_q(d)$$

is even, and hence $v_q(n)$ is even. Thus n is a sum of two squares. $\square$

It is remarkable, and quite rare for equations of this type, that the conditions on n for the equation $x^2 + y^2 = n$ to have either integer solutions or rational solutions are identical. We have a lot of ground to cover before returning to this relationship for a general Diophantine equation in Chapter 8.

Unique factorization in $\mathbb{Z}[i]$ provides still more additional interesting facets to our story. Let us take, for example, $n = 65$. Factoring n in $\mathbb{Z}[i]$, we have

$$65 = 5 \cdot 13 = (2 + i)(2 - i)(3 + 2i)(3 - 2i)$$

factoring into a product of 4 Gaussian primes. Note that if we take *either* factor of norm 5, multiplied by *either* factor of norm 13, the result is a Gaussian integer of norm 65, and hence a way of writing 65 as a sum of two squares. So, on one hand, we could write

$$65 = (2 + i)(3 + 2i) \cdot (2 - i)(3 - 2i) = (4 + 7i)(4 - 7i) = 4^2 + 7^2,$$

but, on the other hand,

$$65 = (2 + i)(3 - 2i) \cdot (2 - i)(3 + 2i) = (8 - i)(8 + i) = 8^2 + 1^2.$$

That is, we get different ways of writing n as a sum of two squares by rearranging the factors of n into different products of two complex conjugates. The exercises will have you explore some consequences of this observation.

Application 2: Pythagorean Triples

The next result should look familiar from Chapter 2. We will tackle the classification of Pythagorean triples again through an algebraic lens, replacing the use of the Diophantus Chord method with the Fundamental Theorem of Gaussian Arithmetic.

> **Theorem 5.7.3 (Classification of Pythagorean Triples)**
> Every primitive Pythagorean triple (a, b, c) (with $a, b, c > 0$) has the form
> $(u^2 - v^2, 2uv, u^2 + v^2)$ for some $u > v \in \mathbb{N}$ with $\gcd(u, v) = 1$.

First, a preliminary philosophical remark. Why do Gaussian integers have anything to do with Pythagorean triples? We have previously mentioned a connection through the factorization $a^2 + b^2 = (a + bi)(a - bi)$, but it's not immediately clear how that helps matters. One argument is that it replaces an additive relationship with a multiplicative one, at the expense of having to work in $\mathbb{Z}[i]$ instead of $\mathbb{Z}$, and this means we can bring the notions of primes and unique factorization into play. Once we have the Fundamental Theorem of Gaussian Arithmetic under our belts, we know number theory works in $\mathbb{Z}[i]$ the way it's supposed to, and the solution becomes almost routine.

The super-brief sketch of the proof is that given $c^2 = (a + bi)(a - bi)$, we can show that $(a + bi)$ and $(a - bi)$ are relatively prime, and whenever a product of relatively prime factors is a square, each factor is itself a square. Writing $a + bi = (u + iv)^2$ gives the result. Let us first formally prove that intermediate step, the analog of Lemma 3.4.10 for $\mathbb{Z}[i]$.

Lemma 5.7.4 (Gaussian Power Lemma)

Let α and β be relatively prime Gaussian integers. If $\alpha\beta$ is an n-th power in $\mathbb{Z}[i]$, then α and β are themselves n-th powers (up to units) in $\mathbb{Z}[i]$. If n is odd, then α and β are themselves n-th powers in $\mathbb{Z}[i]$. ◄

Proof If $\alpha\beta$ is an n-th power, then

$$\alpha\beta = (\varepsilon\pi_1^{a_1} \cdots \pi_k^{a_k})^n = \varepsilon^n \pi_1^{na_1} \pi_2^{na_2} \cdots \pi_k^{na_k},$$

where $\varepsilon \in \mathbb{Z}[i]^\times = \{\pm 1, \pm i\}$ and each π_j is a Gaussian prime. Given our assumption that α and β are relatively prime, each $\pi_j^{na_j}$ divides either α or β but not both. Thus α and β take the form $\alpha = \varepsilon_1 \pi_{j_1}^{na_{j_1}} \cdots \pi_{j_l}^{na_{j_l}}$ and $\beta = \varepsilon_2 \pi_{k_1}^{na_{k_1}} \cdots \pi_{k_m}^{na_{k_m}}$, where $\varepsilon_1, \varepsilon_2 \in \mathbb{Z}[i]^\times$. Thus α and β are n-th powers up to units. When n is odd, each unit of $\mathbb{Z}[i]$ is itself an n-th power in $\mathbb{Z}[i]$, so we can drop the "up to units" clause. □

Note that the "up to units" qualifier is needed in $\mathbb{Z}$ as well: $36 = (-4)(-9)$ is an example of two relatively prime numbers multiplying to give a square, neither of which is a square. Now, on to the classification:

Proof If (a, b, c) is a primitive Pythagorean triple, then no prime p can divide any two of the three (since then by $a^2 + b^2 = c^2$, it would also divide the third). Therefore, we can assume $\gcd(a, b) = 1$, and taking $p = 2$, we see we can assume that a is odd and b is even without loss of generality. This also forces c to be odd.

We claim that the Gaussian integers $a + bi$ and $a - bi$ are necessarily relatively prime. Indeed, letting $\gamma = \gcd(a + bi, a - bi)$, we see that γ also divides their sum, $2a$, and their difference, $2bi$. That γ is a unit (and hence that α and β are relatively prime) will then follow from the claim that $\gcd(\gamma, 2) = 1$, as then Euclid's Lemma shows that $\gamma \mid a$ and $\gamma \mid b$, but $\gcd(a, b) = 1$ (see Corollary 5.5.6). To check that $\gcd(\gamma, 2) = 1$, we note that if the only Gaussian prime divisor of 2, namely, $1 + i$, were to divide γ, then γ would have to have even norm (in $\mathbb{Z}$). But $N(\gamma) \mid N(a + bi) = a^2 + b^2 = c^2$ and c^2 is odd, giving a contradiction.

We can now deduce from Lemma 5.7.4 that $a + bi$ is a square in $\mathbb{Z}[i]$ (up to units), i.e.,

$$a + bi = \varepsilon(u + vi)^2 = \varepsilon\left((u^2 - v^2) + 2uvi\right)$$

for some $u, v \in \mathbb{Z}$ and unit $\varepsilon \in \mathbb{Z}[i]$. If we restrict our attention to positive solutions for a and b and recall that we assumed a to be odd, the only possible values for ε are ± 1 (Exercise 5.10). Equating real and imaginary parts gives $a = \pm(u^2 - v^2)$, $b = \pm 2uv$, and the identity $a^2 + b^2 = c^2$ finishes the result. (We may choose $u > v$ and $\varepsilon = 1$, and note that u and v must be relatively prime, else (a, b, c) would not be primitive.) $\square$

Application 3: Tackling an Elliptic Curve

The technique of using $\mathbb{Z}[i]$ to factor Diophantine equations is not limited to the case of Pythagorean triples, as any sum of two squares admits a factorization in $\mathbb{Z}[i]$. For example, consider the question

Are there any cubes in $\mathbb{Z}$ that are one more than a square?

Translating to an algebraic expression, we are asking for the set $E(\mathbb{Z})$ of integer-coordinate points on the elliptic curve E defined by $y^2 + 1 = x^3$. Immediately re-writing the equation as

$$x^3 = y^2 + 1 = (y + i)(y - i)$$

yet again thrusts us toward $\mathbb{Z}[i]$ in search of solutions. And indeed, the technique is almost identical to before. We check (Exercise 5.9) that $(y + i)$ and $(y - i)$ must be relatively prime, so by Lemma 5.7.4, both $y + i$ and $y - i$ are cubes. Writing $y + i = (u + vi)^3$ and collecting real and imaginary parts gives

$$y = u^3 - 3uv^2 = u(u^2 - 3v^2) \quad \text{and} \quad 1 = 3u^2v - v^3 = v(3u^2 - v^2).$$

Thus v is an integer divisor of 1, implying $v = \pm 1$. If $v = 1$, then $1 = 3u^2 - 1$, which is impossible for $u \in \mathbb{Z}$. If $v = -1$, then $1 = -(3u^2 - 1)$, so $u = 0$, and indeed the pair $(u, v) = (0, -1)$ works! Re-substituting, we find $y = u^3 - 2uv^2 = 0$ and $x^3 = y^2 + 1 = 1$, giving the unique solution $(x, y) = (1, 0)$.

Theorem 5.7.5
In $\mathbb{Z}$, the only cube which is one more than a square is 1 (being one more than 0).

5.8 Exercises

Calculation and Short Answer

Exercise 5.1 Pause and reflect on what happened in this chapter. What parts of moving from $\mathbb{Z}$ to $\mathbb{Z}[i]$ seem easy to you? Which seem hard? Find some questions that you have and ask your instructor. With any luck, they won't know the answer, and you'll know you really do get what's going on.

Exercise 5.2 Factor 17 and 53 into products of Gaussian primes.

Exercise 5.3 Give a factorization of $37 + 3i$ into a product of Gaussian primes.

Exercise 5.4 In the context of the Division Algorithm for $\mathbb{Z}[i]$, describe geometrically how it could be that the quotient and remainder could fail to be unique (as they are in $\mathbb{Z}$), even after fixing a system of coset representatives. Give an explicit example of an α and a β with $N(\beta) > 2$ such that we can write $\alpha = \chi\beta + \rho$ with $N(\rho) < N(\beta)$ in two different ways.

Exercise 5.5 Find a complete set of representatives modulo each given Gaussian integer. Choose representatives that will satisfy the conditions of the Division Algorithm.

(a) $4 + 2i$,　　　　　　　　　　(c) $3 - i$,
(b) $2 + 4i$,　　　　　　　　　　(d) $-2 + 5i$.

Exercise 5.6 For each modulus, find a complete set of representatives compatible with the Division Algorithm, and then compute the reductions:

(a) $6 - 3i \bmod 2 - i$,
(b) $4 + 7i \bmod 3 + 2i$,
(c) $37 - 14i \bmod 1 + i$.

Exercise 5.7 Prove that a rational number $\frac{c}{d}$ is a sum of two rational squares if and only if cd is. Use this and the results of the chapter to decide which of the following rational numbers are a sum of two rational squares:

$$\frac{5}{7} \qquad\qquad \frac{13}{29} \qquad\qquad \frac{43}{39}.$$

Choose one that is and write it as the sum of two rational squares.

Exercise 5.8 The Extended Euclidean Algorithm in $\mathbb{Z}[i]$:

(a) Find a gcd δ of $11 + 3i$ and $1 + 8i$.
(b) Suppose one can purchase Chicken i'Nuggets in boxes of either $11 + 3i$ or $1 + 8i$ nuggets. Show that you can purchase δ (from above) i'Nuggets, assuming that you are allowed to buy and sell Gaussian integer numbers of boxes. That is, find Gaussian integers μ and ν satisfying

$$\mu(11 + 3i) + \nu(1 + 8i) = \delta.$$

Exercise 5.9 Work mod 8 to show that if x, $y \in \mathbb{Z}$ satisfy $x^3 = y^2 + 1$, then y must be even. Now fill in the missing step in the proof of Theorem 5.7.5 by showing that (up to associates), we have

$$\gcd(y + i, y - i) = \begin{cases} 1 + i & \text{if } y \text{ is odd} \\ 1 & \text{if } y \text{ is even.} \end{cases}$$

Exercise 5.10 Finish the missing step in the classification of Pythagorean triples: if $a + bi = \varepsilon(u + vi)^2$ with the conventions on a, b in the theorem, then $\varepsilon = \pm 1$.

Exercise 5.11 Gaussian integer True/False. See how little work you can do to justify each answer:

(a) $5 + 3i \mid 15 + 9i$,   (d) $7 - 2i \mid 37 - 3i$,
(b) $5 + 13i \mid 5 + 14i$,   (e) $7 + 2i \mid 37 - 3i$,
(c) $i \mid 7$,   (f) $3 + 2i \mid 65$.

Exercise 5.12 Generalize Example 5.2.9. That is, find a way of constructing examples of Gaussian integers α, β such that $N(\alpha) \mid N(\beta)$ but $\alpha \nmid \beta$.

Exercise 5.13 Find an integer that is a sum of two squares in at least four different ways and describe a procedure for finding an integer that is a sum of two squares in at least 2^n different ways.

Exercise 5.14 Building off the previous problem, given an integer $k \geq 1$, construct a number that can be expressed as a sum of squares in exactly k different ways.

Exercise 5.15 Fermat conjectured that all numbers of the form $2^{2^n} + 1$ were prime. The first four are indeed prime, but the identity

$$2^{2^5} + 1 = 62264^2 + 20449^2,$$

discovered by Euler, implies that it is not. How?

Exercise 5.16 Follow the proof of Lagrange's Theorem to find α and β in $\mathbb{Z}[i]$ such that $101 \mid \alpha\beta$ but $101 \nmid \alpha$ and $101 \nmid \beta$.

Formal Proofs

Exercise 5.17 Prove that for $a, b \in \mathbb{Z}$, we have $a \mid b$ in $\mathbb{Z}$ if and only if $a \mid b$ in $\mathbb{Z}[i]$. Further, for all $n \in \mathbb{N}$, $a \equiv b$ in $\mathbb{Z}/(n)$ if and only if $a \equiv b$ in $\mathbb{Z}[i]/(n)$.

Exercise 5.18 Let S be a subring of a ring R. Prove that for all $a, b \in S$, if $a \mid b$ in S then $a \mid b$ in R. Find an example of a ring R and subring S showing the converse is not generally true: there exist $a, b \in S$ such that $a \mid b$ in R but $a \nmid b$ in S.

Exercise 5.19 Prove that for $\alpha, \beta \in \mathbb{Z}[i]$, we have

$$\alpha \mid \beta \iff \overline{\alpha} \mid \overline{\beta}.$$

Deduce that for $n \in \mathbb{Z}$ and $\alpha \in \mathbb{Z}[i]$, $\alpha \mid n \iff \overline{\alpha} \mid n$.

Exercise 5.20 Prove that if $(a + bi)$ is a Gaussian prime, then so too are $(a - bi)$, $(-a + bi)$, and $(-a - bi)$.

Exercise 5.21 Let $\alpha, \beta \in \mathbb{Z}[i]$. Show that δ is a gcd of α and β if and only if (1) δ is a common divisor of α and β and (2) for all $\gamma \in \mathbb{Z}[i]$, if γ is a common divisor of α and β, then $\gamma \mid \delta$.

Exercise 5.22 Prove that if γ and δ are both gcds of α and β, then γ and δ are associates.

Exercise 5.23 Give both an algebraic and a geometric proof of the following: For $a, b \in \mathbb{Z}$, we have $(1 + i) \mid (a + ib)$ if and only if a and b are both even or both odd. Because of the "2-like" properties of $1 + i$, we may refer to multiples of $1 + i$ as "$i$ven" numbers.

Exercise 5.24 Prove Euclid's Lemma in $\mathbb{Z}[i]$: Given $\alpha, \beta, \gamma \in \mathbb{Z}[i]$, if $\alpha \mid \beta\gamma$ and α and β are relatively prime, then $\alpha \mid \gamma$.

Exercise 5.25 Prove that if π is a Gaussian prime, then for all $\alpha \in \mathbb{Z}[i]$ with $\pi \nmid \alpha$ we have $\gcd(\alpha, \pi) = 1$. Conclude that there exists a Gaussian integer β such that $\alpha\beta \equiv 1 \bmod \pi$, i.e., that α is a unit mod π.

Computation and Experimentation

Exercise 5.26 Out of the first million natural numbers, how many satisfy the conditions of Theorem 5.7.1 and can hence be written as a sum of two squares?

As we move to more sophisticated algebraic structures, it becomes increasingly desirable to move from raw programming languages like Python to a full computer algebra system. The system SageMath, in particular, allows for work in the Gaussian integers in a way that generalizes nicely to rings we will see in the upcoming chapter.

Exercise 5.27 SageMath uses the notation ZZ[i] for the Gaussian integers. Try running the following code in SageMath:

```
z=ZZ[i](5)
z.is_prime().
```

The code above sets z to be the Gaussian integer 5 (as opposed to the *integer* 5—while they are the same *number*, whether or not it is prime depends on the ambient ring). Experiment with other functions you can do with Gaussian integers—norms, factorizations, conjugates, etc.

Out of the first 100 natural numbers n, for how many such n is the Gaussian integer $n + 0i$ prime? Compare your observations to the theoretical results of the section.

Exercise 5.28 Figure out how to find the gcd of two Gaussian integers in SageMath. Generate a large number of "random" Gaussian integers using the command ZZ[i].random_element() and approximate the probability that two randomly chosen Gaussian integers are relatively prime. Compare the result to the similar problem for randomly chosen integers. It may be worth researching the algorithm SageMath uses to generate random elements, and/or repeating with a different algorithm.

General Number Theory Awareness

Exercise 5.29 Look up Pick's Theorem for counting lattice points inside of a polygon. Apply Pick's Theorem to the fundamental domain defined by the multiples of β to compute $|\mathbb{Z}[i]/(\beta)|$.

Exercise 5.30 Propose a definition for the Euler totient $\phi(\beta)$ of a non-zero Gaussian integer β. How would you compute it for a prime β? Make a prediction for a general formula for $\phi(\beta)$ given a prime factorization. Make a prediction for a Gaussian analog of Euler's Theorem.

Exercise 5.31 Look up Gauss' circle problem. What is it? How would you rephrase the question in terms of norms of Gaussian integers? What is the status of the problem (i.e., what solutions are known, are all solutions known, etc.)?

Exercise 5.32 How many Gaussian primes are there with norm up to 100? Of these, how many are of each of the three types from Theorem 5.6.6? How does this observation influence the Prime Number Theorem for Gaussian primes? And what of the "Gaussian Moat"?

Exercise 5.33 In Exercise 4.68 we argued that $\mathbb{R}[x]/(x^2 + 1)$ and $\mathbb{C}$ had essentially the same algebraic structure. Algebraists would say that these two rings are isomorphic, and write $\mathbb{R}[x]/(x^2 + 1) \cong \mathbb{C}$. One can argue similarly that $\mathbb{Z}[x]/(x^2 + 1) \cong \mathbb{Z}[i]$, and the modular version is even more striking:

$$\mathbb{Z}/(p)[x]/(x^2 + 1) \cong \mathbb{Z}[i]/(p)$$

since both are isomorphic to $(\mathbb{Z}[x]/(x^2 + 1))/(p)$. Explore how much of the chapter can be explained by these relationships, e.g., the number of elements of $\mathbb{Z}[i]/(p)$, and the distinction between primes that are 1 or 3 mod 4.

Exercise 5.34 Setting aside $p = 2$, the chapter has largely focused on the distinction between primes that are 1 mod 4 versus primes that are 3 mod 4. Are there the same number of each? Look up prime races, and Dirichlet's Theorem on Arithmetic Progressions.

Number Theory, from Where We $\mathbb{R}$ to Across the $\mathbb{C}$

6

...wherein i realizes it's not all that special.

6.1 From -1 to $-d$

Despite the central role it has played in the book thus far, the seemingly unique significance of the number $i = \sqrt{-1}$ is somewhat illusory. For example, consider the closing success of the last chapter, using the arithmetic of $\mathbb{Z}[i]$ to find the integral points $E(\mathbb{Z})$ of the elliptic curve E defined by

$$x^3 = y^2 + 1,$$

via the factorization $y^2 + 1 = (y + i)(y - i)$. While this seems at first glance a miracle special to $\mathbb{Z}[i]$ we can just as well tackle the general *Mordell equation*

$$x^3 = y^2 + d$$

via the factorization $y^2 + d = (y+\sqrt{-d})(y-\sqrt{-d})$ in the analogous ring $\mathbb{Z}[\sqrt{-d}] = \{a + b\sqrt{-d} : a, b \in \mathbb{Z}\}$. There is much to verify here, including the claim that this is indeed a ring. But for this introductory section, let's assume everything will work out nicely and take a stab at a second Mordell equation. Here's one that could arise via the question "What numbers, if any, are one more than a square and one less than a cube?" Since such an instance would necessitate the cube being precisely two more than the square, we can rephrase this question as the following Diophantine equation.

▶ **Example 6.1.1** Find all integer solutions to $x^3 = y^2 + 2$.

We hope to mirror the argument for $x^3 = y^2 + 1$, i.e., factor

$$x^3 = (y + \sqrt{-2})(y - \sqrt{-2}),$$

© Springer Nature Switzerland AG 2022
C. McLeman et al., *Explorations in Number Theory*, Undergraduate Texts in Mathematics,
https://doi.org/10.1007/978-3-030-98931-6_6

and then use arithmetic in $\mathbb{Z}[\sqrt{-2}]$ to show that $(y + \sqrt{-2})$ and $(y - \sqrt{-2})$ must be relatively prime. From this we conclude by an analog of Lemma 5.7.4 that each of these factors is itself a cube in $\mathbb{Z}[\sqrt{-2}]$, allowing us to write

$$\begin{aligned} y + \sqrt{-2} &= (m + n\sqrt{-2})^3 \\ &= m^3 + 3m^2 n\sqrt{-2} + 3m(-2n^2) - 2n^3\sqrt{-2} \\ &= (m^3 - 6mn^2) + n(3m^2 - 2n^2)\sqrt{-2}. \end{aligned}$$

This shows that $y = m^3 - 6mn^2$ and $1 = n(3m^2 - 2n^2)$, forcing $n = \pm 1$. From here, it is elementary to finish off the solution: if $n = -1$, our equation becomes $3m^2 - 2 = -1$, which has no integer solutions. Thus, $n = 1$ and $m = \pm 1$, in turn providing $y = \pm 5$ and $x = 3$; a quick check shows that these are indeed solutions. The number 26, between the square 25 and the cube 27, is thus the *only* positive integer which is one more than a square and one less than a cube.

In spite of our many assumptions, we have indeed found all integer solutions for $x^3 = y^2 + 2$. Score one for reckless abandon! Number theory is easier than we thought! Emboldened by our swashbuckling success, let's tackle yet another Mordell equation.

▶ **Example 6.1.2** Find all integer solutions to $x^3 = y^2 + 26$.

Let us mindlessly copy the argument: write

$$x^3 = y^2 + 26 = (y + \sqrt{-26})(y - \sqrt{-26})$$

in the ring $\mathbb{Z}[\sqrt{-26}]$, check again that the two factors are relatively prime and hence must be cubes, and calculate

$$y + \sqrt{-26} = (m + n\sqrt{-26})^3 = (m^3 - 78mn^2) + (3m^2 n - 26n^3)\sqrt{-26}.$$

Equating real and imaginary parts, we get $y = m^3 - 78mn^2$ and $1 = n(3m^2 - 26n^2)$, again giving $n = \pm 1$. Since $-1 = (3m^2 - 26)$ has no solutions, n must equal 1. We solve to find $m = \pm 3$ and substitute to then get $y = m^3 - 78mn^2 = \pm 207$ and $x^3 = 207^2 + 26 = 42875 = 35^3$. We conclude that there are two and only two solutions, $(35, \pm 207)$, to the equation $x^3 = y^2 + 26$.
Fantastic! ...

> *...pausing for dramatic effect...*

...but...wait. We have indeed found two solutions (and pleasantly non-obvious solutions, at that) to this equation, but our proof technique claims to have found *all* the solutions, as it did for the previous two examples. And yet, there is an obvious solution that we missed: 27 is a cube which is 26 more than the square 1, so we have somehow missed the solutions $(3, \pm 1)$. Worse than merely an incompleteness of the process, there must be a fatal theoretical error somewhere along the way, and indeed there is.

So what happened? Before reading on, it is worth reading through the purported solution again, pondering in which of the many unjustified steps the error could lie. We offer reassurances for some previously unchecked facts:

- The set $\mathbb{Z}[\sqrt{-26}]$ is indeed a ring.
- The two elements $y \pm \sqrt{-26}$ are indeed relatively prime.
- The algebra after naming m and n is all correct.

Where else could our error be hiding? There's not much left! The culprit is the claim that a product of two relatively prime factors producing a cube implies that each of the factors is itself a cube. This was a theorem in $\mathbb{Z}[i]$ (and, as we'll see, in $\mathbb{Z}[\sqrt{-2}]$), but is not a theorem in $\mathbb{Z}[\sqrt{-26}]$. Indeed, our missing solution $(x, y) = (3, \pm 1)$ furnishes explicit counter-examples: the numbers $1 \pm \sqrt{-26}$ can be shown to be relatively prime, and are not themselves cubes, but their product is:

$$(1 + \sqrt{-26})(1 - \sqrt{-26}) = 27 = 3^3.$$

In fact, arithmetic in $\mathbb{Z}[\sqrt{-26}]$ is all kinds of messed up, as the above formula *also* represents a failure of our holy grail of unique factorization: the number 27 has the prime factorization $27 = (1 + \sqrt{-26})(1 - \sqrt{-26})$ and a completely separate prime factorization $27 = 3^3$. All of these claims require checking, and we will develop tools to do so in this chapter.

To diagnose where things started going awry in $\mathbb{Z}[\sqrt{-26}]$, it will help to recall *The Path* to unique factorization in $\mathbb{Z}$ and $\mathbb{Z}[i]$ (Figure 6.1). The principal goal of this chapter is to codify this path as a theorem, which we dub the *Fundamental Meta-Theorem of Arithmetic*, and explore rings to which it applies (and to which it doesn't!). Assuming the steps of the path, we can work backward to see what must have gone wrong in $\mathbb{Z}[\sqrt{-26}]$.

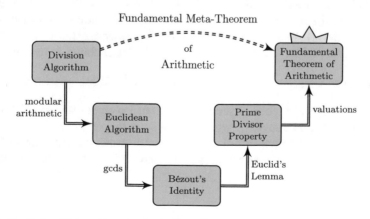

Fig. 6.1 The Path to Unique Factorization in $\mathbb{Z}$ and $\mathbb{Z}[i]$

In $\mathbb{Z}[i]$, the Fundamental Theorem (Theorem 5.5.9) was in turn a consequence of the prime divisor property (Lemma 5.5.8), and hence of Euclid's Lemma (Lemma 5.5.7), and hence of Bézout's Identity (Theorem 5.5.5), and hence of the Euclidean Algorithm (Theorem 5.5.3), and hence, finally, of the Division Algorithm (Theorem 5.4.1). And so the remarkable synthesis of this story is that the failure of our approach to correctly solve $x^3 = y^2 + 26$ can trace its way back to the lack of a functioning Division Algorithm in $\mathbb{Z}[\sqrt{-26}]$. The failure of unique factorization in rings of the form $\mathbb{Z}[\sqrt{d}]$ is potentially demoralizing, and it's okay if you need to go look at pictures of puppies before moving forward. Indeed, the history of mathematics has notable examples (see Exercise 6.36) in which Diophantine equations were erroneously thought solved before someone kindly (or unkindly) pointed out that the purported proof had inappropriately relied on unique factorization in some ring.

The goal of this chapter is now clear: we need to establish that *The Path* holds at a level of generality more than just $\mathbb{Z}$ and $\mathbb{Z}[i]$. In particular, $\mathbb{Z}[\sqrt{-2}]$ seems a promising candidate—are there others? To be explicit, we hope to state and prove a more precise version of the following critical result[1] :

Theorem 6.1.3 (The Fundamental Meta-Theorem of Arithmetic)
A ring with a Division Algorithm has a Fundamental Theorem of Arithmetic.

The precise "non-meta" version of this is the upcoming Theorem 6.5.3.

6.2 Algebraic Numbers and Rings of Integers

To generalize from $\mathbb{Z}[i]$ to other worlds of numbers, it will be useful to identify the properties of i that made $\mathbb{Z}[i]$ a convenient home for doing number theory. A first observation involves a notational subtlety—the reader may have noticed that the square brackets in $\mathbb{Z}[x]$ and $\mathbb{Z}[i]$ seem to do very different things to the ring $\mathbb{Z}$:

$$\mathbb{Z}[x] = \{a_0 + a_1 + \cdots + a_n x^n : a_k \in \mathbb{Z}\} \quad \text{vs.} \quad \mathbb{Z}[i] = \{a_0 + a_1 i : a_k \in \mathbb{Z}\}.$$

Why does $\mathbb{Z}[x]$ involve arbitrary sums of powers of x whereas we limit $\mathbb{Z}[i]$ to only linear expressions? The answer, and thus the relevant property of i, is that i satisfies the clean algebraic relation $i^2 + 1 = 0$. This means that an expanded definition of $\mathbb{Z}[i]$ where you incorporate arbitrary "polynomials in i," we would not actually

---

[1] One author wanted to call this the "Funda-Meta Theorem of Arithmetic," but the other authors would not allow it. This footnote is the resulting compromise. Strangely, we all claim to be the one author.

change $\mathbb{Z}[i]$, as all higher degree terms collapse, e.g.

$$3i^3 + 4i^2 + 5i + 5 = 3i(i^2 + 1) + 4(i^2 + 1) + 2i + 1 = 3i(0) + 4(0) + 2i + 1 = 2i + 1.$$

It is this property, that i satisfies a polynomial algebraic relationship, that merits generalization. Recall that a polynomial $f \in \mathbb{Q}[x]$ is said to be **monic** if its leading coefficient is 1.

Definition 6.2.1

A number $\alpha \in \mathbb{C}$ is **algebraic** if there exists a monic polynomial $f \in \mathbb{Q}[x]$ such that $f(\alpha) = 0$. The **degree** of an algebraic number is the smallest degree of any such polynomial, which is called a **minimal polynomial** for α. A number that is not algebraic is called **transcendental**. ◄

For example, i is algebraic of degree 2 since it is a root of the degree 2 monic polynomial $x^2 + 1 \in \mathbb{Q}[x]$ (and is not the root of any degree 1 polynomial $f \in \mathbb{Q}[x]$). One preliminary observation before further examples:

Lemma 6.2.2

If $\alpha \in \mathbb{C}$ is an algebraic number, then its minimal polynomial is unique, so we can reference *the* minimal polynomial of α. ◄

Proof Suppose α is of degree d, and suppose f and g are distinct monic polynomials of degree d such that $f(\alpha) = g(\alpha) = 0$. Let $h = f - g$, a non-zero polynomial of degree less than d such that $h(\alpha) = 0$. Dividing h by its leading coefficient provides a monic polynomial of degree less than d of which α is a root, contradicting that α was of degree d. □

▶ **Example 6.2.3** Every rational number is algebraic of degree 1. For example, $\alpha = \frac{7}{5}$ is a root of the monic polynomial $x - \frac{7}{5} \in \mathbb{Q}[x]$.

▶ **Example 6.2.4** Radical expressions: $\sqrt{-3}$ and $\sqrt{2}$ are algebraic of degree 2 with respective minimal polynomials $x^2 + 3$ and $x^2 - 2$. The number $\sqrt[3]{5}$ is algebraic, using $f(x) = x^3 - 5$, and is of degree 3 since the output of the quadratic formula shows there is no polynomial of degree 2 or smaller for which $\sqrt[3]{5}$ is a root. Of course, $\sqrt{49}$ and $\sqrt[3]{27}$ are algebraic of degree 1, using the obvious polynomials $x - 7$ and $x - 3$, but in general, if an integer n has no k-th power divisors, then $\sqrt[k]{n}$ is algebraic of degree k via the polynomial $f(x) = x^k - n$.

Our discussion of algebraic numbers will continue to focus on degree 2 examples, so we will not pause to prove carefully statements about higher degree numbers (like the last sentence in the previous example), but higher degree examples do motivate some interesting discussion points.

▶ **Example 6.2.5** Let α be the largest real root of the polynomial $f(x) = x^5 - x + 1$ (the polynomial is of odd degree, so has at least one real root, and hence a well-defined largest one). Despite being perfectly well-defined, it is hard[2] to write α down in terms of radical and standard arithmetic operations—there is no formula like the quadratic formula for expressing roots of fifth-degree polynomials as sums and products of fifth roots. And yet, the number is tautologically algebraic, as f is its minimal polynomial.

▶ **Remark 6.2.6** A philosophical point to make is that we have an unhealthy obsession with trying to write down exact formulas for solutions to equations, often without benefit. For example, to work with the number α in the preceding example, even if a complicated expression in terms of radicals existed, it would likely be just as easy to leave it simply as "α" and make use of its defining algebraic expression when possible. After all, we pretend we have solved a calculus problem when we write down that the answer to a problem is $\sqrt{2}$, but we could just as well be answering that we don't know the answer[3] but know that it is positive and satisfies the polynomial $x^2 - 2 = 0$.

▶ **Example 6.2.7** The numbers $\sqrt{2}$ and $\sqrt{3}$ are algebraic, but what about the number $\alpha = \sqrt{2} + \sqrt{3}$? It is fun to try to come up with a minimal polynomial for α as, unlike the numbers in Example 6.2.4, no power of α is an integer. Here is a nifty approach: We begin by observing that

$$\alpha^2 = (\sqrt{2} + \sqrt{3})(\sqrt{2} + \sqrt{3}) = 2 + 2\sqrt{6} + 3 = 5 + 2\sqrt{6}.$$

This shows that $\alpha^2 - 5 = 2\sqrt{6}$, so $(\alpha^2 - 5)^2 = 24$. Taking

$$f(x) = (x^2 - 5)^2 - 24 = x^4 - 10x^2 + 1 \in \mathbb{Q}[x]$$

thus provides a minimal polynomial for α of degree at most 4 (in fact exactly 4, although verifying that there is no lower-degree monic polynomial for $\sqrt{2} + \sqrt{3}$ requires a further check).

You may have begun to wonder if there even exist transcendental numbers; i.e., complex numbers that are not the root of *any* polynomial with rational coefficients. An appealingly indirect approach to this question makes use of infinite cardinalities: one argues that there are countably many algebraic numbers (there are only countably many polynomials in $\mathbb{Q}[x]$ of which they could be roots, each with finitely many roots), whereas there are uncountably many complex numbers. So somewhat ironically, while most of the numbers we encounter are algebraic, transcendental numbers vastly outnumber the algebraic ones—and yet, it is quite difficult to explicitly point out even one of them! We mention three examples without proof, starting

---

[2] Technically, literally impossible, but we argue that still counts as "hard."

[3] Answering "Solve $x^2 - 2 = 0$" with "$x = \pm\sqrt{2}$" feels particularly circular in this sense.

with one of historical significance as the first number explicitly demonstrated to be transcendental. The other two examples are also of periodic interest.

> **Theorem 6.2.8 (Liouville, 1844)**
> The number
>
> $$0.1010010000001000000000000000000000000100\ldots,$$
>
> whose decimal expansion is formed by interspersing $n!$ zeroes after the n-th one, is transcendental.

> **Theorem 6.2.9 (Hermite, 1873)**
> The number e is transcendental.

> **Theorem 6.2.10 (Lindemann, 1882)**
> The number π is transcendental.

That we are able to say with certainty that there is *no* polynomial with rational coefficients having π as a root is not only remarkable, it is key to solving several famous geometry problems from Greek antiquity, e.g., that it is impossible to "square the circle."

Now, moving on to integers: the phrase "Rings of Integers" in the section header might induce some eyebrow-raising—after all, isn't there the one and only $\mathbb{Z}$, *the* ring of integers? An important perspective to begin adopting is that being an integer (being "integral") is a property that a number can either possess or not possess, much like a number can be even, or prime, or greater than 7, or a square, etc. The viewpoint that emerges is that starting with the rational numbers, $\mathbb{Q}$, the set $\mathbb{Z}$ (*the* integers) is simply "the set of elements of $\mathbb{Q}$ that are integral." We would like to generalize this to other contexts, so that it will make sense to say things like "$\mathbb{Z}[i]$ consists of the integral elements of $\mathbb{Q}[i]$."

What sort of definition would bring this dream to fruition? A tempting but ultimately fruitless answer is to say that "integers are rational numbers with a denominator of 1." There are plenty of mundane objections to this answer (Should we allow a denominator of -1 as well? What about i or other units? What if we run into a non-reduced fraction?) that are easily rectified by a more precise version of the statement, but the general objection is that an arbitrary algebraic number has no notion of its denominator. As a particularly compelling argument, consider the number α from Example 6.2.5. What is its denominator? Is it 1? Is it even a meaningful question?

Fortunately, an elegant fix to this suggestion exists, making use of the fact that we have already identified $\mathbb{Z}$ as an important subset of $\mathbb{Q}$.

Definition 6.2.11

An algebraic number α is an **algebraic integer** (or is **integral**) if its minimal polynomial has all integer coefficients. ◄

For example, $\frac{\sqrt[3]{7}}{5}$ is an algebraic number that is not an algebraic integer since its minimal polynomial $x^3 - \frac{7}{125}$ does not have all integer coefficients. Thus this definition catches something of our "has a denominator" intuition while enabling the study of numbers beyond the comfort of our intuitive understanding. We will experiment with this idea in the upcoming Exploration, but first a brief discussion of these new worlds.

Theorem 6.2.12
The set $\overline{\mathbb{Q}}$ of all algebraic numbers forms a field. The set $\overline{\mathbb{Z}}$ of all algebraic integers forms a ring.

Since both $\overline{\mathbb{Z}}$ and $\overline{\mathbb{Q}}$ are subsets of $\mathbb{C}$, most of the ring and field axioms (associativity, distributivity, etc.) are inherited from those properties in $\mathbb{C}$. The more substantive claims of the theorem are that if α and β are algebraic, then so are $\alpha \pm \beta$, $\alpha\beta$, and $\frac{\alpha}{\beta}$ (for $\beta \neq 0$), and likewise that sums and products of algebraic integers are again algebraic integers. A full proof of this theorem would be formalized through a somewhat lengthy detour into linear algebra, but we have seen and will continue to see the main ingredients in special cases. For example, a proof that $\alpha + \beta$ is algebraic when both α and β are of degree 2 would look much like Example 6.2.7, and an illustrative example where α and β are of different degrees is taken up in Exercise 6.33, where you use linear algebra to find a minimal polynomial for $\sqrt{3} + \sqrt[3]{2}$. We will prove special cases of Theorem 6.2.12 for families of algebraic numbers of particular interest as we encounter them.

▶ **Remark 6.2.13** A principal theme of recent sections has been to extend $\mathbb{Z}$ by the adjoining of algebraic integers like i and $\sqrt{-2}$. Armed with the new ring $\overline{\mathbb{Z}}$, it is initially tempting to bypass such "small" extensions $\mathbb{Z}[i]$ and $\mathbb{Z}[\sqrt{-2}]$ and try to work directly in $\overline{\mathbb{Z}}$. But it is worth noting that $\overline{\mathbb{Z}}$ and $\overline{\mathbb{Q}}$ really are quite exotic worlds to work in. For example, it is not hard to show that if $\alpha \in \overline{\mathbb{Z}}$, then $\sqrt{\alpha} \in \overline{\mathbb{Z}}$ as well (Exercise 6.28), so the identity

$$\alpha = \sqrt{\alpha} \cdot \sqrt{\alpha} = \sqrt[4]{\alpha} \cdot \sqrt[4]{\alpha} \cdot \sqrt[4]{\alpha} \cdot \sqrt[4]{\alpha} = \cdots$$

shows that there are no prime elements at all in $\overline{\mathbb{Z}}$, and hence no notion of a prime factorization.

> **Exploration K**

Algebraic Integers and their Homes ◄

One of the most significant implications of the algebraic approach to number theory is that we should take care when choosing which ring to work in. If we need to work with i, for example, we need to enlarge from $\mathbb{Z}$ to a ring that includes i, but not all the way to $\mathbb{C}$, where notions of primeness and factorization are essentially vacuous. The ring $\mathbb{Z}[i]$ arises as a perfect compromise. For other algebraic numbers, it is a little less clear what that perfect home ring should be. Let's explore.

K.1 Find minimal polynomials for each of the following and decide if they are algebraic integers:

$$1 + \sqrt{5} \qquad 1 + \sqrt{7} \qquad \frac{1 + \sqrt{5}}{2} \qquad \frac{1 + \sqrt{7}}{2} \qquad \frac{1 + \sqrt{5}}{4} \qquad \frac{1 + \sqrt{7}}{4}.$$

For a number $\alpha \in \mathbb{C}$ we can consider the ring

$$\mathbb{Z}[\alpha] = \{a_0 + a_1\alpha + a_2\alpha^2 + \cdots + a_n\alpha^n : a_k \in \mathbb{Z}, n \geq 0\}.$$

We have seen that for $\alpha = i$, we have more simply $\mathbb{Z}[\alpha] = \{a + b\alpha : a, b \in \mathbb{Z}\}$.

K.2 For each of the following rings, decide whether $\mathbb{Z}[\alpha] = \{a + b\alpha : a, b \in \mathbb{Z}\}$:

$$\mathbb{Z}\left[\sqrt{-3}\right] \qquad \mathbb{Z}\left[\frac{1 + \sqrt{-3}}{2}\right] \qquad \mathbb{Z}\left[\frac{1 + \sqrt{-3}}{4}\right].$$

Begin by computing α^2 and α^3 for each of the above.

K.3 Much like $\mathbb{Z}[i]$, the elements of $\mathbb{Z}[\sqrt{-3}] = \{a + b\sqrt{-3} : a, b \in \mathbb{Z}\}$ form a rectangular lattice in the complex plane (though no longer a *square* lattice). Give rough sketches of each of the following sets inside $\mathbb{C}$:

$$\mathbb{Z}\left[\sqrt{-3}\right] \qquad \mathbb{Z}\left[\frac{1 + \sqrt{-3}}{2}\right] \qquad \mathbb{Z}\left[\frac{1 + \sqrt{-3}}{4}\right].$$

Explore any connections with your answers to the previous exercise.

K.4 Taking $\alpha = \pi$ provides a curious example. Get a feel for doing algebra in the ring $\mathbb{Z}[\pi]$. It should feel pretty similar to doing algebra in another ring. Which one? Why?

6.3 Quadratic Fields: Integers, Norms, and Units

The realization that the infamous golden ratio of ancient Greek mathematics,

$$\varphi = \frac{1+\sqrt{5}}{2},$$

is an algebraic *integer* (being the root of the integer polynomial x^2-x-1), despite "having a denominator," calls into question some of our deep-seated prejudices about which numbers should be considered an (algebraic) integer and further demonstrates why a "no denominators" rule is not the way to go. Instead, we begin by allowing any and all denominators (that is, we choose to work in a *field*), and then prove theorems that describe the integral elements in that field. For example, we will prove that the algebraic integers in the field $\mathbb{Q}$ are precisely $\mathbb{Z}$, and the algebraic integers in the field $\mathbb{Q}[i]$ are precisely $\mathbb{Z}[i]$. In other examples, e.g., $\mathbb{Q}[\sqrt{5}] = \{a + b\sqrt{5} \colon a, b \in \mathbb{Q}\}$, we are forced to admit elements like $\frac{1+\sqrt{5}}{2}$.

To set up a general statement, we introduce the notion of a quadratic field and then pose the question of determining the ring of integers of that field. One precaution about introducing rings of the form $\mathbb{Q}[\sqrt{d}]$ is that we should be slightly picky about which d we allow—for example,

$$\mathbb{Q}[\sqrt{4}] = \{a + b\sqrt{4} \colon a, b \in \mathbb{Q}\} = \{a + 2b \colon a, b \in \mathbb{Q}\} = \mathbb{Q},$$

and similarly, $\mathbb{Q}[\sqrt{12}] = \mathbb{Q}[\sqrt{3}]$, $\mathbb{Q}[\sqrt{-25}] = \mathbb{Q}[i]$, etc. In general, if $d = a^2 b$, then $\mathbb{Q}[\sqrt{d}] = \mathbb{Q}[\sqrt{b}]$, and so it is often convenient to suppose that d has no non-trivial square factors. We say that an integer d is **square-free** if for all primes p we have $p^2 \nmid d$ (or equivalently, $v_p(d) \leq 1$ for all primes p). Only one special case leaves us in a bit of a pickle:

Pickle 6.3.1

Should 1 be considered square-free? If so, is it the unique square-free square? If not, why not? What about 0? Choose a side and pick a fight with someone on these very topics[4]. ◄

Definition 6.3.2

A **quadratic field** is a field of the form

$$\mathbb{Q}[\sqrt{d}] = \{a + b\sqrt{d} \colon a, b \in \mathbb{Q}\}$$

for some square-free integer $d \in \mathbb{Z}$ ($d \neq 0, 1$). We say K is a **complex quadratic field** if $d < 0$ and a **real quadratic field** if $d > 0$. ◄

---

[4] or have a lovely reasoned conversation over coffee.

We have already had experience with complex quadratic fields (taking $d = -1$ gives $\mathbb{Q}[i]$) and can leverage this understanding when exploring new examples. First and foremost, it is routine to verify that the definition of $\mathbb{Q}[\sqrt{d}]$ above does indeed agree with its definition as "$\mathbb{Q}[x]$ with $x = \sqrt{d}$ plugged in" and that this set defines a field. Only the calculation that every non-zero element has a multiplicative inverse merits spelling out, and that follows from the identity

$$\frac{1}{a + b\sqrt{d}} = \frac{1}{a + b\sqrt{d}} \cdot \frac{a - b\sqrt{d}}{a - b\sqrt{d}} = \frac{a}{a^2 - db^2} + \frac{-b}{a^2 - db^2}\sqrt{d} \in \mathbb{Q}[\sqrt{d}].$$

Here, note that the denominator $a^2 - db^2$ is never zero, as $a^2 - db^2 = 0$ would imply that $d = \frac{a^2}{b^2}$ was a square, contradicting its square-freeness.

Definition 6.3.3

Given a quadratic field[5] $\mathbb{Q}[\sqrt{d}]$, if $\alpha = a + b\sqrt{d} \in \mathbb{Q}[\sqrt{d}]$, then the **conjugate** of α is

$$\overline{\alpha} = \overline{a + b\sqrt{d}} = a - b\sqrt{d},$$

and its **norm** is given by

$$N(\alpha) = \alpha\overline{\alpha} = (a + b\sqrt{d})(a - b\sqrt{d}) = a^2 - db^2.$$ ◀

Again, when $d = -1$, we recover the norm function studied extensively in the previous chapter, and in general, for $d < 0$, this is just the complex norm—it gives the square of the distance from the element to 0 in the complex plane. When $d > 0$, however, this represents a new construction. In $\mathbb{Q}[\sqrt{2}]$, for example, the norm of the element $7 + 5\sqrt{2}$ is given by

$$N(7 + 5\sqrt{2}) = (7 + 5\sqrt{2})(7 - 5\sqrt{2}) = 49 - 50 = -1. \qquad (6.1)$$

This norm of -1 for this element certainly doesn't reflect the distance from $7 + 5\sqrt{2}$ to the origin in either the real line or the complex plane, and it is hard to interpret this number as a key geometric property of any sort. On the other hand, however, from an algebraic perspective this norm continues to exhibit many of the nice properties that the complex norm did in our study of the Gaussian integers (recall that the norm was our principal tool for finding primes in $\mathbb{Z}[i]$). To generalize this to an arbitrary quadratic field, we will need to finish the story started by the Golden Ratio above to characterize the algebraic integers in a quadratic field.

---

[5] The notation $\mathbb{Q}[\sqrt{d}]$ will for us always reference a quadratic field. In particular, we assume d is a square-free integer and $d \neq 0, 1$.

Definition 6.3.4

The **ring of integers** of a quadratic field is the subset of elements of the field that are algebraic integers. ◀

Of course, you can't define a set into being a ring. That these sets are rings is a small part of the following theorem, which gives an explicit description of the algebraic integers of a quadratic field.

Theorem 6.3.5

Let $d \neq 0, 1$ be a square-free integer. Then:

- If $d \not\equiv 1 \pmod 4$, then the ring of integers of $\mathbb{Q}[\sqrt{d}]$ is

$$\mathbb{Z}[\sqrt{d}] = \{a + b\sqrt{d} : a, b \in \mathbb{Z}\}$$

- If $d \equiv 1 \pmod 4$, then the ring of integers of $\mathbb{Q}[\sqrt{d}]$ is

$$\mathbb{Z}\left[\frac{1 + \sqrt{d}}{2}\right] = \left\{a + b\left(\frac{1+\sqrt{d}}{2}\right) : a, b \in \mathbb{Z}\right\}$$

Proof First, it is routine to verify that the described numbers are algebraic integers in both cases: the quadratic formula verifies that the number $a + b\sqrt{d}$ has the minimal polynomial

$$(x - (a + b\sqrt{d}))(x - (a - b\sqrt{d})) = x^2 - 2ax + (a^2 - db^2) \in \mathbb{Z}[x],$$

and the number $a + b\left(\frac{1+\sqrt{d}}{2}\right)$ has minimal polynomial

$$x^2 - (2a + b)x + a^2 + ab - b^2\left(\frac{d-1}{4}\right),$$

which is in $\mathbb{Z}[x]$ since $d \equiv 1 \mod 4$.

It remains to show the reverse direction, that if $\alpha \in \mathbb{Q}[\sqrt{d}]$ is an algebraic integer then it is in $\mathbb{Z}[\frac{1+\sqrt{d}}{2}]$ when $d \equiv 1 \pmod 4$ and is in $\mathbb{Z}[\sqrt{d}]$ otherwise.

Suppose $\alpha = a + b\sqrt{d} \in \mathbb{Q}[\sqrt{d}]$ is an algebraic integer, so its minimal polynomial $x^2 - 2ax + (a^2 - db^2)$ has integer coefficients. Let $e = 2a$ and $f = a^2 - db^2$. Write $b = \frac{m}{n}$ for $m, n \in \mathbb{Z}$ with $\gcd(m, n) = 1$. Then substituting $a = \frac{e}{2}$ and $b = \frac{m}{n}$ into the expression for f and clearing denominators gives

$$e^2 n^2 - 4dm^2 = 4fn^2. \tag{$*$}$$

This shows that e^2n^2 must be a multiple of 4, and so at least one of e or n is even. We break into these two cases.

First, if e is even, then $a = \frac{e}{2}$ is an integer, and hence $a^2 - f = db^2 = \frac{dm^2}{n^2}$ must be an integer as well, so $n^2 \mid dm^2$. Since $\gcd(m, n) = 1$, Euclid's Lemma shows that $n^2 \mid d$, which since d is square-free implies that $n = \pm 1$. Since $b = \frac{m}{n}$, this shows that b, along with a, must be an integer.

Otherwise e is odd and n is even, and since $\gcd(m, n) = 1$, m must be odd. Substituting $n = 2t$ into $(*)$ and canceling a factor of 4 gives

$$e^2t^2 - dm^2 = 4ft^2.$$

Since e and m are odd, we have $e^2 \equiv m^2 \equiv 1 \bmod 4$ and so reducing this equation mod 4 gives $t^2 \equiv d \bmod 4$. Since d is a square mod 4, d is either 0 or 1 mod 4, but since d is square-free, we know $4 \nmid d$ and hence $d \equiv 1 \bmod 4$. Substituting $m = nb = 2tb$ into $e^2t^2 - dm^2 = 4ft^2$, we find $e^2t^2 - d(2tb)^2 = 4ft^2$, so we see that $t^2 \mid d(2tb)^2$. Say $d(2tb)^2 = t^2k$. (We stress that b is not necessarily an integer, but $2tb$ is.)

Since d is square-free, it must be the case that $t^2 \mid (2tb)^2$, and so $(2b)^2 \in \mathbb{Z}$. It thus makes sense to cancel a t^2 and reduce mod 4:

$$e^2t^2 - d(2tb)^2 = 4ft^2$$
$$e^2 - d(2b)^2 = 4f$$
$$1 - 1(2b)^2 \equiv 0 \bmod 4,$$

showing that $2b$ must be odd.

Combining the two cases, we conclude that for $a + b\sqrt{d}$ ($a, b \in \mathbb{Q}$) to be an algebraic integer we must either have a and b both integers, or $2a$ and $2b$ both odd integers and $d \equiv 1 \bmod 4$. That is, if $d \not\equiv 1 \pmod 4$, then $a + b\sqrt{d}$ is an algebraic integer if and only if $a, b \in \mathbb{Z}$. If $d \equiv 1 \pmod 4$, then those same numbers are algebraic integers, and so are the $a + b\sqrt{d}$ for which a and b are both half-integers.

Finally, given our explicit descriptions of these sets, it is easy to verify that they are rings. The only ring axioms that need checking are that the two sets are closed under addition and multiplication. Of these, addition is trivial and multiplication follows from the identities

$$(a + b\sqrt{d})(m + n\sqrt{d}) = (am + bdn) + (an + bm)\sqrt{d} \in \mathbb{Z}[\sqrt{d}]$$

and, writing $\gamma = \frac{1+\sqrt{d}}{2}$ when $d \equiv 1 \bmod 4$,

$$(a + b\gamma)(m + n\gamma) = \left(am + bn \cdot \frac{d-1}{4}\right) + (an + bn + bm)\gamma \in \mathbb{Z}\left[\frac{1 + \sqrt{d}}{2}\right],$$

both verifiable via direct computation. $\qquad\square$

The conclusion of this train of thought is worth pausing to appreciate. While we may have been taken aback by the appearance of "denominators" in our algebraic integers, Theorem 6.3.5 spells out exactly how far this can go. So as we temper any adverse reactions to denominators, we stress that it is the ring of integers of $\mathbb{Q}[\sqrt{d}]$ (as opposed to always using $\mathbb{Z}[\sqrt{d}]$) that is the "right" place to develop number theory for that field.

Let us connect the notions of norms and integers. For an algebraic integer of the form $\alpha = a + b\sqrt{d}$, it's clear that its norm $N(\alpha) = a^2 - db^2$ is also an integer. When $d \equiv 1 \bmod 4$, however, we need the further computation

$$N\left(a + b\left(\frac{1 + \sqrt{d}}{2}\right)\right) = \left(a + \frac{b}{2} + \frac{b}{2}\sqrt{d}\right)\left(a + \frac{b}{2} - \frac{b}{2}\sqrt{d}\right)$$

$$= \left(a + \frac{b}{2}\right)^2 - \left(\frac{b\sqrt{d}}{2}\right)^2$$

$$= a^2 + ab + b^2\left(\frac{1 - d}{4}\right).$$

Since $a, b, d \in \mathbb{Z}$ and $d \equiv 1 \pmod 4$, we see that $N(\alpha)$ is still an integer. As with the Gaussian integers, we can now piggyback on our understanding of $\mathbb{Z}$ to develop number theory in these larger rings of integers.

Lemma 6.3.6

Let R be the ring of integers of $\mathbb{Q}[\sqrt{d}]$ and let $\alpha, \beta, \gamma \in R$. Then $N(\alpha\beta) = N(\alpha)N(\beta)$, and thus if $\alpha \mid \gamma$ in R, then $N(\alpha) \mid N(\gamma)$ in $\mathbb{Z}$. ◀

Proof For the first claim, we compute

$$N(\alpha\beta) = (\alpha\beta)(\overline{\alpha\beta}) = \alpha\overline{\alpha}\beta\overline{\beta} = N(\alpha)N(\beta).$$

For the second claim, if $\alpha \mid \gamma$, then $\gamma = \alpha\beta$ for some $\beta \in R$, so $N(\alpha)N(\beta) = N(\gamma)$ by the first claim. Thus $N(\alpha) \mid N(\gamma)$ in $\mathbb{Z}$. □

Lemma 6.3.7

Let R be the ring of integers of $\mathbb{Q}[\sqrt{d}]$. Then for all $\mu \in R$, μ is a unit in R if and only if $N(\mu)$ is a unit in $\mathbb{Z}$. That is,

$$\mu \in R^\times \iff N(\mu) \in \mathbb{Z}^\times.$$ ◀

Proof Suppose $\mu \in R$ is a unit. Then there exists $\nu \in R$ such that $\mu\nu = 1$. Then by Lemma 6.3.6,

$$1 = N(1) = N(\mu\nu) = N(\mu)N(\nu),$$

and since $N(\mu)$ and $N(\nu)$ are integers that multiply to 1, we see they must be units. Conversely, if $N(\mu)$ is a unit, then $N(u) = \mu\bar{\mu} = \pm 1$, so either $\bar{\mu}$ or $-\bar{\mu}$ is a multiplicative inverse of μ in R. (Note that from the explicit form given in Theorem 6.3.5, μ is an algebraic integer if and only if $\bar{\mu}$ is.)   □

For complex quadratic fields, the story of units is particularly simple. Namely, we already know that when $d = -1$, we have units of $\{\pm 1, \pm i\}$, and there is only one other case, $d = -3$ (of sufficient significance that we will deal with it again in Section 6.6), with units other than ± 1.

Corollary 6.3.8

Let $d < 0$, and let R be the ring of integers of $\mathbb{Q}[\sqrt{d}]$. Then with the exceptions of $d = -1$ and $d = -3$, we have $R^{\times} = \{\pm 1\}$.

For $d = -1$ we have $R^{\times} = \{\pm 1, \pm i\}$, and for $d = -3$, $R^{\times} = \{\pm 1, \pm \frac{1 \pm \sqrt{-3}}{2}\}$. ◀

Proof By Lemma 6.3.7, an element $\mu \in R$ is a unit if and only if it has norm ± 1, but norms of complex numbers are always positive, so we need only solve the equation $N(\mu) = +1$. If $d \not\equiv 1 \bmod 4$, this is the equation

$$1 = a^2 - db^2.$$

If $d = -1$, we quickly find the only solutions to be $(a, b) = (\pm 1, 0)$ and $(a, b) = (0, \pm 1)$, corresponding to the units ± 1 and $\pm i$ respectively. When $d < -1$, the term $-db^2$ is greater than 1 unless $b = 0$, so has no other solutions than $(a, b) = (\pm 1, 0)$, corresponding to the units $\mu = \pm 1$.

For the $d \equiv 1 \bmod 4$ case, we need to solve the Diophantine equation

$$1 = a^2 + ab + b^2 \left(\frac{1 - d}{4} \right),$$

or after a little algebra, the equivalent equation

$$(2a + b)^2 - db^2 = 4.$$

Again, both $(2a + b)^2$ and $-db^2$ are positive. If $b = 0$, then $a = \pm 1$ and we retrieve the units $\mu = \pm 1$ for any d. Now we look for any extra solutions: if $d = -3$, we find $(a, b) = (0, \pm 1), \pm(1, -1)$ corresponding to the units

$$\mu = \pm \left(\frac{1}{2} \pm \frac{\sqrt{-3}}{2} \right),$$

But for any other $d \equiv 1 \bmod 4$, d is at most -7, and so $-db^2 > 4$ whenever $b \neq 0$, providing no more solutions.   □

For real quadratic fields, the situation is significantly wilder, as units can now have norm ± 1, and can be vastly greater in number. In $\mathbb{Q}[\sqrt{2}]$, for example, Equation (6.1) provides the unit $\mu = 7 + 5\sqrt{2}$ of norm -1. Recalling that the product of two units in any ring is again necessarily a unit, it must be that μ^2 is also a unit, and a computation gives

$$(7 + 5\sqrt{2})^2 = 99 + 70\sqrt{2}$$

as another unit (check: $99^2 - 2 \cdot 70^2 = 9801 - 9800 = 1$). In fact, μ too is of this form, as it turns out to be the cube of another unit: $7 + 5\sqrt{2} = (1 + \sqrt{2})^3$.

Corollary 6.3.9

The ring $\mathbb{Z}[\sqrt{2}]$ has infinitely many units, including at least the family $(1 + \sqrt{2})^n$ for each $n \in \mathbb{Z}$, and their negatives. ◄

Though the example highlights the general phenomenon, we will take up a more nuanced discussion of the units of real quadratic fields in Section 9.4.

6.4 Euclidean Domains

Recall the goal of this chapter: to follow *The Path* and prove the Fundamental Meta-Theorem of Arithmetic, that rings having some notion of a Division Algorithm also have unique factorization into primes. As we uncover in Exploration L, the presence of zero divisors is enough to dash our hopes of unique factorization, so we start by limiting our attention to integral domains, i.e., commutative rings with unity in which the product of non-zero elements is always non-zero.

Most of the rings we have encountered (e.g., $\mathbb{Z}$, $\mathbb{Q}$, $\mathbb{R}$, $\mathbb{C}$, $\mathbb{Z}[i]$, $\mathbb{R}[x]$) are integral domains, since in none of these rings is it possible to multiply two non-zero elements and get zero. The modular rings $\mathbb{Z}/(n)$ make an enlightening set of contrasting examples. Here, if n is composite, then $\mathbb{Z}/(n)$ is *not* an integral domain—for example, in $\mathbb{Z}/(6)$ we have $[2] \cdot [3] = [0]$. If n is prime, on the other hand, we have seen that $\mathbb{Z}/(n)$ is always an integral domain (and in fact, a field). Among integral domains, the ones in which we have hope of defining a Division Algorithm are precisely those where we have some notion of the *size* of an element, allowing us to make sense of the claim that the remainder is "smaller" than the divisor.

Definition 6.4.1

A **Euclidean Domain** is an integral domain R with a **Euclidean norm**, which is a function $\mathcal{N}$ from $R - \{0\}$ to the non-negative integers satisfying the following two properties:

(i) For all $a, b \in R - \{0\}$, if $a \mid b$ then $\mathcal{N}(a) \leq \mathcal{N}(b)$.

(ii) For all $a, b \in R - \{0\}$, there exist elements $q, r \in R$ such that $a = bq + r$ and either $r = 0$ or $\mathcal{N}(r) < \mathcal{N}(b)$.

◄

In principle, it is possible to have many different Euclidean norms on a given integral domain, so an integral domain can be a Euclidean domain in multiple ways. In addition, it may be difficult in principle to decide if an integral domain is Euclidean or not, as it's difficult to rule out the existence of a possibly complicated Euclidean norm function defined on it. That said, we will be almost exclusively focused on viewing rings as Euclidean domains under reasonably standard norm maps. Finally, note that we do not insist on giving $0 \in R$ a norm, though in most of our examples, it will be clear how to define $\mathcal{N}(0)$ as well.

▶ **Example 6.4.2** The ring $\mathbb{Z}$ is a Euclidean domain via the absolute value function $\mathcal{N}(n) = |n|$.

▶ **Example 6.4.3** The ring $\mathbb{Z}[i]$ is a Euclidean domain via the norm function $\mathcal{N}(a + bi) = a^2 + b^2$.

▶ **Example 6.4.4** The ring $\mathbb{R}[x]$ is a Euclidean domain via the norm function $\mathcal{N}(f) = \deg(f)$.

The proofs of these three claims follow from the existence of a Division Algorithm in these rings, with the remainder guaranteed to have a smaller value than the divisor under the given norm.

▶ **Example 6.4.5** Any field (e.g., $\mathbb{Q}$ or $\mathbb{R}$ or $\mathbb{C}$) is a Euclidean domain via the trivial function $\mathcal{N}(x) = 1$ (Exercise 6.24).

The previous section on rings of integers in quadratic fields provided another potentially rich source of Euclidean domains. For now, let us see how much the number theory we developed in $\mathbb{Z}$ and $\mathbb{Z}[i]$ continues to hold in an arbitrary Euclidean ring (spoiler alert: like all of it).

Definition 6.4.6

Given elements a, b, not both 0, of a Euclidean domain R with Euclidean norm $\mathcal{N}$, a **gcd** of a and b is an element $g \in R$ such that:

- $g \mid a$ and $g \mid b$; and
- If $d \mid a$ and $d \mid b$, then $\mathcal{N}(d) \leq \mathcal{N}(g)$.

That is, g is a common divisor of a and b of greatest Euclidean norm. If 1 is a gcd of a and b, we write $\gcd(a, b) = 1$ and say a and b are **relatively prime**. ◄

Note that both definitions extend their counterparts from Chapter 3.

Lemma 6.4.7

Let R be a Euclidean domain with Euclidean norm $\mathcal{N}$ and $a \neq 0 \in R$. If $\mathcal{N}(a)=0$, then $a \in R^{\times}$. ◄

Proof Suppose that $\mathcal{N}(a) = 0$. Then applying the Division Algorithm to divide 1 by a, we find $q, r \in R$ such that $1 = aq + r$ with $r = 0$ or $\mathcal{N}(r) < \mathcal{N}(a)$. Since $\mathcal{N}(a) = 0$, it must be the case that $r = 0$, so $aq = 1$, and thus a is a unit.   □

This is a good moment to check your close mathematical reading skills. How does Lemma 6.4.7 not contradict Lemma 6.3.7? Continuing along *The Path*, our next step is to give Bézout a clearly deserved promotion.

Theorem 6.4.8 (General Bézout Identity)

Suppose R is a Euclidean domain with Euclidean norm $\mathcal{N}$. Let $a, b \in R$ not both be zero, and let g be any gcd of a and b. Then there exist $m, n \in R$ such that

$$am + bn = g.$$

Proof This is trivial if a is a unit, since then $a(a^{-1}g) + b(0) = g$, so assume without loss of generality that a is not a unit. Then $\mathcal{N}(a) > 0$ by the lemma, and we mimic prior proofs of Bézout's Identity: let $S = \{as + bt : s, t \in R$ and $\mathcal{N}(as + bt) > 0\}$, and let $c = as + bt \in S$ be any element of least norm (since $a = a(1) + b(0) \in S$, S is non-empty). We will show $c \mid g$, which is enough to imply g also lies in S.

By the Division Algorithm, we can write $a = qc + r$ with either $r = 0$ or $\mathcal{N}(r) < \mathcal{N}(c)$. But since

$$r = a - qc = a - q(as + bt) = a(1 - qs) + b(-qt)$$

is a linear combination of a and b, $r \in S$ unless $r = 0$; thus we cannot have $\mathcal{N}(r) < \mathcal{N}(c)$ by minimality of $\mathcal{N}(c)$. Hence $r = 0$ and so $c \mid a$. The analogous argument proves that $c \mid b$.

Now write $g = cq' + r'$ for some $q', r' \in R$ with $r' = 0$ or $N(r') < N(c)$. Then $r' = g - cq' = g - (as + bt)q'$, and since g divides g, a, and b, $g \mid r'$ as well, which in turn implies $N(g) \leq N(r')$. But $N(r') < N(c) \leq N(g)$ since g is a gcd of a and b, so we have a contradiction unless $r' = 0$. Therefore, $c \mid g$, as desired.   □

To reach the culmination of our path—a statement about unique prime factorizations—we need to identify the prime elements in an arbitrary Euclidean domain. An interesting dichotomy presents itself here. In the integers, we have two fundamental properties of primeness: the intuitive "no non-trivial divisors" characterization, and the Prime Divisor Property, that if p is a prime dividing a product, then p must divide one of the factors. For primes of $\mathbb{Z}$ (and $\mathbb{Z}[i]$), these two notions are equivalent. Moving forward, we will see this is not always the case, and thus we need more precise terminology to distinguish these properties.

Definition 6.4.9

Let R be a ring and let $p \in R$ be neither 0 nor a unit.

- We say p is **irreducible** if and only if its only divisors are units and associates of p.
- We say p is **prime** if and only if for all $a, b \in R$, if $p \mid ab$, then $p \mid a$ or $p \mid b$.

◀

Many of the consequences of irreducibility/primeness in $\mathbb{Z}$ extend to other rings, even ones where the two notions do not coincide. As an example of this, the induction argument from $\mathbb{Z}$ generalizes to show that if $p \in R$ is prime and $p \mid a_1 \cdots a_k$, then $p \mid a_j$ for some $1 \leq j \leq k$.

Turning to quadratic rings, we again find the one-way irreducibility test (Theorem 5.2.11) we had for $\mathbb{Z}[i]$:

Lemma 6.4.10

Suppose R is the ring of integers of $\mathbb{Q}[\sqrt{d}]$ and $\alpha \in R$. If $N(\alpha)$ is irreducible in $\mathbb{Z}$, then α is irreducible in R. ◀

Proof This follows immediately from Lemma 6.3.6. □

As the following lemma attests, we can specify the relationship between primeness and irreducibility for a general class of algebraic structures.

Lemma 6.4.11

In any integral domain, every prime element is irreducible. ◀

Proof Suppose that p is a prime element of the integral domain R, and suppose that $p = ab$ for some $a, b \in R$. Then $p \mid ab$, so $p \mid a$ or $p \mid b$ by the definition of prime. Without loss of generality, suppose $p \mid a$, so $a = pc$ for some $c \in R$. Now $p = ab = pcb$, so $p(1 - cb) = 0$ and since R is an integral domain, this implies that $1 = cb$, proving that b is a unit. □

The Prime Divisor Property in $\mathbb{Z}$ and $\mathbb{Z}[i]$ (Lemmas 3.3.7 and 5.5.8, respectively) shows the converse, that irreducible elements are always prime in *these* two rings, but crucially, this is not in general true. Let's explore.

As in so many other instances, the norm map is a principal tool in establishing primeness and irreducibility for algebraic integers. As an illustrative example, let's return to the aforefactored

$$3^3 = 27 = (1 + \sqrt{-26})(1 - \sqrt{-26}).$$

First, we claim that both 3 and $1 \pm \sqrt{-26}$ are irreducible in the ring $\mathbb{Z}[\sqrt{-26}]$. Indeed, suppose $1 + \sqrt{-26} = \alpha\beta$ for non-units $\alpha, \beta \in \mathbb{Z}[\sqrt{-26}]$. Then taking norms of both sides, we get $N(\alpha)N(\beta) = 27$, and by Lemma 6.3.7, it must be that one of these two factors has norm 3 and the other has norm 9. But there are no elements of $\mathbb{Z}[\sqrt{-26}]$ of norm 3 since the Diophantine equation $a^2 + 26b^2 = 3$ has no solutions, and so no such factorization can exist. Nearly identical arguments work for both 3 and $1 - \sqrt{-26}$. On the other hand, none of these elements are prime: We have $1 + \sqrt{-26} \mid 3 \cdot 3 \cdot 3$, but $1 + \sqrt{-26} \nmid 3$ (again by considering norms).

That primes and irreducibles become distinct concepts in certain rings of integers is a curious phenomenon and, echoing our discussion from Chapter 3, reinforces the assertion that the Fundamental Theorem of Arithmetic is not as trivial as it may first seem. Indeed, we will soon see that the analog of the Fundamental Theorem of Arithmetic cannot hold in rings where primes and irreducible elements are not one and the same, so any claim that unique factorization is *obvious* contains the subclaim that in $\mathbb{Z}$, the notions of primeness and irreducibility coincide. Regardless, our inexorable march down *The Path* continues.

Theorem 6.4.12 (Euclid's Lemma in Euclidean Domains)
Let R be a Euclidean Domain, and $a, b, c \in R$ satisfy $a \mid bc$ and $\gcd(a, b) = 1$. Then $a \mid c$.

Proof Once again, the proof completely mirrors that of $\mathbb{Z}$, making use of the general Bézout Identity. We leave the details to Exercise 6.30.   □

Corollary 6.4.13 (Prime Divisor Property in a Euclidean Domain)

In a Euclidean domain R, every irreducible element is prime. ◄

Proof Suppose p is irreducible and $p \mid bc$ for some $b, c \in R$. If $p \mid b$, we're done, so suppose $p \nmid b$. A common divisor of p and b cannot then be an associate of p, so since p is irreducible, must be a unit, giving $\gcd(p, b) = 1$. By Euclid's Lemma above, $p \mid c$, showing that p is prime.   □

Exploration L

Euclidean Norms ◄

L.1 Let's examine why the "integral domain" condition is necessary for a sensible notion of unique factorization. Which elements of the ring $\mathbb{Z}/(6)$ are prime? Which are irreducible? Show by example that you can factor elements of $\mathbb{Z}/(6)$ into distinct products of primes; that is, express an element as a product of primes in at least two different ways that differ by more than ordering and associates.

L.2 Show that $\mathbb{R}[x]$ is indeed a Euclidean domain using the degree function as its Euclidean norm. This norm differs from those we are used to: for $\mathbb{Z}$ and $\mathbb{Z}[i]$, we have $\mathcal{N}(ab) = \mathcal{N}(a)\mathcal{N}(b)$, and $\mathcal{N}(u) = 1$ if and only if u is a unit. Show that both of these properties fail for the degree norm on $\mathbb{R}[x]$.

A key tool in proofs involving a general Euclidean domain is to divide elements (often units) by other elements and examine the quotient and remainder that appear from the Division Algorithm. The next exercise gives some practice with this idea, and "fixes" the observation about units in the previous exercise.

L.3 Suppose R is a Euclidean domain with norm $\mathcal{N}$, and let

$$d = \min\{\mathcal{N}(r) : r \in R\}.$$

Prove that

$$\mathcal{N}(r) = d \iff r \text{ is a unit}.$$

Keep in mind that all you have to work with are the two properties defining a Euclidean norm.

L.4 The argument that $\mathbb{R}[x]$ is a Euclidean domain generalizes to the ring $F[x]$ for any field F. However, this is not true for a general integral domain. Show that the degree function does *not* define a Euclidean norm on $\mathbb{Z}[x]$. In fact, only slightly harder is to show that $\mathbb{Z}[x]$ is not a Euclidean domain. Why is that second sentence not redundant with the first?

6.5 Unique Factorization Domains

Euclid's Lemma and the Prime Divisor Property, established for general Euclidean domains in the previous section, are the penultimate step along *The Path* to unique factorization. All that remains is to formally define that last step and show the implication. As we saw in $\mathbb{Z}$ and $\mathbb{Z}[i]$ it takes some care to properly state what we mean by "unique factorization."

Definition 6.5.1

A **Unique Factorization Domain** (or **UFD** for short) is an integral domain in which every non-zero element can be written uniquely, up to ordering and associates, as a product of irreducible elements. ◄

Chapter 3 showed that $\mathbb{Z}$ is a UFD, and Chapter 5 likewise for $\mathbb{Z}[i]$. But we know that $\mathbb{Z}[\sqrt{-26}]$ is not a UFD from our recurring factorization of 27 in that ring into two different products of irreducibles, $27 = (1+\sqrt{-26})(1-\sqrt{-26}) = 3\cdot3\cdot3$, which we showed differ by more than ordering and associates. Recall that as a consequence, our attempt to find integral points on the elliptic curve $x^3 = y^2 + 27$ went awry. In short, the recognition that a ring is a UFD often drastically simplifies the practice of doing number theory in that ring. As an example, we show that the equivalence of primeness and irreducibility that we found in Euclidean Domains via clever use of the General Bézout Identity admits a much more succinct proof when we have unique factorization at our disposal.

Theorem 6.5.2
An element of a UFD is prime if and only if it is irreducible.

Proof Since UFDs are integral domains, we already know that prime elements are irreducible, so it only remains to show that irreducible elements are also prime. Accordingly, assume that p is irreducible, and suppose that $p \mid ab$ for some $a, b \in R$. Then there exists some $c \in R$ such that $pc = ab$, and since R is a UFD, the elements a and b have unique factorizations into irreducible elements: say

$$a = a_1 \cdots a_k \qquad \text{and} \qquad b = b_1 \cdots b_\ell.$$

Then $pc = ab = a_1 \cdots a_k b_1 \cdots b_\ell$. Because p is irreducible and the factorization is unique, p (or an associate of p) must appear among the irreducibles $a_1, \ldots, a_k, b_1, \ldots, b_\ell$. That is, $\epsilon p = a_j$ or $\epsilon p = b_j$ for some unit ϵ and some j. Therefore, either $p \mid a$ or $p \mid b$, so p is prime. □

Other consequences of unique factorization abound, as is illustrated by the number of times we appeal to the existence of the prime factorization of an element of $\mathbb{Z}$ in order to solve a problem. For example, we can find the gcd of two elements in a UFD by the analog of Lemma 3.4.4, simply comparing powers appearing in the prime factorizations of those elements. Most significantly for elliptic curve problems, like those at the start of this chapter, is the analog of the Power Lemma. In either case, it behooves us to streamline the process of demonstrating a ring is a UFD, and so our principal goal is the following.

> **Theorem 6.5.3 (The Fundamental Meta-Theorem of Arithmetic)**
> Every Euclidean Domain is a Unique Factorization Domain.

We establish two lemmas first, based on the intuitive understanding gained in Exploration L. For both lemmas, suppose that R is a Euclidean domain with Euclidean norm $\mathcal{N}$.

> **Lemma 6.5.4**
>
> If u is a unit of R, then $\mathcal{N}(u) = \mathcal{N}(1)$. If d is a non-zero non-unit of R, then $\mathcal{N}(d) > \mathcal{N}(1)$. ◄

Proof Let u be a unit of R, so $uv = 1$ for some $v \in R$. Then $u \mid 1$, so $\mathcal{N}(u) \leq \mathcal{N}(1)$. Of course, $1 \mid u$ as well, so $\mathcal{N}(1) \leq \mathcal{N}(u)$, and thus $\mathcal{N}(u) = \mathcal{N}(1)$. Now let d be a non-zero non-unit of R, and divide 1 by it: there exist $q, r \in R$ such that $1 = dq + r$, where $r = 0$ or $\mathcal{N}(r) < \mathcal{N}(d)$. Since d is not a unit, $r \neq 0$, so we must have $\mathcal{N}(r) < \mathcal{N}(d)$. But as $1 \mid r$, $\mathcal{N}(1) \leq \mathcal{N}(r) < \mathcal{N}(d)$, as desired. □

> **Lemma 6.5.5**
>
> If $d = ab$ is non-zero and a is not a unit, then $\mathcal{N}(b) < \mathcal{N}(d)$. ◄

Proof Dividing b by d, there exist $q, r \in R$ such that $b = dq + r$ and $r = 0$ or $\mathcal{N}(r) < \mathcal{N}(d)$. If $r = 0$, we find $b = dq = (ab)q \Rightarrow aq = 1$, contradicting the assumption that a is not a unit. Thus $r \neq 0$, and $\mathcal{N}(r) < \mathcal{N}(d)$. Now $b = dq + r$, so $b = abq + r$. Thus $b \mid r$ (by the Linear Combination Lemma), so $\mathcal{N}(b) \leq \mathcal{N}(r) < \mathcal{N}(d)$, again as desired. □

Proof (of Theorem 6.5.3) We prove existence and uniqueness of factorizations into irreducibles. For existence, note that any unit can be written in the desired form (using that unit and an empty product of irreducibles), and every unit has the same norm as 1. Now we proceed as in $\mathbb{Z}$ or $\mathbb{Z}[i]$, by strong induction on the Euclidean norm: Fix $n > \mathcal{N}(1)$ and suppose each element of R with Euclidean norm less than n is expressible as a product of irreducible elements. Now suppose $d \in R$ has Euclidean norm $\mathcal{N}(d) = n$. We need to show that d can be written as a product of irreducible elements. If d is irreducible, then we are done. Else, $d = ab$ with neither a nor b a unit. By Lemma 6.5.5, we have $\mathcal{N}(a) < n$ and $\mathcal{N}(b) < n$, so by the induction hypothesis, both a and b can be written as products of irreducibles. Multiplying these products together shows that d can be so written as well.

For uniqueness, we proceed identically to Theorem 5.5.9. Suppose

$$a_1 \cdots a_k = d = b_1 \cdots b_\ell$$

are two factorizations of an element of d into irreducibles $a_1, \ldots a_k$ and $b_1, \ldots, b_\ell$. Since R is a Euclidean Domain, the Prime Divisor Property (Theorem 6.4.12) tells us that each of the a_i and b_j is also a prime element. Since each a_i divides the product $b_1 \cdots b_\ell$, we must have $a_i \mid b_j$ for some j, and since b_j is irreducible, it must be that a_i and b_j are associates. This shows that, up to associates, the irreducibles in each of the two factorizations are the same.   □

▶ **Remark 6.5.6** Since the notions of prime and irreducible are synonymous in UFDs, we can interchangeably speak either of unique prime factorizations or of unique factorizations into irreducibles.

And there we have it, our long-awaited, hard-won, and much-deserved prize: the Fundamental Meta-Theorem of Arithmetic. It is worth pausing for a moment to mention the principles of mathematical practice that historically served to develop the theory we've just pursued. While no doubt the actual content of the Fundamental Theorem of Arithmetic for ℤ was known well before even Euclid, the first proofs of this statement that stand up to modern scrutiny only came thousands of years later. But if we knew the theorem for thousands of years, what can really be said about the significance of having a formal proof? Beyond the obvious (we are sometimes mistaken about things "everyone knows"), we have just uncovered a second tremendous benefit. The analysis of any proof reveals how and when each hypothesis of the theorem was used. In the case of the Theorem "If $R = \mathbb{Z}$, then R is a UFD," the hypothesis that $R = \mathbb{Z}$—that the elements of R are literally the standard integers—is never used! Rather it is the existence of a Division Algorithm that does all the heavy lifting. Once this has been identified, we realize that a vast generalization is at hand, and that *any* ring with a Division Algorithm is destined to enjoy unique factorization as well.

▶ **Example 6.5.7** By Exploration H.2, the rings $\mathbb{R}[x]$ and $\mathbb{C}[x]$ are Euclidean domains and hence, by Theorem 6.5.3, unique factorization domains. This, along with a description of the primes in these rings—linear polynomials or quadratic polynomials in $\mathbb{R}[x]$ with no real roots—is precisely the Fundamental Theorem of Algebra, providing an interesting connection between these two "Fundamental" Theorems. Note that an analogous description classifying all the primes in ℤ is completely missing!

As a consequence of our hard-earned victory, while we don't pause to call each of them out explicitly, we can now make use of any corollary of unique factorization (e.g., the existence of well-defined valuations as in Definitions 5.6.9 and 3.4.1, the analogous Power Lemma, etc.) when working in any Euclidean Domain. As our victory lap in celebration of this result, let's return to Euclidean Rings to see what we've won.

6.6 Euclidean Rings of Integers

We now have an idea of how to approach any particular Diophantine equation whose solution necessitates the existence of a number $\sqrt{d}$. We move to the quadratic field $\mathbb{Q}[\sqrt{d}]$, find its ring of integers, and attempt to prove that this ring is a UFD. While this won't typically be the case, in this section, we tackle in turn three explicit case studies where we can make progress, namely those of $d = -2$, $d = 2$, and $d = -3$.

Case Study 1: $d = -2$

We began the chapter (Example 6.1.1) by finding the integral points on the elliptic curve defined by

$$y^2 = x^3 - 2,$$

assuming some calculations that relied on the ring $\mathbb{Z}[\sqrt{-2}]$ being a Unique Factorization Domain (though we did not use that language at the time). Our milestone result, Theorem 6.5.3, shows that to demonstrate $\mathbb{Z}[\sqrt{-2}]$ is a UFD, we need only demonstrate the existence of a Euclidean norm on $\mathbb{Z}[\sqrt{-2}]$. To do so, we recall what worked so well for $\mathbb{Z}[i]$. Here we used the geometry of the lattice $\mathbb{Z}[i]$ in the complex plane to reinterpret the division algorithm as a statement about diagonals of squares. We approach $\mathbb{Z}[\sqrt{-2}]$ in a similar fashion, with the adjustment that this lattice in the complex plane is comprised not of squares, but rather of 1-by-$\sqrt{2}$ rectangles. For example, the analog of Figure 5.4 in $\mathbb{Z}[\sqrt{-2}]$ is Figure 6.2, consisting of the (arbitrarily chosen) number $\beta = 2 + \sqrt{-2}$ and all of its $\mathbb{Z}[\sqrt{-2}]$-multiples.

   Again, the norm on $\mathbb{Z}[\sqrt{-2}]$ (which, again, is just the complex norm) plays an important geometric role in this diagram, as for $\beta \in \mathbb{Z}[\sqrt{-2}]$, the integer $N(\beta)$ reflects the square of the distance from 0 to β in the complex plane. We get a rectangular lattice since β and $\sqrt{-2}\beta$ are perpendicular: if $\beta = a + b\sqrt{-2} = (a, b\sqrt{2})$, then $\sqrt{-2}\beta = -2b + a\sqrt{-2} = (-2b, a\sqrt{2})$, and the dot product of the vector representations is 0. Reusing the concept from Section 5.3 of the *fundamental domain* for the shaded rectangle with vertices at 0, β, $\sqrt{-2}\beta$, and $(1 + \sqrt{-2})\beta$, we see that for any given $\beta \in \mathbb{Z}[\sqrt{-2}]$, the complex plane can be partitioned into translates of that fundamental domain. That is, every element of $\mathbb{Z}[\sqrt{-2}]$ can be represented as the sum of a multiple of any given β plus a remainder term. The condition for this norm to be Euclidean is precisely the claim that this remainder can always be chosen to have norm smaller than $N(\beta)$.

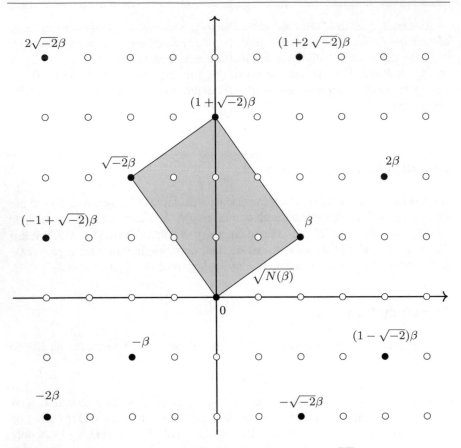

Fig. 6.2 The rectangular lattice and fundamental domain for $\beta = 2 + \sqrt{-2}$

Theorem 6.6.1
The complex norm $N(a + b\sqrt{-2}) = a^2 + 2b^2$ on $\mathbb{Z}[\sqrt{-2}]$ is a Euclidean norm, so $\mathbb{Z}[\sqrt{-2}]$ is a Euclidean domain, and hence a UFD.

Proof From Definition 6.4.1, we have two things to check. First is that whenever $\alpha \mid \beta$, we must have $N(\alpha) \le N(\beta)$, but this is an immediate consequence of Lemma 6.3.6, as $N(\alpha)$ and $N(\beta)$ are positive integers with $N(\alpha) \mid N(\beta)$.

The second thing to check is the Division Algorithm: for all $\alpha, \beta \in \mathbb{Z}[\sqrt{-2}]$, there exist $\chi, \rho \in \mathbb{Z}[\sqrt{-2}]$, with $0 \le N(\rho) < N(\beta)$, such that $\alpha = \chi\beta + \rho$. The lattice reduces this to a geometric argument: given a rectangle of side lengths $\sqrt{N}$ and $\sqrt{2N}$ units, no point in the interior is farther than $\frac{\sqrt{3N}}{2}$ (the length of half a diagonal) away from one of the vertices. Applying this to the fundamental domain generated by β, this says that every number $\alpha \in \mathbb{Z}[\sqrt{-2}]$ is within $\frac{\sqrt{3N(\beta)}}{2}$ of some multiple $\chi\beta$ of β. Thus, if we let $\rho = \alpha - \chi\beta$, then since the norm is the square of the distance, $N(\rho) \leq \frac{3}{4}N(\beta) < N(\beta)$, showing that our norm N satisfies both conditions required to be a Euclidean norm.

To clean up shop a little, let's complete our solution to $x^3 = y^2 + 2$ by proving that $(y + \sqrt{-2})$ and $(y - \sqrt{-2})$ are relatively prime. The rest of our argument in Section 5.7 then holds.

▶ **Example 6.6.2** As elements of the ring $\mathbb{Z}[\sqrt{-2}]$, the numbers $(y + \sqrt{-2})$ and $(y - \sqrt{-2})$ are relatively prime.

Solution We begin by noting that if y is even then $x^3 = y^2 + 2 \equiv 2 \pmod{4}$. However 2 is not a cube in $\mathbb{Z}/(4)$. Thus, y must be odd and so must x. Suppose δ is a gcd of $(y + \sqrt{-2})$ and $(y - \sqrt{-2})$. By the Linear Combination Lemma, δ divides their difference $2\sqrt{-2}$, and so $N(\delta) \mid N(2\sqrt{-2}) = 8$. If $N(\delta) \neq 1$ then it must be even. However, since $\delta \mid (y + \sqrt{-2})$ it also divides x^3 and thus $N(\delta) \mid N(x^3) = x^6$, which is odd since x is. Thus $N(\delta)$ cannot be even and must be 1. It follows that δ is a unit. ☐

Thereby completing our original argument, we definitively conclude that the *only* integer solutions for (x, y) satisfying $x^3 = y^2 + 2$ are $(3, \pm 5)$. This solution was observed by Diophantus, and Fermat claimed there were no other natural number solutions (though naturally did not write that proof down anywhere). A proof was finally given by Euler a little over 100 years later, though his proof assumed unique factorization without remark.

Case Study 2: $d = 2$

The difference between real quadratic fields and complex quadratic fields is probably nowhere more clear than in the geometric significance of its norm. For $\alpha = a + b\sqrt{2} \in \mathbb{Z}[\sqrt{2}]$, the norm $N(\alpha) = a^2 - 2b^2$ has no obvious interpretation as a distance in the complex plane—in fact, the value of this norm is just as often negative as positive. One profound consequence of this observation occurred in our discussion of units: when $d < 0$, the positivity of the norm forced there to be finitely many units, as there are only finitely many lattice points within one unit from the origin. But in real quadratic fields, the divorce from any geometric significance permitted the existence of infinitely many units, as in Corollary 6.3.9. Nevertheless, we can develop unique factorization in these rings with this norm (or rather, its absolute value):

Theorem 6.6.3
The function $|N|$ defined by

$$|N(a + b\sqrt{2})| = |a^2 - 2b^2|$$

is a Euclidean norm, so $\mathbb{Z}[\sqrt{2}]$ is a Euclidean Domain and hence a UFD.

Proof We proceed as in the other cases. First, note that the norm $|N|$ is multiplicative, so if $\alpha \mid \beta$, then $|N(\alpha)| \mid |N(\beta)|$ and thus $|N(\alpha)| \leq |N(\beta)|$. Now we need to consider division. Let $\alpha = a + b\sqrt{2}$ and $\beta = c + d\sqrt{2}$, so

$$\frac{\alpha}{\beta} = \frac{a + b\sqrt{2}}{c + d\sqrt{2}} \cdot \frac{c - d\sqrt{2}}{c - d\sqrt{2}} = \frac{ac - 2bd}{c^2 - 2d^2} + \frac{bc - ad}{c^2 - 2d^2}\sqrt{2}.$$

Like any real number, both $\dfrac{ac - 2bd}{c^2 - 2d^2}$ and $\dfrac{bc - ad}{c^2 - 2d^2}$ are within $\frac{1}{2}$ of an integer, say r and s, respectively. Let $\gamma = r + s\sqrt{2}$, and let $\rho = \alpha - \beta\gamma$. Then

$$\frac{\rho}{\beta} = \frac{\alpha}{\beta} - \gamma = \left(\frac{ac - 2bd}{c^2 - 2d^2} - r\right) + \left(\frac{bc - ad}{c^2 - 2d^2} - s\right)\sqrt{2}.$$

Therefore, by the triangle inequality,

$$\left|N\left(\frac{\rho}{\beta}\right)\right| = \left|\left(\frac{ac - 2bd}{c^2 - 2d^2} - r\right)^2 - 2\left(\frac{bc - ad}{c^2 - 2d^2} - s\right)^2\right| \leq \frac{1}{4} + 2 \cdot \frac{1}{4} < 1.$$

Thus $|N(\rho)| < |N(\beta)|$, as desired. $\qquad\qquad\square$

The integral points on the elliptic curve $y^2 = x^3 + 2$ can now be tackled via the factorization $x^3 = (y + \sqrt{2})(y - \sqrt{2})$ and using the full range of number theory following from unique factorization. See Exercise 6.2.

Case Study 3: $d = -3$

The case $d = -3$ turns out to be of significant interest in the story we are developing so far, as we have already learned that the ring $\mathbb{Z}[\sqrt{-3}]$ is *not* the full ring of integers of $\mathbb{Q}[\sqrt{-3}]$. Let's see what happens if we try the same thing we did for $d = -2$. We have $N(a + b\sqrt{-3}) = a^2 + 3b^2$, the square of the distance from $a + b\sqrt{-3}$ to the origin in the complex plane. As always, we can consider the rectangular lattice generated by an algebraic integer β and ask whether it is necessarily the case that every element of $\mathbb{Z}[\sqrt{-3}]$ is close enough to a vertex of the fundamental domain to make the division algorithm work. In Figure 6.36.36, we see the analog of Figure 5.6,

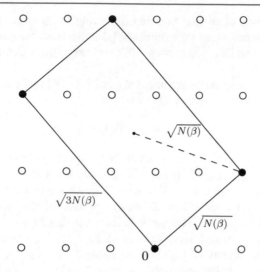

Fig. 6.3 A diagonal too long

used for the Gaussian integers. There we showed that every lattice point was a distance less than $\sqrt{N(\beta)}$ from a vertex of the fundamental domain, since every point in the entire plane is within $\sqrt{N(\beta)}$ from one of these vertices.

This argument fails for $\mathbb{Z}[\sqrt{-3}]$ since the half-diagonal (the distance from β to the center of the fundamental domain) has length

$$\sqrt{\left(\frac{\sqrt{N(\beta)}}{2}\right)^2 + \left(\frac{\sqrt{3N(\beta)}}{2}\right)^2} = \sqrt{N(\beta)}.$$

In terms of the Division Algorithm, this tells us that we cannot guarantee ourselves that the remainder when a number $\alpha \in \mathbb{Z}[\sqrt{-3}]$ is divided by β will have a *strictly* smaller norm than β has, as they may be equal. This prevents the Euclidean Algorithm from functioning as we cannot guarantee that the remainders of the successive divisions will have norms tending toward zero. It is tempting to declare in light of this calculation that $\mathbb{Z}[\sqrt{-3}]$ is not a Euclidean domain, but two impediments stand in our way:

- You may object that the center of the rectangle in Figure 6.3 is not actually a lattice point, and in fact every *lattice point* in that Figure is indeed with $\sqrt{N(\beta)}$ of one of the four vertices.
- You may also object that while *this* norm is not a Euclidean norm, that does not prove that there is *no* norm on $\mathbb{Z}[\sqrt{-3}]$ that shows it to be a Euclidean domain.

The first objection is dealt with in Exercise 6.8, where you will argue that using $\beta = 2 + 2\sqrt{-3}$ provides us a suitable counter-example (i.e., the center of a fundamental domain *is* a lattice point). The second objection is excellent, and you should be proud of thinking of it! It *is* possible for a ring of integers to be a Euclidean Domain via a norm other than the natural one! But this is not the case here, as we can prove by showing that $\mathbb{Z}[\sqrt{-3}]$ is not a UFD, and hence not a Euclidean Domain by Theorem 6.5.3.

We proceed similarly to our argument about $\mathbb{Z}[\sqrt{-26}]$ at the end of Section 6.4. Consider the following identity in $\mathbb{Z}[\sqrt{-3}]$:

$$2 \cdot 2 = (1 + \sqrt{-3})(1 - \sqrt{-3}).$$

We will show that the number 2 is irreducible but not prime in this ring, which implies by Theorem 6.5.2 that $\mathbb{Z}[\sqrt{-3}]$ is not a UFD. First we check that 2 is irreducible: Indeed, recalling that $N(a + b\sqrt{-3}) = a^2 + 3b^2$, we see that 2 has norm 4, and so any non-trivial factorization of 2 would be as a product of two elements of norm 2. But since $a^2 + 3b^2 = 2$ has no integer solutions, the ring $\mathbb{Z}[\sqrt{-3}]$ has no elements of norm 2, and hence no such factors can exist. Last, it is easy to check that 2 is not prime, i.e., that we can have $2 \mid ab$ without either $2 \mid a$ or $2 \mid b$. The factorization above furnishes the requisite example; we have $2 \mid (1 + \sqrt{-3})(1 - \sqrt{-3})$, but 2 divides neither of the individual factors as we know $\frac{1 \pm \sqrt{-3}}{2} \notin \mathbb{Z}[\sqrt{-3}]$.

This is, intriguingly, the second appearance of the number $\frac{1 + \sqrt{-3}}{2}$ in this chapter. Indeed, if we recall some of the lessons learned earlier in this chapter, then we shouldn't have been investigating unique factorization in the ring $\mathbb{Z}[\sqrt{-3}]$ in the first place—for $d = -3$, the rightful home of number theory is instead the full ring of integers of $\mathbb{Q}[\sqrt{-3}]$. Working here instead, as it turns out, restores unique factorization. We also use this opportunity to give this specific quadratic ring a name.

Theorem 6.6.4
Under the complex norm, the **ring of Eisenstein integers** $\mathbb{Z}\left[\frac{1 + \sqrt{-3}}{2}\right]$ is a Euclidean Domain and hence a UFD.

Before completing this argument, let us recall from Section 6.3 some facts about the Eisenstein integers, abbreviating

$$\gamma = \frac{1 + \sqrt{-3}}{2}.$$

We recall that the complex norm is given by $N(a + b\gamma) = a^2 + ab + b^2$ and that the units of this ring (the elements of norm 1) are the six elements $\{\pm 1, \pm \gamma, \pm \overline{\gamma}\}$.

It is reasonable to be skeptical that moving to the full ring of integers will fix anything. In particular, don't we still have the following two distinct factorizations of 4 into irreducible components?

$$(1 + \sqrt{-3})(1 - \sqrt{-3}) = 2 \cdot 2.$$

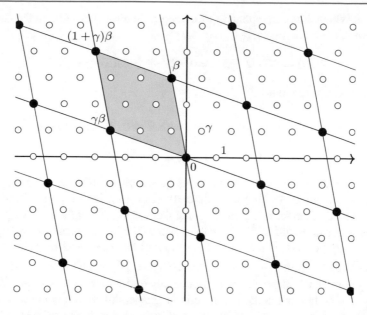

Fig. 6.4 The Eisenstein Lattice and fundamental domain for $\beta = -2 + 3\gamma$

The algebraic identity itself is still valid, of course, but recall that the precise statement of "unique factorization" is careful to point out that factorizations are unique only up to rearrangements and associates. And here is the point: Since $\frac{1+\sqrt{-3}}{2}$ is a *unit* of our ring, the identity

$$1 + \sqrt{-3} = \frac{1 + \sqrt{-3}}{2} \cdot 2$$

shows that the numbers $1 + \sqrt{-3}$ and 2 are in fact *associates* in the Eisenstein integers (even though they are *not* in $\mathbb{Z}[\sqrt{-3}]$!). It is a subtle but important realization— inserting more units into a ring connects more elements as associates. In this case, working in the full ring of integers has clarified the situation—the factorization above is not two distinct factorizations, but rather the same one twice, disguised by units. Of course, we still need to prove that no *other* failures of unique factorization arise in the Eisenstein integers, and we turn to this now. The key change is the geometric shape of the lattice from rectangular to parallelogramular.

The white dots in Figure 6.4 represent the elements $a + b\gamma$ for $a, b \in \mathbb{Z}$. The black dots represent the $\mathbb{Z}[\gamma]$-multiples of $\beta = -2 + 3\gamma$, and the shaded parallelogram is a fundamental domain. We see that every element of $\mathbb{Z}[\gamma]$ is either a multiple of β or is in one of six other possible congruence classes modulo β: Every $\alpha \in \mathbb{Z}[\gamma]$ is ρ "more" than some multiple of β for some

$$\rho \in \{0, \pm 1, \pm 2, \pm \gamma\}.$$

(Exercise: Which dots correspond to which representative?). Similar to the Gaussian integer case, there are precisely

$$N(\beta) = N(-2 + 3\gamma) = (-2)^2 + (-2)(3) + 3^2 = 7$$

congruence classes modulo β.

Proof (of Theorem 6.6.4) Let $\alpha, \beta \in \mathbb{Z}[\gamma]$ with $\beta \neq 0$. We need to show that we can write

$$\alpha = \chi\beta + \rho$$

with $N(\rho) < N(\beta)$. As with our previous geometric proofs, we need only argue that every element $\alpha \in \mathbb{Z}[\gamma]$ is sufficiently close to a lattice point. Subdividing the fundamental domain using the diagonal from β to $\gamma\beta$ results in two equilateral triangles with side length $\sqrt{N(\beta)}$, and any α must be in one of the two triangles.

Elementary geometry shows that the *center* of an equilateral triangle with side length $\sqrt{N(\beta)}$ is within $\frac{\sqrt{3N(\beta)}}{4}$ of a vertex (Figure 6.5), which is less than $\sqrt{N(\beta)}$. Thus *any* point in the plane, and in particular any lattice point α, is within $\sqrt{N(\beta)}$ of a multiple of β. In our standard arithmetic language, this says that the remainder ρ (the difference between α and the nearest multiple $\chi\beta$ of β) has norm $N(\rho) < N(\beta)$.

$\square$

This case of $d = -3$ is significant in one more surprising aspect that will play a major role in the next chapter, obtained by considering the powers of γ. Since the group $\mathbb{Z}[\gamma]^\times$ of units of $\mathbb{Z}[\gamma]$ has finitely many units, and γ is one of them ($N(\gamma) = 1$), some power of γ must equal 1. Indeed, we compute

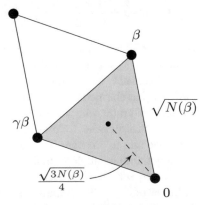

Fig. 6.5 Shaded: a fundamental half-domain in $\mathbb{Z}[\frac{1+\sqrt{-3}}{2}]$

$$\gamma = \frac{1 + \sqrt{-3}}{2} \qquad \gamma^2 = \frac{-1 + \sqrt{-3}}{2} \qquad \gamma^3 = -1$$

$$\gamma^4 = -\gamma \qquad\qquad \gamma^5 = -\gamma^2 \qquad\qquad \gamma^6 = 1$$

In the words of the next chapter, γ is a *6-th root of unity*, since its sixth power is 1. We will eventually label this number ζ_6 for this reason, and likewise label its square

$$\zeta_3 = \gamma^2 = \frac{-1 + \sqrt{-3}}{2}$$

as a *third root of unity* ζ_3 as it satisfies $\zeta_3^3 = 1$. (Note we have already discovered a *4-th root of unity* in the number i.) Viewing certain elements as powers of roots of unity helps simplify calculation, e.g., that ζ_3^2 is the conjugate of ζ_3. In fact, since $\mathbb{Z}[\gamma] = \mathbb{Z}[\zeta_3]$ (Exercise 6.23), this point is compelling enough that it is much more common to describe this ring as $\mathbb{Z}[\zeta_3]$ instead of $\mathbb{Z}[\gamma]$.

A word of caution is in order: we have changed from $\mathbb{Z}[\gamma]$ to $\mathbb{Z}[\zeta_3]$. While these are the same ring, we observe that $N(a + b\gamma) = a^2 + ab + b^2$, while $N(a + b\zeta_3) = a^2 - ab + b^2$. (check this!).

▶ **Example 6.6.5**   Use the Division Algorithm to divide $-2 + 2\zeta_3$ by $1 + 3\zeta_3$.

Solution   We can use *divide and round*:

$$\frac{-2 + 2\zeta_3}{1 + 3\zeta_3} = \frac{-2 + 2\zeta_3}{1 + 3\zeta_3} \cdot \frac{1 + 3\zeta_3^2}{1 + 3\zeta_3^2} = \frac{10 + 8\zeta_3}{7} = \frac{10}{7} + \frac{8}{7}\zeta_3.$$

Rounding gives a nearest Eisenstein integer $\chi = 1 + \zeta_3$, and a remainder of

$$\rho = -2 + 2\zeta_3 - (1 + \zeta_3)(1 + 3\zeta_3) = \zeta_3,$$

which is admissible since $N(\zeta_3) = 1$ is less than $N(1 + 3\zeta_3) = 7$.   □

As we saw for $\mathbb{Z}[i]$, since every $\dfrac{\alpha}{\beta} \in \mathbb{Q}[\zeta_3]$ is within 1 unit of some $\chi \in \mathbb{Z}[\zeta_3]$, there will be an appropriate χ and ρ in $\mathbb{Z}[\zeta_3]$ to satisfy the conclusion of the Division Algorithm. In essence, the question of whether the usual complex norm gives a Euclidean norm for our ring of integers becomes the question of whether its lattice is "tightly packed" enough. If all complex numbers are within one unit of a lattice point, the complex norm gives us a Euclidean domain.

Finally, we see that working in the full ring of integers, rectifying unique factorization at the cost of introducing "denominators," allows us to solve Diophantine equations via the traditional path.

▶ **Example 6.6.6**   Find all integer solutions to $x^3 = y^2 + 3$.

Solution Suppose that x and y are integers such that $x^3 = y^2 + 3$. If y is odd, then $x^3 \equiv 4 \pmod{8}$, which has no solutions, so y must be even, which makes x odd. We now factor in $\mathbb{Z}[\zeta_3]$, using that $\sqrt{-3} = 1 + 2\zeta_3$:

$$x^3 = y^2 + 3 = (y + \sqrt{-3})(y - \sqrt{-3}) = (y + (1 + 2\zeta_3))(y - (1 + 2\zeta_3)).$$

We claim these factors are relatively prime: suppose $\delta \mid (y \pm (1 + 2\zeta_3))$ is a common prime divisor (and thus also a divisor of x). Then δ divides their sum, $2y$, and their difference, $2(1 + 2\zeta_3)$. Now $N(1 + 2\zeta_3) = 1^2 - 1 \cdot 2 + 2^2 = 3$, so $1 + 2\zeta_3 = \sqrt{-3}$ is prime in $\mathbb{Z}[\zeta_3]$. By considering norms, we see that 2 is also prime (a nontrivial factor of 2 would be an element of $\mathbb{Z}[\zeta_3]$ of norm 2, none of which exist). Thus $\delta \mid 2$ or $\delta \mid (1 + 2\zeta_3)$. The first case would imply that δ is an associate of 2, which would imply that x is even—a contradiction. Therefore, we must have $\delta \mid (1 + 2\zeta_3)$, so δ is an associate of $\sqrt{-3}$ and has norm 3, and $\delta \mid y$ (since $\delta \mid 2y$ and $\delta \nmid 2$). Taking norms gives $3 \mid N(y) = y^2$ and so $3 \mid y^2 + 3 = x^3$. Recall that x and y are integers, and so since 3 is prime in $\mathbb{Z}$, it follows that 3 divides both x and y. Write $y = 3z$ and $x = 3w$. Then

$$x^3 = y^2 + 3 \implies 27w^3 = 9z^2 + 3 \implies 9w^3 = 3z^2 + 1,$$

but this is impossible since 3 divides the left-hand side but not the right-hand side. Thus $y \pm (1 + 2\zeta_3)$ are relatively prime, and since $\mathbb{Z}[\zeta_3]$ is a UFD, the Power Lemma tells us that $y + (1 + 2\zeta_3)$ must be a cube:

$$\begin{aligned}
(y + 1) + 2\zeta_3 &= (a + b\zeta_3)^3 \\
&= a^3 + 3a^2 b\zeta_3 + 3ab^2\zeta_3^2 + b^3\zeta_3^3 \\
&= (a^3 + b^3 - 3ab^2) + 3ab(a - b)\zeta_3.
\end{aligned}$$

This forces $2 = 3ab(a - b)$, which is impossible for integers a and b. Thus, the equation $x^3 = y^2 + 3$ has no solutions in integers. □

A final application of our newfound ability to work in this ring comes from an unlikely place, the Law of Cosines. Recall that our classification of Pythagorean Triples was redone in the context of the Gaussian integers via the norm function $N(a + bi) = a^2 + b^2$. An Eisensteinian analog occurs if we consider triangles not with a right angle, but an angle of 60 degrees. Since $\cos(60) = \frac{1}{2}$, the Law of Cosines relates the side lengths a, b, c of such a triangle by the formula

$$c^2 = a^2 + b^2 - ab,$$

and the right-hand side of the expression is precisely $N(a + b\zeta_3)$. If we call (a, b, c) an *Eisensteinian triple* whenever it satisfies that identity, then we can find all Eisensteinian triangles by our knowledge of primes and factorization in $\mathbb{Z}[\zeta_3]$. For example, we computed above that $N(1 + 3\zeta_3) = 7$, so $(1 + 3\zeta_3)^2 = -8 - 3\zeta_3$ has norm $7^2 = 49$, i.e.,

$$8^2 - (-3)(-8) + 3^2 = 7^2.$$

There thus exists an Eisensteinian triangle with side lengths $(3, 8, 7)$. See Exercise 6.40 for the rest of this story.

6.7 Exercises

Calculation & Short Answer

Exercise 6.1 Show that $3 \in \mathbb{Z}[\sqrt{10}]$ is irreducible but not prime.

Exercise 6.2 Use that $\mathbb{Z}[\sqrt{2}]$ is a UFD to show that there are no integer points on the elliptic curve $y^2 = x^3 + 2$.

Exercise 6.3 Find a simple formula for $N(a + b\zeta_3)$. Use your formula to show that $5 - 4\zeta_3$ is prime.

Exercise 6.4 Find some rational primes p that remain prime in $\mathbb{Z}[\zeta_3]$ and some rational primes p that factor in $\mathbb{Z}[\zeta_3]$. What happens for $p = 3$?

Exercise 6.5 Use the Eisenstein Lattice to find a quotient q and remainder r for the division of $5 - \zeta_3$ by $3 + 2\zeta_3$.

Exercise 6.6 Show that if $\omega = \frac{-1+\sqrt{-5}}{2}$, the set $\{a + b\omega : a, b \in \mathbb{Z}\}$ isn't even a ring (under the usual operations).

Exercise 6.7 Show that if $\omega = \frac{m+n\sqrt{d}}{3}$, where $m, n \in \mathbb{Z}, n \neq 0$, m and n are not both multiples of 3, and d is square-free, then $\{a + b\omega : a, b \in \mathbb{Z}\}$ is never a ring (under the usual operations).

Exercise 6.8 Consider the ring $R = \mathbb{Z}[\sqrt{-3}]$ with its natural norm function N. Use a visual argument to show that for the element $\beta = 2 + 2\sqrt{-3}$, there exists an $\alpha \in R$ for which the Division Algorithm fails to provide any remainder of norm smaller than $N(\beta)$.

Exercise 6.9 For $d = -3$, we used the lattice $\mathbb{Z}[\frac{1+\sqrt{d}}{2}]$ to show that the Eisenstein integers form a UFD. What is the least $d \equiv 1 \bmod 4$ for which that same distance argument works?

Exercise 6.10 For each of the following numbers α, decide if it is an algebraic number and/or algebraic integer, and if so, find a minimal polynomial for α.

- $\sqrt{13}$
- $\frac{5+7\sqrt{13}}{2}$
- $\frac{5+6\sqrt{13}}{2}$
- $\frac{1}{7}$

- $\sqrt[3]{17}$
- $\frac{1+\sqrt{7}}{2}$
- $\sqrt{5} + \sqrt{7}$
- The largest root of $x^5 - x - 1$

Exercise 6.11 Use factorization in the ring of integers of $\mathbb{Q}[\sqrt{-67}]$ to find a pair of integer points on the elliptic curve $y^2 = x^3 - 67$.

Exercise 6.12 By factoring the left-hand side of $x^2 - y^2 = 1$, show that it has only two integer solutions. Then tackle $x^2 - y^2 = 57$.

For $d > 1$ square-free, *Brahmagupta-Bhaskara's Equation* is the equation $x^2 - dy^2 = 1$, for which we are trying to find integer solutions.

Exercise 6.13 Show that $x^2 - ny^2 = 1$ has only two integer solutions when n is a square positive integer.

Exercise 6.14 Find eight units of $\mathbb{Z}[\sqrt{17}]$: Two are ± 1; find one more by inspection of the norm equation, and then think structurally to find some more.

Exercise 6.15 Find four positive integer solutions to $x^2 - 122y^2 = 1$.

Exercise 6.16 Generalize the previous two exercises. What's special about 17 and 122 that makes this calculation easier?

Exercise 6.17 Find three non-trivial solutions of $x^2 - 7y^2 = 1$ with both x and y positive integers.

Exercise 6.18 Find three non-trivial solutions of $x^2 - 37y^2 = 1$ with both x and y positive integers.

Exercise 6.19 Provide solutions to Brahmagupta-Bhaskara's equation for the following special values of d:

- $d = n^2 + 1$.
- $d = n^2 - 1$.
- $d = n(n + 1)$.

Exercise 6.20 Show that for $d \leq -3$, the element $2 \in \mathbb{Z}[\sqrt{d}]$ is irreducible but not prime, and hence $\mathbb{Z}[\sqrt{-d}]$ is neither a UFD nor a Euclidean Domain.

Exercise 6.21 A common tool for prime factorization in early grades is the use of "factor trees." Explore using factor trees for prime factorization in non-UFDs.

Exercise 6.22 A *divide-and-round* method, introduced in Chapter 5 for $\mathbb{Z}[i]$, would propose to enact the Division Algorithm for $\alpha, \beta \in \mathbb{Z}[\sqrt{-2}]$ by writing $\frac{\alpha}{\beta} \in \mathbb{Q}[\sqrt{-2}]$ in the form $a + b\sqrt{-2}$ and then rounding the rational numbers a and b to the nearest integers to find the quotient. Why does this work? Why doesn't it work for $\mathbb{Z}[\sqrt{-3}]$?

Formal Proofs

Exercise 6.23 With the notation as before Example 6.6.5, verify that $\zeta_3 = \gamma - 1$. Use this to prove that
$$\mathbb{Z}[\gamma] = \mathbb{Z}[\zeta_3].$$

Exercise 6.24 Prove that the function $\mathcal{N}(x) = 1$ is a Euclidean norm on any field.

Exercise 6.25 For d square-free and $R = \mathbb{Z}$ or $\mathbb{Q}$, consider the ring

$$R(\sqrt{d}) = \left\{ \frac{a + b\sqrt{d}}{m + n\sqrt{d}} : a, b, m, n \in R, m + n\sqrt{d} \neq 0 \right\}.$$

Prove that $\mathbb{Q}[\sqrt{d}] = \mathbb{Q}(\sqrt{d})$ but that $\mathbb{Z}[\sqrt{d}] \neq \mathbb{Z}(\sqrt{d})$.

Exercise 6.26 For $d > 0$ an integer, prove that if there are at least three solutions to $x^2 - dy^2 = 1$, then there are infinitely many.

Exercise 6.27 Find all integer points on the elliptic curve

$$y^2 = x^3 + 4$$

with $x, y \in \mathbb{Z}$. Hint: Assume x is odd first.

Exercise 6.28 Prove that if $\alpha \in \mathbb{C}$ is an algebraic number, then so is $\sqrt{\alpha}$, and likewise for being an algebraic integer.

Exercise 6.29 The ring of integers of $\mathbb{Q}[\sqrt{-67}]$ is a UFD. Use this to prove that the two points found in Exercise 6.11 are the only integer points on that elliptic curve.

Exercise 6.30 Prove Theorem 6.4.12, that Euclid's Lemma holds in an arbitrary Euclidean ring.

Exercise 6.31 Let R be an integral domain, and let $a \in R$. Define congruence modulo a by $x \equiv y \pmod{a}$ if and only if $x - y = az$ for some $z \in R$.

1. Prove that congruence modulo a is an equivalence relation on R.
2. Prove that addition and multiplication of equivalence classes (as defined in $\mathbb{Z}/(n)$) is well defined for this equivalence relation on R. Conclude that it is reasonable to talk about the *ring $R/(a)$*.
3. Describe the set of equivalence classes $R/(a)$ for $R = \mathbb{Z}[\zeta_3]$ and $a = \sqrt{-3}$.

Exercise 6.32 Prove that the set of units with norm equal to $+1$ in $\mathbb{Z}[\sqrt{d}]$ for a square-free integer d forms a subgroup of the group of units, but that this is false for units with norm equal to -1.

Computation and Experimentation

Exercise 6.33 Find a minimal polynomial for $\alpha = \sqrt{3} + \sqrt[3]{2}$. To achieve this, for each $0 \leq n \leq 6$, write $(\sqrt{3} + \sqrt[3]{2})^n$ in the form

$$a + b\sqrt{3} + c\sqrt[3]{2} + d\sqrt[3]{4} + e\sqrt{3}\sqrt[3]{2} + f\sqrt{3}\sqrt[3]{4}.$$

Then use SageMath to set up and solve a system of equations in the variables $c_0, \ldots, c_6$ representing the identity

$$c_0 + c_1\alpha + c_2\alpha^2 + \cdots + c_6\alpha^6 = 0.$$

Exercise 6.34 A powerful generalization of the previous exercise is to have Sage-Math do the first step as well. Find out how to ask SageMath to describe more general algebraic numbers and compute their minimal polynomials.

In addition to the Gaussian Integers, SageMath has built-in functions for defining the Eisenstein integers.

Exercise 6.35 Begin with the following code for establishing 7 as an Eisenstein integer and asking for its prime factorization.

```
R=EisensteinIntegers()
z=R(7)
z.factor()
```

Experiment with this construction and make some conjecture about Eisensteinian primality.

General Number Theory Awareness

Exercise 6.36 Unique factorization was not always addressed as conscientiously as we do today. Find some places in history where unjustified assumptions about unique factorization led to some faulty proofs. As is often the case, a good place to start is Fermat's Last Theorem.

Exercise 6.37 The chapter provides some explicit examples of transcendental numbers. Describe our current state of understanding—can we prove that most "obviously transcendental" numbers like π^e are transcendental? For results that aren't known, what are the leading conjectures in this area?

Exercise 6.38 Look up the Gelfond–Schneider theorem on Wikipedia. Use the result to write down a transcendental number that no one has likely ever written down before.

Exercise 6.39 In the rings of integers considered in this chapter, it has always been the function $N(x+y\sqrt{d}) = x^2 - dy^2$ that has served as our candidate for a Euclidean norm. It happens, however, that a ring of integers could fail to be a Euclidean domain using this norm but still be a Euclidean domain via some other norm. Investigate.

Exercise 6.40 Replicate for Eisensteinian triples all the principal results we developed in the previous chapter for Pythagorean triples. To the extent possible, connect to prime factorization in $\mathbb{Z}[\zeta_3]$ and culminate in a complete classification/parameterization of such triples.

Cyclotomic Number Theory: Roots and Reciprocity

...wherein we take minding our *p*'s and *q*'s to its logical limit.

7.1 Introduction

One of the biggest remaining generalizations in moving from the Gaussian integers to arbitrary quadratic fields is the series of results that concluded Chapter 5, and in particular Section 5.6, in which we classified the Gaussian primes and how rational primes p behave in $\mathbb{Z}[i]$. We recall Theorem 5.6.4:

Theorem 7.1.1 (aka Theorem 5.6.4)
The following are equivalent for all rational primes p:

  (i) p factors in $\mathbb{Z}[i]$; i.e., p is not a Gaussian prime.
  (ii) $p = a^2 + b^2$ for (essentially unique) integers a, b.
  (iii) -1 is a square modulo p.
  (iv) $p \equiv 1$ or $2 \pmod 4$.

Candidates for how to generalize *some* of these bullets are relatively easy to come by. To replace the $p = a^2 + b^2$ condition, for example, we recognize the close tie between the primes that factor in $\mathbb{Z}[\sqrt{d}]$ and the primes expressible in the form $p = a^2 - db^2$ through the norm identity

$$p = a^2 - db^2 = (a + b\sqrt{d})(a - b\sqrt{d}) = N(a + b\sqrt{d}).$$

Supplementary Information The online version contains supplementary material available at https://doi.org/10.1007/978-3-030-98931-6_7.

We saw expressions of this type repeatedly in the previous chapter. In particular, if $\mathbb{Z}[\sqrt{d}]$ is a UFD, then the above will represent the unique prime factorization of p in this ring, and the analog of the equivalence (i) $\Longleftrightarrow$ (ii) will carry over. Similarly, if $p = a^2 - db^2$, then reduction mod p gives

$$a^2 - db^2 \equiv 0 \bmod p,$$

and so d is a square mod p (one shows that we can take $b \not\equiv 0 \bmod p$ and then $d \equiv (ab^{-1})^2$). In the context of the Gaussian integers, where $d = -1$, this is precisely condition (iii) that -1 is a square mod p.

This brings us to the last bullet. Whereas bullets (i)–(iii) represent the properties of a prime that are of tremendous interest but are hard to mentally evaluate from first principles, bullet (iv) permits an almost trivial calculation. It is not much of an overstatement to say that the inclusion of bullet (iv) in that list, that such a direct method exists for testing if a prime satisfies any of (i)–(iii), is the true meat of the theorem. In the case of Gaussian integers, we used Lagrange's Lemma and Fermat's Two-Square Lemma to deduce that the condition that -1 is a square mod p is equivalent to $p \equiv 1$ or $2 \bmod 4$. These methods were somewhat *ad hoc*, and so it is this final part of the story that we aim to generalize in this chapter. Can we find something as simple as a congruence condition to generalize bullet (iv) from the list? That is, can we find a list of modular arithmetic conditions on a prime p that will dictate whether or not d is a square mod p? We motivate the ensuing discussion with the following questions about primes p and q:

- For fixed q, for which p does $x^2 \equiv p \bmod q$ have a solution?
- For fixed p, for which q does $x^2 \equiv p \bmod q$ have a solution?

Despite differing in only one single character, there is a world of difference between these two questions. The first of them, when q is fixed, can be answered by brute force: we square all the values mod q and the results are the equivalence classes mod q that have a square root. The second question, on the other hand, is in principle an infinite calculation.

The remarkable punchline of this story is a completely algorithmic approach to figuring out which numbers have square roots modulo which other numbers. The upcoming *Law of Quadratic Reciprocity* tells us that despite all appearances, the two questions above are of exactly the same difficulty—anyone capable of efficiently answering the first question can also efficiently answer the second, and vice versa.

In proving this theorem, we will find cause to depart substantially from the world of *quadratic* fields to peek at algebraic numbers of higher degree. The equation

$$x^n = 1$$

is rather innocuous as far as equations go, but its solutions (in any ring) are of fundamental importance. In particular, its solutions in the complex plane are the *n-th roots of unity*, and they have already manifested themselves several times in the book. When $n = 2$, the solutions are ± 1, numbers of such significance as not to need introduction. The solutions for the cases $n = 3$ and $n = 4$, respectively, give rise to the Eisenstein and Gaussian integers. We can chalk up to Gauss the realization that a more methodical study of these numbers is needed for all n. We will alternate between developing these two seemingly disparate studies—the study of square roots in $\mathbb{Z}/(p)$ and the study of p-th roots of unity in $\mathbb{C}$—before arriving at Gauss's conclusion that the two are inextricably linked.

7.2 Quadratic Residues and Legendre Symbols

Definition 7.2.1

A **quadratic residue modulo** n is an integer $a \in \mathbb{Z}$ (or sometimes $[a] \in \mathbb{Z}/(n)$) for which the equation $x^2 \equiv a \pmod{n}$ has a solution. If there is no solution, we say a is a **quadratic non-residue modulo** n. ◄

The word "residue" here is a near-synonym for "remainder" (an amount left over), and so a quadratic residue mod n is an integer whose remainder mod n (or rather, its congruence class) is a square in $\mathbb{Z}/(n)$. We will continue our use of slightly less formal language, saying that a is (or is not) "a square mod n." For example, we have already seen that the squares mod 4 are 0 and 1, as we can find solutions to $x^2 \equiv 0$ (mod 4) and $x^2 \equiv 1 \pmod{4}$, but not to $x^2 \equiv 2 \pmod{4}$ or $x^2 \equiv 3 \pmod{4}$. There is a straight-forward algorithm for identifying the squares mod n—we simply compute $[x]^2$ for each $[x] \in \mathbb{Z}/(n)$—and looking for patterns in the results while doing so consumed countless hours for generations of pre-Gaussian mathematicians. As primes will be of primary interest in our investigation, let us focus on the case where n is prime.

▶ **Example 7.2.2** List all of the quadratic residues modulo 3, 5, and 7.

Solution We simply square all of the elements in $\mathbb{Z}/(p)$.

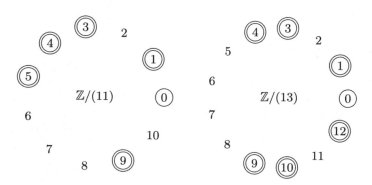

Fig. 7.1 Our hopes and dreams made manifest: the quadratic residues modulo 11 and 13

(a) Modulo 3, we have $0^2 = 0$, $1^2 = 1$, and $2^2 = 4 \equiv 1 \pmod 3$.
   Thus the squares in $\mathbb{Z}/(3)$ are $[0]$ and $[1]$.
(b) The squares modulo 5 are $0^2 = 0$, $1^2 = 1$, $2^2 = 4$, $3^2 = 9 \equiv 4$, and $4^2 = 16 \equiv 1$ $\pmod 5$. Thus in $\mathbb{Z}/(5)$, we have the squares $[0]$, $[1]$, and $[4]$.
(c) Modulo 7, we have $0^2 = 0$, $1^2 = 1$, $2^2 = 4$, $3^2 \equiv 2$, $4^2 \equiv 2$, $5^2 \equiv 4$, and $6^2 \equiv 1$ $\pmod 7$. Thus in $\mathbb{Z}/(7)$, $[0]$, $[1]$, $[2]$, and $[4]$ are the only squares. $\square$

The following notational convention is not only a convenient shorthand for discussion but also sets up the "right" way of thinking about modular square roots.

Definition 7.2.3

For $a \in \mathbb{Z}$ and an odd prime p, we define the **Legendre symbol** $\left(\frac{a}{p}\right)$ by

$$\left(\frac{a}{p}\right) = \begin{cases} 0 & \text{if } p \mid a. \\ +1 & \text{if } a \text{ is a non-zero square mod } p. \\ -1 & \text{if } a \text{ is not a square mod } p. \end{cases}$$

◀

Be sure not to interpret a Legendre symbol as a fraction—the symbol $\left(\frac{\cdot}{\cdot}\right)$ is a function of two variables which spits out ± 1 (or 0) according to whether the top input is a square modulo the bottom input. That said, as a matter of convenience and maybe a pinch of masochism, we will typically make reference to the "numerator" and "denominator" of the Legendre symbol instead of the awkward terms "top input" and "bottom input." It is not obvious yet why we should have to pull out the case $p \mid a$ as a special case in the definition, but for now note that the case that $a \equiv 0$ $\pmod p$ is the case that a has only *one* square root modulo p, whereas there will always be two solutions (if any) when $a \not\equiv 0 \pmod p$ (since p is odd, square roots come in distinct $\pm$ pairs; see Exercise 7.14).

Example 7.2.2(b) can be encoded in this notation as follows:

$$\left(\frac{0}{5}\right) = 0, \quad \left(\frac{1}{5}\right) = \left(\frac{4}{5}\right) = 1, \quad \text{and} \quad \left(\frac{2}{5}\right) = \left(\frac{3}{5}\right) = -1.$$

We promise opportunities to explore patterns in Legendre symbols momentarily, but first an aside: there is probably no more evocative use of the word *ring* for our beloved algebraic structures than the organization of $\mathbb{Z}/(n)$ in a circular diagram. As it turns out, this picture also sheds a little geometric light on our search for patterns among Legendre symbols. We begin our journey with the following pair of diagrams, consisting of the rings $\mathbb{Z}/(11)$ and $\mathbb{Z}/(13)$, arranged ringularly, in which each element has been circled once for each square root it has in that ring (Figure 7.1).

What conjectures do these tantalizing images elicit? What shrouded symmetries seethe secretly sub the seemingly serene surface? All in good time, dear readers, all in good time.

Exploration M

Collecting Quadratic Data ◄

[1] Some preliminary experimentation with Legendre symbols:

M.1 Solve the following two problems:

- Find the first five primes p satisfying $\left(\dfrac{p}{3}\right) = 1$.
- Find the first five primes p satisfying $\left(\dfrac{3}{p}\right) = 1$.

Then discuss the relative difficulty of the two problems.

M.2 Fill in the following tables with the data and look for patterns.

p	3	7	11	13	17
$\left(\dfrac{5}{p}\right)$					
$\left(\dfrac{p}{5}\right)$					

p	3	7	11	17
$\left(\dfrac{13}{p}\right)$				
$\left(\dfrac{p}{13}\right)$				

p	3	7	13	17
$\left(\dfrac{11}{p}\right)$				
$\left(\dfrac{p}{11}\right)$				

M.3 Step back and discern patterns both (a) within and (b) *among* the tables for 5, 13, and 11. Make some predictions for what might emerge in the analogous $\left(\frac{17}{p}\right)$ or $\left(\frac{19}{p}\right)$ tables.

---

[1] See the Python worksheet "Legendre Symbols" for a programming approach.

7.3 Quadratic Residues and Non-Residues Mod p

We will soon see that the heart of the matter lies in focusing on Legendre symbols $\left(\frac{a}{p}\right)$ for p a positive odd prime and a an arbitrary integer. Currently, our approach to evaluating any particular one of these symbols is via exhaustion. For example, we can compute that $\left(\frac{11}{17}\right) = -1$ by squaring each element of $\mathbb{Z}/(17)$ and observing that none of the results are 11. But this algorithm does not scale well to larger denominators and, ultimately, it feels somewhat disappointing to have an algorithm rather than a formula. Fortunately, we can rectify both of these dilemmas at once.

Recall from Section 4.8 that a primitive root modulo p is an element $g \in \mathbb{Z}/(p)^\times$ of order $p - 1$ and that Theorem 4.8.5 guarantees the existence of such an element modulo any prime. A pleasant consequence of the existence of a primitive root is a systematic organization of the elements of $\mathbb{Z}/(p)^\times$ as increasing powers of such a root. For example, 2 is a primitive root mod 13, and the first 12 powers of 2 arrange the elements of $\mathbb{Z}/(13)^\times$ as

$$(2^0, 2^1, 2^2, 2^3, 2^4, 2^5, 2^6, 2^7, 2^8, 2^9, 2^{10}, 2^{11}) \equiv (1, 2, 4, 8, 3, 6, 12, 11, 9, 5, 10, 7) \bmod 13.$$

Choosing a different primitive root will produce a different ordering, and so the ordering itself is not particularly special. Indeed, 6 is another primitive root mod 13, and arranging $\mathbb{Z}/(13)^\times$ as powers of 6 provides the ordering

$$(6^0, 6^1, 6^2, 6^3, 6^4, 6^5, 6^6, 6^7, 6^8, 6^9, 6^{10}, 6^{11}) \equiv (1, 6, 10, 8, 9, 2, 12, 7, 3, 5, 4, 11) \bmod 13.$$

While the specific ordering isn't of particular significance, there is a potent commonality between the two orderings: we can identify the squares in $(\mathbb{Z}/13)^\times$ very quickly from either list! The observation is simply that regardless of their reduction mod 13, the elements $\{2^0, 2^2, 2^4, 2^6, 2^8, 2^{10}\}$ are clearly all squares mod 13 (being the respective squares of 2^0 through 2^5), and likewise for the even powers of 6. In the lists above, this manifests as the observation that the set of values in the even-indexed slots are the same in both lists—$\{1, 4, 3, 12, 9, 10\}$ vs $\{1, 10, 9, 12, 3, 4\}$—and are all squares mod 13. Proving that this is the *complete* set of squares modulo 13 is not difficult.

Lemma 7.3.1

Let g be a primitive root mod p. Then the non-zero quadratic residues mod p are precisely the even powers of g. ◄

Proof We begin by observing that g itself is a non-square modulo p. Indeed, if $g = h^2$, then by Fermat's Little Theorem, $1 \equiv h^{p-1} \equiv g^{(p-1)/2} \bmod p$, contradicting the fact that g has order $p - 1$. Now, every $a \in \mathbb{Z}/(p)^\times$ can be written in the form g^k for some $1 \leq k \leq p-1$. We show that a is a square if and only if k is even. First, if $k = 2\ell$ is even, then $a = g^k = g^{2\ell} = (g^\ell)^2$ is a square. Second, if $k = 2\ell + 1$ is odd and $a = b^2$ were a square, then $b^2 \equiv (g^\ell)^2 \cdot g$, so $g \equiv (bg^{-\ell})^2$ is a square, a contradiction.   □

Corollary 7.3.2

For p odd, exactly half of the non-zero elements of $\mathbb{Z}/(p)$ are squares, and half are non-squares. ◄

The lemma leads to a succinct algebraic encoding of the question of whether a is a square mod p, called *Euler's Criterion*. It is interesting to note that while it makes use of the above lemma, and thus hinges on the *existence* of a primitive root, it is not necessary to actually *find* one in order to use it to compute Legendre symbols.

Theorem 7.3.3 (Euler's Criterion)
For all $a \in \mathbb{Z}$,

$$\left(\frac{a}{p}\right) \equiv a^{(p-1)/2} \quad (\bmod\ p).$$

Proof If $p \mid a$, both sides of the equation are 0. Otherwise, let g be a primitive root mod p. If $\left(\frac{a}{p}\right) = 1$, then by Lemma 7.3.1, $a = g^{2\ell}$ for some ℓ, and then

$$a^{(p-1)/2} \equiv \left(g^{2\ell}\right)^{(p-1)/2} \equiv \left(g^{p-1}\right)^\ell \equiv 1^\ell \equiv 1 \bmod p$$

by Fermat's Little Theorem, so both sides of the claimed equality agree. Similarly, if $\left(\frac{a}{p}\right) = -1$, then $a = g^{2\ell+1}$ for some ℓ, and then

$$a^{(p-1)/2} \equiv \left(g^{p-1}\right)^\ell \cdot g^{(p-1)/2} \equiv 1^\ell \cdot (-1) \bmod p.$$

Here we have used that $h = g^{(p-1)/2}$ is either ± 1 since $h^2 \equiv g^{p-1} \equiv 1$, but h is not 1 since g has order $p - 1$ (and not $(p - 1)/2$).   □

It is hard to overstate how valuable an algebraic resolution to the question of the existence of square roots is. As a first example, we see that the Legendre symbols is multiplicative in its top input.

Corollary 7.3.4

For all integers a and b and any odd prime p, we have

$$\left(\frac{ab}{p}\right) = \left(\frac{a}{p}\right)\left(\frac{b}{p}\right).$$

◄

Proof If p divides either a or b, both sides are zero. Otherwise,

$$\left(\frac{ab}{p}\right) \equiv (ab)^{(p-1)/2} \equiv a^{(p-1)/2}b^{(p-1)/2} \equiv \left(\frac{a}{p}\right)\left(\frac{b}{p}\right) \quad (\mathrm{mod}\ p).$$

Since both sides of this expression are integers equal to ± 1 and p is odd, they must be equal. □

It is remarkable how closely this situation mirrors that of $\mathbb{R}$. In the context of non-zero real numbers, note that the notions of being positive and of being a square are completely synonymous: the non-zero real numbers with square roots are precisely the positive ones and it is thus not unreasonable to say, as in Corollary 7.3.2, that half of real numbers are squares, and half are non-squares. In $\mathbb{R}$, the product of two negative real numbers is a positive real number, i.e., the product of two non-squares is a square, much like Corollary 7.3.4. This provides some validation that the notation $\left(\frac{\cdot}{\cdot}\right)$ is more than just a convenient shorthand, but actually clarifies the structure—being a quadratic residue or non-residue is analogous to being positive or negative in the reals, so associating these characteristics with the values ± 1 is eminently reasonable.

7.4 Application: Counting Points on Curves

We will use the algebraification of quadratic residueness to study algebraic curves, specifically for counting the number of points on such a curve defined over $\mathbb{Z}/(p)$. We conclude the section with one concrete computation which will provide a crucial step in our eventual proof of the main result of the chapter, the Quadratic Reciprocity Law.

We consider an elliptic curve E defined over $\mathbb{Z}/(p)$; that is, an equation of the form $y^2 = f(x) = x^3 + sx + t$ where $s, t \in \mathbb{Z}/(p)$ and $4s^3 - 27t^2 \neq 0$ (Definition 2.3.3). We would like to know how many points are on this curve, i.e., the number $|E(\mathbb{Z}/(p))|$ of pairs $(x, y) \in \mathbb{Z}/(p) \times \mathbb{Z}/(p)$ that satisfy the given equation. Figure 7.2 shows the curve $y^2 = x^3 + x + 1$ (so $s = t = 1$) over each of $\mathbb{Z}/(13)$ and $\mathbb{Z}/(101)$. Note that we need to quite drastically update our expectation as to what a "curve" should look like in this setting!

The graphs reveal several tantalizing patterns. Perhaps most glaring is the apparent horizontal line of symmetry halfway up the y-axis— this is explained by the observation that $(p - y)^2 \equiv y^2 \bmod p$. That is, whenever (x, y) is on the curve, so

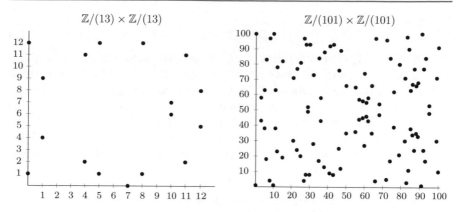

Fig. 7.2 The Elliptic Curve $y^2 = x^3 + x + 1$ over $\mathbb{Z}/(13)$ and over $\mathbb{Z}/(101)$

is $(x, p - y)$. In fact, if we had chosen coset representatives for our y-axis to range from -6 to 6 instead of 0 to 12 in the first graph, we would have seen a horizontal line of symmetry precisely at the x-axis, much like it is for our standard pictures of elliptic curves drawn in $\mathbb{R}^2$ (e.g., Figure 2.4). A second observation pertains more directly to counting points: as with the picture over $\mathbb{R}$, any vertical line $x = a$ for $a \in \mathbb{Z}/(p)$ intersects the curve in 0, 1, or 2 points. How many? It depends on the Legendre symbol $\left(\frac{f(a)}{p}\right)$!

- If $f(a)$ is a non-zero square mod p — say, $f(a) = b^2$ —then both (a, b) and $(a, -b)$ satisfy the equation $y^2 = f(x)$.
- If $f(a)$ is a non-square mod p, then $y^2 = f(a)$ is unsolvable and so there is no point on E with x-coordinate a.
- Finally, each time $f(a)$ is 0, we get the single point $(a, 0)$ on the curve.

   This observation permits a remarkable approach to counting the total number of points on the curve: We simply run from $a = 0$ up to $a = p - 1$ and add up $1 + \left(\frac{f(a)}{p}\right)$, an expression which contributes a summand of 2 if $f(a)$ is a square, 1 if $f(a) = 0$, and 0 if $f(a)$ is a non-square. That is,

$$|E(\mathbb{F}_p)| = \sum_{a=0}^{p-1} 1 + \left(\frac{f(a)}{p}\right) = p + \sum_{a=0}^{p-1} \left(\frac{f(a)}{p}\right).$$

In addition to being a remarkably concise formula, this also permits a relatively easy first estimate for its value. Recall that by Corollary 7.3.2, half of the non-zero elements of $\mathbb{Z}/(p)$ are squares and half are non-squares. It is reasonable to predict that the summands in the right-most expression above are thus $+1$ for about half the possible values of a and -1 for the other about-half of the possible values of a (there are at most 3 values of a for which $f(a) = 0$, so let's neglect them for estimation purposes). Adding roughly the same number of $+1$'s and -1's in a sum is like not having them at all, and so

$$|E(\mathbb{F}_p)| \approx p. \tag{$*$}$$

Of course, nothing *forces* the values of $f(x)$ to take on square values precisely half the time. However, while its proof lies beyond the scope of this book, we mention in passing the renowned Hasse-Weil Theorem, a remarkable theorem of arithmetic geometry that says while any given $f(x)$ might skew towards square (or non-square) values more often than a strict 50/50 split, they can't stray *too* far. Specifically, the estimate $(*)$ can be off by at most roughly $2\sqrt{p}$. For example, given the elliptic curve E defined by $y^2 = x^3 + x + 1$ and $p = 15,485,863$ (the one millionth prime number), the theorem gives that

$$15,477,992 < |E(\mathbb{F}_p)| < 15,493,734.$$

This is a remarkably tight range given the Herculean task[2] of enumerating all such points[3] . We refer the reader to Exercise 7.30 to find a more precise statement of the theorem.

Conics and the Return of Diophantus

While the arguments above to count points on elliptic curves lead to some pretty high-end algebraic geometry, the exact same arguments provide fruitful, concrete results when applied nearly verbatim to conics. Let us consider a curve C of the form

$$y^2 = f(x) = x^2 + sx + t, \qquad s, t \in \mathbb{Z}/(p).$$

Using the same reasoning as for elliptic curves, the number of points on C over $\mathbb{Z}/(p)$ is given by

$$|C(\mathbb{Z}/(p))| = \sum_{a=0}^{p-1} 1 + \left(\frac{f(a)}{p}\right) = p + \sum_{a=0}^{p-1} \left(\frac{f(a)}{p}\right).$$

Now unlike the case of elliptic curves, for these conics, we can obtain an exact count of the number of points using simple factorization.

Theorem 7.4.1
Let p be an odd prime, and $f(x) = x^2 + sx + t$ a quadratic polynomial over $\mathbb{Z}/(p)$ with non-zero discriminant ($s^2 - 4t \neq 0$). Then the conic $y^2 = f(x)$ has precisely $p - 1$ points on it.

---

[2] Counting points on elliptic curves is one of his lesser-known labors.
[3] In case it was going to keep you awake at night, there are exactly 15,486,474 of them.

Proof First, since p is odd, 2 is invertible modulo p, so we can "complete the square" to re-write the conic as

$$y^2 = x^2 + sx + t = \left(x + \frac{s}{2}\right)^2 + u$$

where $u = t - \frac{s^2}{4} = \frac{4t-s^2}{4} \neq 0$ by the discriminant hypothesis. (Note that we are writing $\frac{s}{2}$ as a shorthand for $s \cdot 2^{-1}$). Rearranging gives

$$u = y^2 - \left(x + \frac{s}{2}\right)^2 = \left(y - x - \frac{s}{2}\right)\left(y + x + \frac{s}{2}\right).$$

Thus every point on the conic provides a factorization of u in $\mathbb{Z}/(p)$. Conversely, given a factorization $u = u_1 u_2$, then the system

$$u_1 = y - x - \frac{s}{2}$$

$$u_2 = y + x + \frac{s}{2}$$

provides the following solution to the conic:

$$x = \frac{u_2 - u_1 - s}{2} \qquad y = \frac{u_1 + u_2}{2}.$$

We thus have a bijection between points on the conic and (ordered) factorizations of u into two factors $u = u_1 u_2$ in $\mathbb{Z}/(p)$. This establishes the result, as there are exactly $p - 1$ such factorizations: for any $u_1 \neq 0$ in $\mathbb{Z}/(p)$, there is a unique u_2, namely $u_2 = u u_1^{-1}$, that gives a factorization. $\qquad\square$

It is intriguing to think of this proof as a multiplicative analog of the *Diophantus Chord Method* of Chapter 2. Recall that given a point (x_0, y_0) on a conic over $\mathbb{Q}$, we found all other points by intersecting the conic with the lines $y = m(x - x_0) + y_0$ as m varied over all rational slopes. Similarly, the above proof has us enumerate the factorizations $u = u_1 \cdot u_2$ of a given non-zero $u \in \mathbb{Z}/(p)$. To do so, we begin with the initial solution $(u_1, u_2) = (u, 1)$ and form every other factorization as $(u \cdot m, 1 \cdot m^{-1})$ as m varies over all non-zero elements of $\mathbb{Z}/(p)$.

In fact, even the regular version of the chord method (that is, not a multiplicative analog) can be made to work for conics over $\mathbb{Z}/(p)$, a fact made all the more surprising since the geometric nature of that proof seems to crumble under the discreteness of modular geometry (e.g., Figure 7.2). A counting argument (Exercise 7.17) shows that every conic over $\mathbb{Z}/(p)$ has at least one point and that, roughly speaking, one can account for every other point on the conic as above, by intersecting it with the appropriate line of slope m. Pursuing this approach carefully, however, would take us far afield into the world of projective geometry and modular algebraic geometry (Pickle: what, for example, do "tangent lines" look like mod p?), and so we will leave this as a tantalizing avenue for exploration, except to note that Exercise 7.18 presents a case where everything works rather nicely.

Corollary 7.4.2

For f as in Theorem 7.4.1, we have

$$\sum_{a=0}^{p-1} \left(\frac{f(a)}{p} \right) = -1.$$

◀

Proof Let C be the curve $y^2 = f(x)$. Then by the theorem,

$$p - 1 = |C(\mathbb{F}_p)| = \sum_{a=0}^{p-1} \left(1 + \left(\frac{f(a)}{p} \right) \right) = p + \sum_{a=0}^{p-1} \left(\frac{f(a)}{p} \right),$$

giving the result. □

7.5 The Quadratic Reciprocity Law: Statement and Use

The key problem remaining is the efficient computation of Legendre symbols. While we have several feasible algorithms (brute-force squaring, organizing $\mathbb{Z}/(p)$ as powers of a primitive root, or computation using Euler's Criterion), these all pale in efficiency when compared with the one example we've developed a complete answer for. When $a = -1$, Theorem 7.1.1 shows that computing $\left(\frac{a}{p} \right)$ is as simple as reducing p mod 4. Of course, this also follows by Euler's Criterion, as the expression $(-1)^{(p-1)/2}$ is very easily evaluated. This special case will continue to be a special case for some time, and we view it as a supplement to the upcoming main law:

Theorem 7.5.1 (Quadratic Reciprocity, Supplemental Law #1)

For all odd primes p, we have

$$\left(\frac{-1}{p} \right) = (-1)^{(p-1)/2} = \begin{cases} 1 & \text{if } p \equiv 1 \pmod 4 \\ -1 & \text{if } p \equiv 3 \pmod 4 \end{cases}$$

The goal is now to complete the process that this preliminary result begins, systematically evaluating the symbols $\left(\frac{a}{p} \right)$. The proofs of these results will require developing a reasonable amount of new machinery, so the current section serves as an appetizer, establishing the significance of said results and developing fluency in their use. We begin with the other special case, determining when $a = 2$ is a square mod p:

Theorem 7.5.2 (Quadratic Reciprocity, Supplemental Law #2)

For all odd primes p, we have

$$\left(\frac{2}{p}\right) = (-1)^{(p^2-1)/8} = \begin{cases} 1 & \text{if } p \equiv \pm 1 \pmod 8 \\ -1 & \text{if } p \equiv \pm 3 \pmod 8 \end{cases}$$

The proof of the first equality in the theorem statement is the bulk of the claim and will be addressed in Section 7.7 in tandem with the proof of the Quadratic Reciprocity Law. The second equality is simply casework: any odd prime p is congruent to one of 1, 3, 5, or 7 mod 8, and each one of these four values squares to be congruent to 1 mod 8, so $\frac{p^2-1}{8}$ is always an integer. We verify by hand that $(-1)^{(p^2-1)/8}$ evaluates to $+1$ when $p \equiv \pm 1$ mod 8 and -1 when $p \equiv \pm 3$ mod 8.

Now, moving on to arbitrary numerators, the multiplicativity of the Legendre symbol (Corollary 7.3.4) allows us to reduce the calculation of any Legendre symbol to one in which all of the numerators are prime (or -1). For example, if we wanted to decide if -90 were a square modulo a prime $q > 5$, the factorization $-90 = -2 \cdot 3^2 \cdot 5$ implies that

$$\left(\frac{-90}{q}\right) = \left(\frac{-1}{q}\right)\left(\frac{2}{q}\right)\left(\frac{3}{q}\right)^2\left(\frac{5}{q}\right).$$

Of these four factors, the first two are handled by the two Supplemental Laws above, and the third is trivially $+1$ (as regardless of the value of $\left(\frac{3}{q}\right)$, its square is $+1$). This leaves only the problem of evaluating the Legendre symbol $\left(\frac{5}{q}\right)$, and more generally the Legendre symbols $\left(\frac{p}{q}\right)$ for two odd primes p and q. The *Quadratic Reciprocity Law* provides a remarkable mechanism for evaluating these symbols efficiently.

Theorem 7.5.3 (Quadratic Reciprocity Law)
For odd primes p and q,

$$\left(\frac{p}{q}\right) = \begin{cases} +\left(\frac{q}{p}\right) & \text{if either } p \text{ or } q \equiv 1 \pmod 4 \\ -\left(\frac{q}{p}\right) & \text{if both } p \text{ and } q \equiv 3 \pmod 4. \end{cases}$$

There are several equivalent reformulations of this result that one might find in the literature, e.g., that

$$\left(\frac{p}{q}\right)\left(\frac{q}{p}\right) = (-1)^{\frac{p-1}{2}\frac{q-1}{2}}.$$

We refer the reader to the exercises for this and several other equivalencies.

The surprising consequence of the Quadratic Reciprocity Law is that the question of whether p is a square mod q appears to be inexorably tied to that of whether q is a square mod p. Why should those two questions have anything to do with one another? Surprisingly enough, we learn from the theorem that if at least one of p or q is congruent to 1 modulo 4, then not only are the two answers to these questions *related*, they're the *same*! And if both are 3 mod 4, the answers are opposite. The patterns in Exploration M provide further corroboration on the relationship articulated by the theorem, and we take some time now to develop the mechanics of applying the theorem to calculate Legendre symbols. The calculations of $\left(\frac{a}{p}\right)$ below all proceed by way of well-timed application of some collection of the following operations:

(1) Applying the Quadratic Reciprocity Law.
(2) Applying one of the Supplemental Laws in case $a = -1$ or $a = 2$.
(3) Using the multiplicativity of the Legendre symbol.
(4) Removing square factors of a (since $\left(\frac{c^2}{p}\right) = \left(\frac{c}{p}\right)^2 = 1$ whenever $p \nmid c$).
(5) Modular reduction: By definition, we have $\left(\frac{a}{p}\right) = \left(\frac{a \bmod p}{p}\right)$.

▶ **Example 7.5.4** Is 41 a square mod 103? Yes, since

$$\left(\frac{41}{103}\right) = \left(\frac{103}{41}\right) = \left(\frac{21}{41}\right) = \left(\frac{3}{41}\right)\left(\frac{7}{41}\right) = \left(\frac{41}{3}\right)\left(\frac{41}{7}\right) = \left(\frac{-1}{3}\right)\left(\frac{-1}{7}\right) = (-1)\cdot(-1) = 1.$$

▶ **Example 7.5.5** Is 79 a square mod 101? Yes, since

$$\left(\frac{79}{101}\right) = \left(\frac{101}{79}\right) = \left(\frac{22}{79}\right) = \left(\frac{2}{79}\right)\left(\frac{11}{79}\right) = 1\cdot\left(\frac{11}{79}\right) = -\left(\frac{79}{11}\right) = -\left(\frac{2}{11}\right) = -(-1) = 1.$$

Of course, both of the above examples could be in principal done by brute force, squaring all the elements of $\mathbb{Z}/(103)$ and $\mathbb{Z}/(101)$ and seeing if we ever got out 41 or 79, respectively. A more substantial application of quadratic reciprocity allows us to fix the numerator and vary the denominator of the Legendre symbol, in which case a brute-force search is no longer possible.

▶ **Example 7.5.6** For which odd primes p is 5 a square mod p? By Quadratic Reciprocity, $\left(\frac{5}{p}\right) = \left(\frac{p}{5}\right)$, so for odd primes p,

$$\left(\frac{5}{p}\right) = 1 \iff \left(\frac{p}{5}\right) = 1 \iff p \equiv 1 \text{ or } 4 \bmod 5.$$

▶ **Example 7.5.7**   For which odd primes p is 7 a square mod p? This problem differs from the previous example in that the relationship between $\left(\frac{7}{p}\right)$ and $\left(\frac{p}{7}\right)$ depends on the value of p mod 4 so we need to break into these two cases. First, if $p \equiv 1$ (mod 4), then $\left(\frac{7}{p}\right) = \left(\frac{p}{7}\right)$, which is 1 if and only if $p \equiv 1, 2,$ or 4 (mod 7). And if $p \equiv 3$ (mod 4), then $\left(\frac{7}{p}\right) = -\left(\frac{p}{7}\right)$, which is 1 if and only if $p \equiv 3, 5,$ or 6 (mod 7). So, in summary, for odd primes p we have

$$\left(\frac{7}{p}\right) = 1 \quad \Longleftrightarrow \quad \begin{array}{c} (p \equiv 1 \bmod 4) \text{ and } (p \equiv 1, 2, 4 \bmod 7) \\ \text{or} \\ (p \equiv 3 \bmod 4) \text{ and } (p \equiv 3, 5, 6 \bmod 7) \end{array}.$$

In principal, we can stop our answer here, though it's often preferable to report an answer as a single modular congruence rather than a system of congruences with different moduli. Since 4 and 7 are relatively prime, an application of Sunzi's Theorem (Theorem 4.7.5) converts each possible pair into a single condition mod 28. We conclude that for odd primes p,

$$\left(\frac{7}{p}\right) = 1 \quad \Longleftrightarrow \quad p \equiv \pm 1, \pm 3, \text{ or } \pm 9 \quad (\bmod \ 28).$$

As a culminating calculation of this type, we leave it as an exercise (Exercise 7.4) to finish the problem started in the section, finding all primes q for which -90 is a square mod q.

7.6   Some Unexpected Helpers: Roots of Unity

Much of our focus in this book has dealt with "quadratic phenomena," e.g., quadratic Diophantine equations and quadratic fields, and in this chapter an obsession with *square* roots mod p. It may come as a surprise (or perhaps a welcome respite) that we now drastically shift gears—from square roots to n-th roots, and from the modular world to the complex one. Explicitly, despite seeming to have little connection to our Legendre symbols, we will move toward considering n-th roots of the number 1 in the complex plane, which are solutions to the equation

$$z^n = 1, \qquad z \in \mathbb{C}.$$

Fortunately, we have a lot of experience with such numbers, at least for small n: the cases $n = 1$ and $n = 2$ have solution sets of $\{1\}$ and $\{\pm 1\}$, respectively. When $n = 3$, the solution set $\{1, -\frac{1}{2} \pm \frac{\sqrt{-3}}{2}\}$ led us to the Eisenstein integers, and when $n = 4$ our solutions $\{\pm 1, \pm i\}$ led to the Gaussian integers.

Definition 7.6.1

Given a natural number n, an **n-th root of unity** is a complex number z such that $z^n = 1$. An n-th root of unity z is **primitive** if $z^m \neq 1$ for all natural numbers $m < n$. ◄

That is, a primitive n-th root of unity is an n-th root of unity that isn't an m-th root of unity for any $m < n$. For example, each element of $\{\pm 1, \pm i\}$ is a 4th root of unity, but only $\pm i$ are *primitive* 4th roots of unity since $1^1 = 1$ and $(-1)^2 = 1$. We remark that the tools of Section 1.3 show that at least one primitive n-th root of unity exists for any n: as mentioned in that section, if $z = 1 \angle \theta$, then $z^n = 1 \angle (n\theta)$. Indeed, using complex exponential notation, substituting $z = re^{i\theta}$ into $z^n = 1$ gives

$$r^n e^{in\theta} = 1 = 1 e^{i2\pi},$$

showing that the n-th roots of unity must have $r = 1$ and $n\theta = 2\pi$, and so we obtain n distinct solutions by taking $\theta = 2k\pi/n$ for any $0 \leq k \leq n - 1$. Taking $k = 1$ guarantees that the root is primitive, as its powers run through all of the other solutions before returning to 1. The linear combinations of these powers form another interesting ring.

Definition 7.6.2

For a natural number n, we denote by ζ_n the primitive n-th root of unity $\zeta_n = e^{2\pi i/n}$, or just ζ when the n is fixed or unambiguous. The **n-th ring of cyclotomic integers** is the ring

$$\mathbb{Z}[\zeta_n] = \{a_0 + a_1\zeta_n + a_2\zeta_n^2 + \cdots + a_n\zeta_n^{n-1} : a_i \in \mathbb{Z}, 0 \leq i \leq n - 1\}. \quad ◄$$

The word "integer" in the definition needs some justification.

Theorem 7.6.3

Every element of $\mathbb{Z}[\zeta_n]$ is an algebraic integer.

This follows trivially from the observation (Theorem 6.2.12) that sums and products of algebraic integers are again algebraic integers, but since we did not fully justify that claim back in Chapter 6, let us give a careful proof for this special case.

Proof Abbreviate $\zeta = \zeta_n$ and take some $z \in \mathbb{Z}[\zeta]$. Our task is to find a monic polynomial $f \in \mathbb{Z}[x]$ for which $f(z) = 0$. Observe that for any $0 \leq j \leq n - 1$, the number $z\zeta^j$ is also in $\mathbb{Z}[\zeta]$, and so expressible in the form

$$z\zeta^j = \sum_{k=0}^{n-1} a_{jk}\zeta^k$$

for some integers a_{jk}, providing a system of equations:

$$
\begin{array}{rcl}
a_{00} + a_{01}\zeta + \cdots + a_{0,n-1}\zeta^{n-1} &=& z \\
a_{10} + a_{11}\zeta + \cdots + a_{1,n-1}\zeta^{n-1} &=& z\zeta \\
\vdots \qquad \vdots \qquad\qquad \vdots && \vdots \\
a_{n-1,0} + a_{n-1,1}\zeta + \cdots + a_{n-1,n-1}\zeta^{n-1} &=& z\zeta^{n-1}.
\end{array}
$$

Write this system as the matrix equation

$$
\begin{bmatrix}
a_{00} & a_{01} & \cdots & a_{0,n-1} \\
a_{10} & a_{11} & \cdots & a_{1,n-1} \\
\vdots & \vdots & \ddots & \vdots \\
a_{n-1,0} & a_{n-1,1} & \cdots & a_{n-1,n-1}
\end{bmatrix}
\begin{bmatrix}
1 \\ \zeta \\ \vdots \\ \zeta^{n-1}
\end{bmatrix}
=
\begin{bmatrix}
z \\ z\zeta \\ \vdots \\ z\zeta^{n-1}
\end{bmatrix},
$$

or just $Av = zv$, where A is the coefficient matrix on the left and v is the vector whose entries are the powers of ζ. This is all we need: this says that z is an eigenvalue of A corresponding to the eigenvector ζ; i.e., z is a root of the monic polynomial $f(\lambda) = \det(\lambda I - A)$. □

The definition of ζ_n above as $e^{2\pi i/n}$ is one of the last references to explicit complex exponentials in the book. All that will typically matter is that we set ζ_n to be any one particular primitive n-th root of unity, and not $e^{2\pi i/n}$ specifically[4]. In fact, while for select values of n we can express ζ_n in terms of concise algebraic expressions involving radicals (e.g., $\zeta_3 = \frac{-1+\sqrt{-3}}{2}$), these are the exception rather than the rule. Instead, for most values of n, to do arithmetic with ζ_n we need only invoke its principal property, that it is a root of the polynomial $x^n - 1 = 0$. As an exercise in this philosophy, let us deduce some key arithmetic properties of n-th roots of unity directly from the algebraic definition, foreshadowing our goal of finding relationships between n-th roots of 1 and square roots of n.

Here goes. Let ζ be a primitive n-th root of unity ($n > 1$). Then $\zeta^n = 1$, and by taking conjugates, $\overline{\zeta}^n = 1$ as well. Thus $N(\zeta) = \zeta\overline{\zeta}$ is a positive real number satisfying $(\zeta\overline{\zeta})^n = 1$, and so $\zeta\overline{\zeta} = 1$. We conclude that $\overline{\zeta} = \zeta^{-1}$ and thus that $\overline{\zeta^k} = \zeta^{-k} = \zeta^{n-k}$ for all $k \in \mathbb{Z}$. Since the equation $x^n = 1$ has the n complex solutions ζ^a ($0 \le a \le n - 1$), including the real solution $x = 1$, we have the factorization

$$
\begin{aligned}
x^n - 1 &= (x - 1)(x^{n-1} + x^{n-2} + \cdots + x^2 + x + 1) \\
&= (x - 1)(x - \zeta)(x - \zeta^2) \cdots (x - \zeta^{n-1}).
\end{aligned}
$$

----

[4] This brings up an interesting philosophical point: For $n = 4$, for example, this is the statement that we could have developed all of Gaussian arithmetic using $-i$ instead of $+i$. In fact, how do you know which of the two complex solutions to $z^2 = -1$ is the one that should be called $+i$ and which should be called $-i$? Answer: You don't. You can't!

Since $\mathbb{C}[x]$ is a Unique Factorization Domain (see Example 6.5.7), we can cancel $x - 1$ from both factorizations to arrive at

$$x^{n-1} + x^{n-2} + \cdots + x^2 + x + 1 = (x - \zeta)(x - \zeta^2) \cdots (x - \zeta^{n-1}).$$

Evaluating both sides of this identity at different values of x provides useful numerical relationships: first, evaluating at $x = \zeta$ gives the identity

$$1 + \zeta + \zeta^2 + \cdots + \zeta^{n-1} = 0,$$

and second, evaluating at $x = 1$ gives

$$n = (1 - \zeta)(1 - \zeta^2) \cdots (1 - \zeta^{n-1}),$$

a factorization of the integer n in the ring $\mathbb{Z}[\zeta]$.

Assume for simplicity that n is odd (indeed, we mostly care about the case that n is an odd prime), so that $\frac{n-1}{2}$ is an integer. Then splitting the above factorization at the half-way point allows us to write

$$n = (1 - \zeta)(1 - \zeta^2) \cdots (1 - \zeta^{\frac{n-1}{2}})(1 - \zeta^{\frac{n+1}{2}}) \cdots (1 - \zeta^{n-2})(1 - \zeta^{n-1})$$

$$= (1 - \zeta)(1 - \zeta^2) \cdots (1 - \zeta^{\frac{n-1}{2}})\overline{(1 - \zeta^{\frac{n-1}{2}}) \cdots (1 - \zeta^2)(1 - \zeta)}$$

$$= N((1 - \zeta)(1 - \zeta^2) \cdots (1 - \zeta^{\frac{n-1}{2}})).$$

That is, the element $u = (1 - \zeta)(1 - \zeta^2) \cdots (1 - \zeta^{\frac{n-1}{2}}) \in \mathbb{Z}[\zeta]$ has $N(u) = n$ and $|u| = \sqrt{N(u)} = \sqrt{n}$. This bodes well for our desired connection between $\mathbb{Z}[\zeta_n]$ and square roots of n. Foreshadowing accomplished! As it turns out, u is in general neither $\sqrt{n}$ nor $\sqrt{-n}$ (try it for $n = 3$). And yet, it offers the tantalizing possibility that some *other* arithmetic combination of ζ's might succeed in demonstrating for us that $\sqrt{n}$ or $\sqrt{-n}$ is itself in $\mathbb{Z}[\zeta]$, so that the arithmetic of roots of unity could be brought to bear upon understanding Diophantine equations like $p = x^2 - ny^2$. The following exploration continues this quest, attempting to find hints of a $\sqrt{n}$ appearing in this exotic cyclotomic world. For ease of reference in doing so, we collect below the most important of the identities deduced above.

Theorem 7.6.4

Let $n \in \mathbb{N}$ and let $\zeta \in \mathbb{C}$ be an n-th root of unity. Then $\overline{\zeta} = \zeta^{-1}$ and for any $k \in \mathbb{Z}$, we have $\zeta^k = \zeta^{k \bmod n}$. Further, we have

$$1 + \zeta + \zeta^2 + \cdots + \zeta^{n-1} = 0.$$

Exploration N

Playing with 5th Roots of Unity ◄

Let us explore some basic (or at least, basic-looking) sums in $\mathbb{Z}[\zeta]$, where $\zeta = \zeta_5$ is our chosen primitive 5-th root of unity. The powers of ζ are displayed in the complex plane below.

N.1 Mentally place each of the following quantities in the complex plane.

$$\zeta^6 \qquad \zeta^{1234} \qquad \zeta^{-37} \qquad \zeta + \zeta^4 \qquad \zeta + \zeta^2 + \zeta^3 + \zeta^4$$

Prompted by our quest for $\sqrt{n}$, we turn our attention to sums of the form

$$G_\varepsilon = \varepsilon_1 \zeta^1 + \varepsilon_2 \zeta^2 + \varepsilon_3 \zeta^3 + \varepsilon_4 \zeta^4$$

where each $\varepsilon_i = \pm 1$ and $\varepsilon = (\varepsilon_1, \varepsilon_2, \varepsilon_3, \varepsilon_4)$.

N.2 Use Figure 7.3 to decide if there are choices for ε so that G_ε is:

• real and positive   • real and negative   • imaginary

N.3 Use Theorem 7.6.4 (and that $\varepsilon_i^2 = 1$) to find a simple expression for

$$G_\varepsilon^2 = \left(\varepsilon_1 \zeta^1 + \varepsilon_2 \zeta^2 + \varepsilon_3 \zeta^3 + \varepsilon_4 \zeta^4 \right)^2.$$

N.4 Use your results of Problem N.3 to corroborate and further your findings in Problem N.2. In particular, to find the positive real G_ε, figure out how to choose ε so that your expression for G_ε^2 is real.

N.5 Much of our upcoming work centers around generalizing the previous problem. Based on the ε you found there, make a prediction about how to assign values for $\varepsilon_1, \varepsilon_2, \ldots, \varepsilon_{p-1}$, each equal to ± 1, so that

$$\left(\varepsilon_1 \zeta_p + \varepsilon_2 \zeta_p^2 + \varepsilon_3 \zeta_p^3 + \cdots + \varepsilon_{p-1} \zeta_p^{p-1} \right)^2 \text{ is real.}$$

Fig. 7.3 The Power(s) of $\zeta = \zeta_5$

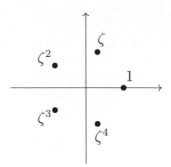

7.7 A Proof of Quadratic Reciprocity

Once per generation a mathematical theorem comes along encoding an observation so breathtakingly beautiful, so eye-openingly insightful, that its proof seems to just write itself. The Law of Quadratic Reciprocity, articulated by Euler and Legendre after lifetimes of pattern-searching not dissimilar to what you yourself have done over the course of this chapter, is not such a result. To be sure, reciprocity is highly non-obvious, and your loving authors, in rare consensus, agree that this is one of the most beautiful results in number theory, leading to its prominence as a hallmark result in almost any number theory book. But the proof of quadratic reciprocity does *not* simply fall out after the observation is made. Indeed, neither Euler nor Legendre managed to put together a complete proof[5], leaving that trophy to be claimed by one nineteen-year-old Carl Friedrich Gauss[6].

Before we move on, while we will not dwell on the history and evolution of the result (though we encourage the readers to peruse some of the historically-oriented research exercises), it is worth pausing for one comment on the interplay of history and notation. The notation we have used throughout the chapter, and indeed the whole book, has been shaped by centuries of a kind of notational natural selection: notation that elucidates key ideas gets picked up and used again by subsequent authors, whereas notation that obscures patterns falls into disuse. Gauss, for example, did not have at his disposal Legendre symbols, a piece of notation we have been positively beaming about for several sections now. Likewise, Legendre did not have Gauss' notation for modular congruence, and through modern eyes it is hard to imagine getting this far in number theory without it. As a consequence, the proof we provide below is significantly less complicated than Gauss' *original* proof (to distinguish it from the several other proofs he devised over his lifetime[7]) and other early proofs. Finally, we note that the result is often cited as one of the "most proved" results of mathematics, as measured by the number of different proofs, so our approach via roots of unity is by no means the only one. In fact, it seems likely that a complete understanding of the result necessitates processing quite a few different proofs, and

---

[5] QED, we are all mere mortals.

[6] Except maybe Gauss.

[7] Sigh.

so we again reference the curious reader to the exercises for more information on alternate proofs.

Moving back into mathematics proper, the previous Exploration motivates further investigation into sums of the form

$$\varepsilon_1 \zeta + \varepsilon_2 \zeta^2 + \cdots + \varepsilon_{p-1} \zeta^{p-1}$$

for $\zeta = \zeta_p$ a primitive p-th root of unity (p odd) and each $\varepsilon = \pm 1$. For example, when $p = 5$, we found that the assignment

$$\varepsilon_1 = \varepsilon_4 = +1 \qquad \varepsilon_2 = \varepsilon_3 = -1$$

gave an interesting value for the sum. While the sample size is still quite small, it is intriguing that the coefficients that gave this value amounted to setting $\varepsilon_a = 1$ if a was a square mod 5 (i.e., when a is 1 or 4) and -1 if not (i.e., when a is 2 or 3), that is, $\varepsilon_a = \left(\frac{a}{p}\right)$. We are led to the following definition.

Definition 7.7.1

Let p be an odd prime and ζ a primitive p-th root of unity. Then the **Gauss sum** for p is the sum

$$G = G_p = \sum_{a=0}^{p-1} \left(\frac{a}{p}\right) \zeta^a.$$

◀

While we will mostly approach this sum algebraically, it is worth keeping in mind that for each p, the Gauss sum G for p is a specific complex number representing an actual point in the complex plane. In fact, the geometry lends itself to some insight.

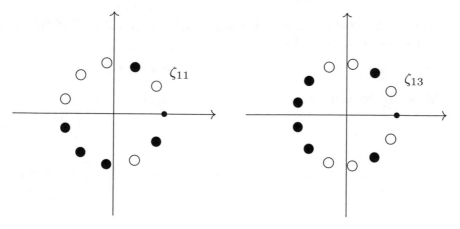

Fig. 7.4 The Non-Residue and Residue Powers of ζ_{11} and ζ_{13} in black and white

Figure 7.4 shows the 11th and 13th roots of unity in the complex plane[8]. A root ζ^a is colored white when $\left(\frac{a}{p}\right) = 1$ and black when $\left(\frac{a}{p}\right) = -1$. One obvious pattern, that the roots appear in conjugate pairs (symmetric over the real axis), is well-known to us, as $\overline{\zeta^a} = \zeta^{-a}$. More tantalizing is the relationship between the colors within a conjugate pair. In the first image, for $p = 11$, a node appears to be colored white if and only if its conjugate is colored black, whereas in the second image, for $p = 13$, conjugate pairs seem to always have the same color. The distinction is that of primes being 1 vs. 3 modulo 4: by the first supplemental law of quadratic reciprocity, if $p \equiv 3 \bmod 4$ then

$$\left(\frac{-a}{p}\right) = \left(\frac{-1}{p}\right)\left(\frac{a}{p}\right) = -\left(\frac{a}{p}\right),$$

explaining the color switch after reflecting. If $p \equiv 1 \bmod 4$, then $\left(\frac{-a}{p}\right) = \left(\frac{a}{p}\right)$ by the same calculation, whence the symmetry.

The Gauss sum G is the complex number formed by adding the white dots and then subtracting the black dots in the figure. We are interested in where the result lands, and we begin this calculation with some familiar friends.

▶ **Example 7.7.2** Let $p = 3$. Since 1 is a square mod 3 and 2 is not, we have

$$G_3 = \zeta_3 - \zeta_3^2 = \zeta_3 - (-1 - \zeta_3) = 2\zeta_3 + 1 = \sqrt{-3}.$$

▶ **Example 7.7.3** For $p = 5$, the residues are 1 and 4 and non-residues are 2 and 3, so

$$\begin{aligned}
G_5 &= \zeta_5 - \zeta_5^2 - \zeta_5^3 + \zeta_5^4 \\
&= (\zeta_5 + \zeta_5^{-1}) - (\zeta_5^2 + \zeta_5^{-2}) \\
&= (\zeta_5 + \zeta_5^{-1}) - (\zeta_5 + \zeta_5^{-1})^2 + 2,
\end{aligned}$$

where we have used in the last step that $(\zeta_5 + \zeta_5^{-1})^2 = \zeta_5^2 + 2 + \zeta_5^{-2}$. Similarly, and relevantly, we have (Theorem 7.6.4)

$$0 = 1 + \zeta_5 + \zeta_5^2 + \zeta_5^3 + \zeta_5^4 = (\zeta_5 + \zeta_5^{-1})^2 + (\zeta_5 + \zeta_5^{-1}) - 1,$$

whence by the quadratic formula (of all things!) we get

$$\zeta_5 + \zeta_5^{-1} = \frac{-1 + \sqrt{5}}{2},$$

where we have chosen the $+\sqrt{5}$ based on the geometry of $\zeta_5 + \zeta_5^{-1}$, which we know to be a positive real number.

---

[8] Exercise to do literally right now: Compare Figure 7.4 to Figure 7.1. Do it. Go!

Substituting this in to continue the above calculation gives

$$G_5 = \left(\frac{-1+\sqrt{5}}{2}\right) - \left(\frac{-1+\sqrt{5}}{2}\right)^2 + 2$$

$$= \left(\frac{-1+\sqrt{5}}{2}\right) - \left(\frac{1-2\sqrt{5}+5}{4}\right) + 2 = \sqrt{5}.$$

This relationship appears to be the much-ballyhooed relationship between p-th roots of 1 and square roots of p. Moving forward, it seems foolhardy to attempt (or even hope for) a generalization wherein we solve algebraically for ζ_p (or even $\zeta_p + \zeta_p^{-1}$) first. Instead, we search more systematically for ways to reduce the algebra. To this end, let us consider the case $p = 7$ and pretend that we are unable to evaluate the Legendre symbols mod 7, so that (writing ζ for ζ_7) we have

$$G_7 = \left(\frac{1}{7}\right)\zeta + \left(\frac{2}{7}\right)\zeta^2 + \left(\frac{3}{7}\right)\zeta^3 + \left(\frac{4}{7}\right)\zeta^4 + \left(\frac{5}{7}\right)\zeta^5 + \left(\frac{6}{7}\right)\zeta^6.$$

Now, squaring hexanomials is not everyone's idea of a good time[9], and so to foreshadow the technique of the proof, let's approach G_7^2 one coefficient at time. Consider the coefficient of ζ^3 in the expansion of G_7^2. Applying the distributive law pairs each $\left(\frac{a}{7}\right)\zeta^a$ from the first sum with a $\left(\frac{b}{7}\right)\zeta^b$ from the second sum, and we get a contribution to the ζ^3 coefficient in the result whenever $a + b \equiv 3 \bmod 7$. That is, the coefficient of ζ^3 in G_7^2 is given by

$$\left(\frac{1}{7}\right)\left(\frac{2}{7}\right) + \left(\frac{2}{7}\right)\left(\frac{1}{7}\right) + \left(\frac{4}{7}\right)\left(\frac{6}{7}\right) + \left(\frac{5}{7}\right)\left(\frac{5}{7}\right) + \left(\frac{6}{7}\right)\left(\frac{4}{7}\right).$$

Finally, note that since we can add on the convenient 0's $\left(\frac{0}{7}\right)\left(\frac{3}{7}\right)$ and $\left(\frac{3}{7}\right)\left(\frac{0}{7}\right)$, this is precisely the sum

$$\sum_{a=0}^{6} \left(\frac{a}{p}\right)\left(\frac{3-a}{p}\right) = \left(\frac{-1}{p}\right)\sum_{a=0}^{6}\left(\frac{a^2-3a}{p}\right).$$

Fortunately, we have seen sums of this form before (do you remember where?), and the following proof thus nicely dispenses of the whole calculation.

---

[9] Can you believe it?

Theorem 7.7.4
Let p be an odd prime. Then

$$G_p^2 = \left(\frac{-1}{p}\right)p.$$

Proof Let $\zeta = \zeta_p$. Write $G = G_p = \sum_{a=1}^{p-1} \left(\frac{a}{p}\right)\zeta^a$, so

$$G^2 = \sum_{a=0}^{p-1}\left(\frac{a}{p}\right)\zeta^a \cdot \sum_{b=0}^{p-1}\left(\frac{b}{p}\right)\zeta^b = \sum_{b=0}^{p-1}\sum_{a=0}^{p-1}\left(\frac{ab}{p}\right)\zeta^{a+b} = \sum_{a=0}^{p-1}\sum_{c=a}^{a+(p-1)}\left(\frac{a(c-a)}{p}\right)\zeta^c,$$

where the last step makes the substitution $c = a + b$. Note that in the last sum, the inner summation takes c through each equivalence class modulo p exactly once, so we can rewrite it as

$$\sum_{a=0}^{p-1}\sum_{c=0}^{p-1}\left(\frac{a(c-a)}{p}\right)\zeta^c = \sum_{c=0}^{p-1}\sum_{a=0}^{p-1}\left(\frac{a(c-a)}{p}\right)\zeta^c.$$

Now we consider the coefficient $\sum \left(\frac{a(c-a)}{p}\right)$ of ζ^c in this expansion. First, if $c = 0$ we get

$$\sum_{a=0}^{p-1}\left(\frac{a(0-a)}{p}\right) = \sum_{a=0}^{p-1}\left(\frac{-1}{p}\right)\left(\frac{a^2}{p}\right) = \sum_{a=1}^{p-1}\left(\frac{-1}{p}\right) = \left(\frac{-1}{p}\right)(p-1),$$

and if $c \neq 0$, then by Corollary 7.4.2 we have

$$\sum_{a=0}^{p-1}\left(\frac{a(c-a)}{p}\right) = \sum_{a=0}^{p-1}\left(\frac{-1}{p}\right)\left(\frac{a^2-ca}{p}\right) = \left(\frac{-1}{p}\right)\sum_{a=1}^{p-1}\left(\frac{f(a)}{p}\right) = -\left(\frac{-1}{p}\right).$$

Now, plugging in these coefficients, we can continue our calculation of G^2:

$$G^2 = \sum_{c=0}^{p-1}\sum_{a=0}^{p-1}\left(\frac{a(c-a)}{p}\right)\zeta^c$$

$$= \left(\frac{-1}{p}\right)\left((p-1) - \zeta - \zeta^2 - \cdots - \zeta^{p-1}\right) = \left(\frac{-1}{p}\right)p,$$

where the last equality follow from the identity $\zeta + \zeta^2 + \cdots + \zeta^{p-1} = -1$. $\square$

Remarkable! We conclude that if $p \equiv 1 \bmod 4$, then $G^2 = p$, and so $G = \pm\sqrt{p}$. Since $G \in \mathbb{Z}[\zeta_p]$ by definition, we conclude that $\sqrt{p} \in \mathbb{Z}[\zeta_p]$. When $p \equiv 3 \bmod 4$, the same argument applies to $\sqrt{-p}$, providing the following corollary.

Corollary 7.7.5

Let $p^* = (-1)^{(p-1)/2}p$, so $p^* = p$ if $p \equiv 1 \bmod 4$ and $p^* = -p$ if $p \equiv 3 \bmod 4$. Then

$$\mathbb{Z}[\sqrt{p^*}] \subseteq \mathbb{Z}[\zeta_p] \quad (\text{and } \mathbb{Q}[\sqrt{p^*}] \subseteq \mathbb{Q}[\zeta_p]). \quad \blacktriangleleft$$

This fundamental algebraic inclusion can be seen as the reason that the study of cyclotomic integers is of any relevance in the study of quadratic ones, and to an algebraic number theorist is often viewed as the most important interpretation of quadratic reciprocity. Notice that we may recast the Law of Quadratic Reciprocity as follows:

Theorem 7.7.6 (Law of Quadratic Reciprocity)
Let p and q be distinct odd primes. Then

$$\left(\frac{q}{p}\right) = \left(\frac{G_p^2}{q}\right).$$

We will prove this version of the result, which is equivalent to the version stated earlier (in Theorem 7.5.3) since $\left(\frac{-1}{p}\right) = (-1)^{(p-1)/2}$, and so

$$\left(\frac{G_p^2}{q}\right) = \left(\frac{\left(\frac{-1}{p}\right)p}{q}\right) = \left(\frac{-1}{q}\right)^{(p-1)/2}\left(\frac{p}{q}\right) = (-1)^{(p-1)(q-1)/4}\left(\frac{p}{q}\right).$$

But first, one important technical warning: a central calculation in the proof centers around the quantity "$G_p^q \pmod{q}$," where we pause to emphasize that since $G_p^q \notin \mathbb{Z}$ this is necessarily a slightly new use of the word "mod." In the upcoming proof, for $a, b \in \mathbb{Z}[\zeta_p]$ the notation $a \equiv b \bmod q$ means that $a - b$ is a multiple of q in $\mathbb{Z}[\zeta_p]$, that there exists $z \in \mathbb{Z}[\zeta_p]$ such that $(a - b) = qz$. Note that since $\mathbb{Z} \subseteq \mathbb{Z}[\zeta_n]$, this new usage of the term extends the usual use of mod, since if $a - b = qn$ for some $n \in \mathbb{Z}$, then certainly $a - b = qz$ for some $z \in \mathbb{Z}[\zeta_p]$ (just taking $z = n$). We require only one modular arithmetic tool in the proof, that the nØØb's Binomial Theorem (Theorem 4.6.25, and specifically that the proof in Remark 4.6.26) continues to hold in this setting: for all $x, y \in \mathbb{Z}[\zeta_p]$, we have the relation

$$(x + y)^q \equiv x^q + y^q \bmod q.$$

Proof The idea is to compute the number G_p^q mod q in two ways. First, using Euler's Criterion (7.3.3):

$$G_p^q = G_p^{q-1} G_p = (G_p^2)^{\frac{q-1}{2}} G_p \equiv \left(\frac{G_p^2}{q}\right) G_p \pmod{q}.$$

Second, we reduce mod q the sum $G_p = \sum_{k=0}^{p-1} \left(\frac{k}{p}\right) \zeta_p^k$ raised to the q-th power by applying Theorem 4.6.25:

$$G_p^q \equiv \sum_{k=0}^{p-1} \left(\frac{k}{p}\right)^q \zeta_p^{qk} \equiv \sum_{k=0}^{p-1} \left(\frac{k}{p}\right) \zeta_p^{qk} \equiv \left(\frac{q}{p}\right) \sum_{k=0}^{p-1} \left(\frac{qk}{p}\right) \zeta_p^{qk} \equiv \left(\frac{q}{p}\right) G_p \pmod{q},$$

where for the second congruence we used that q was odd and in the last step[10] we use that since $\gcd(q, p) = 1$, as k runs over all values mod p, so does qk. Now, setting these two expressions for G_p^q equivalent mod q gives

$$\left(\frac{G_p^2}{q}\right) G_p \equiv \left(\frac{q}{p}\right) G_p \pmod{q} \tag{*}$$

in $\mathbb{Z}[\zeta_p]$. To show the *equality* $\left(\frac{G_p^2}{q}\right) = \left(\frac{q}{p}\right)$, and not just a mod-$q$ congruence in $\mathbb{Z}[\zeta_p]$, we suppose for the sake of contradiction that $\left(\frac{G_p^2}{q}\right) = -\left(\frac{q}{p}\right)$. Then substituting this into (∗) gives $G_p \equiv -G_p$ mod q. Multiplying by another G_p and rearranging gives $2G_p^2 \equiv 0$ mod q, so $\frac{2G_p^2}{q} = z$ for some $z \in \mathbb{Z}[\zeta_p]$. Now

$$z = \frac{2G_p^2}{q} = \left(\frac{-1}{p}\right) \frac{2p}{q}$$

(by Theorem 7.7.4), which shows that z is a non-integer rational number and hence not an algebraic integer, contradicting Theorem 7.6.3. This finishes the proof of the theorem. □

Finally, to settle our tab, we still owe one proof of the Second Supplemental Law. Much like every other encounter with the prime 2, the argument is similar but requires a seemingly unpredictable appearance of a higher power of 2. We leave a well-scaffolded proof of the result below, and encourage the readers to fill in the arguments.

---

[10] Which leaves to you, our astute reader, the justification for the third step.

> **Theorem 7.7.7 (Supplemental Law #2)**
> For p an odd prime, we have
>
> $$\left(\frac{2}{p}\right) = \begin{cases} 1 & \text{if } p \equiv \pm 1 \pmod 8 \\ -1 & \text{if } p \equiv \pm 3 \pmod 8 \end{cases}$$

Proof We work not with ζ_2 but ζ_8. First, we claim that $\zeta = \zeta_8 = \frac{\sqrt{2}}{2} + \frac{\sqrt{-2}}{2}$ is such a primitive 8th root of unity, which mostly amounts to checking that $\zeta^2 = i$ so indeed $\zeta^8 = 1$. Rather than using the Gauss sums G_2 or G_8, the analogous concept for this proof is $G = \zeta + \zeta^{-1}$. Compute that $G^2 = 2$, and so

$$G^{p+1} = G^{p-1} \cdot G^2 = 2^{(p-1)/2} \cdot 2 \equiv 2\left(\frac{2}{p}\right) \pmod p$$

by Euler's Criterion. As in the main proof, Theorem 4.6.25 gives the alternate computation

$$G^{p+1} = G(\zeta + \zeta^{-1})^p \equiv G(\zeta^p + \zeta^{-p}) \pmod p.$$

Now if $p \equiv \pm 1 \pmod 8$, then since $\zeta^8 = 1$ we get $\zeta^p + \zeta^{-p} = \zeta + \zeta^{-1} = G$, so equating the two evaluations of G^{p+1} gives $2\left(\frac{2}{p}\right) \equiv G(G) \equiv 2 \pmod p$, so $\left(\frac{2}{p}\right) = 1$ (with the same argument about mod-p arithmetic in $\mathbb{Z}[\zeta]$ as in the main proof). We leave the remaining case of $p \equiv \pm 3 \pmod 8$ to Exercise 7.19. $\qquad\square$

Application: Quadratic Equations mod p

Our theme of investigating square roots of elements $c \in \mathbb{Z}/(p)$, can be alternatively viewed as finding the $\mathbb{Z}/(p)$-roots of the polynomial $x^2 - c$. It is an instructive application of abstraction to generalize this to the quadratic polynomial $ax^2 + bx + c$ (with $a \neq 0$). Recall the standard derivation of the quadratic formula:

$$ax^2 + bx + c = 0$$

$$x^2 + \frac{b}{a}x + \frac{c}{a} = 0$$

$$\left(x + \frac{b}{2a}\right)^2 + \left(\frac{c}{a} - \frac{b^2}{4a^2}\right) = 0$$

$$\left(x + \frac{b}{2a}\right)^2 = \frac{b^2 - 4ac}{4a^2},$$

and so after rooting and subtracting,

$$x = \frac{-b \pm \sqrt{b^2 - 4ac}}{2a},$$

where we understand that if $b^2 - 4ac$ has no square roots, then no such x exists.

It is remarkable how many facts about algebra in $\mathbb{R}$ are used in this derivation. Of course, we use the ring operations and axioms as we re-arrange and collect terms, but we also use, for example, the commutative law (that $\frac{b}{2a} x = x \frac{b}{2a}$), that every non-zero a has a multiplicative inverse (to divide by a), and our understanding of square roots (that when they exist, they come in $\pm$ pairs).

From an abstract algebra perspective, what this means is that we should be able to get an analogous quadratic formula in any ring where all of these algebra rules hold. For example:

▶ **Example 7.7.8** Find a function f such that

$$\sin(x)f(x)^2 - 3xf(x) + e^x = 0.$$

Solution The quadratic formula applies here to solve for $f(x)$, using $a = \sin(x)$, $b = -3x$, and $c = e^x$ to give

$$f(x) = \frac{3x \pm \sqrt{9x^2 - 4\sin(x)e^x}}{2\sin(x)}.$$ □

While the implementation of the quadratic formula plays out differently in different rings, it works out well in the rings $\mathbb{Z}/(p)$ with p odd (see Exercise 7.13 for $p = 2$). Here, for a polynomial $ax^2 + bx + c$ with $a, b, c \in \mathbb{Z}/(p)$, the condition $a \neq 0$ automatically implies that a is a unit, and since p is odd, 2 is also a unit (and hence so is $2a$). We have already observed that square roots behave somewhat like square roots in $\mathbb{R}$: a number $a \in \mathbb{Z}/(p)$ can have 0, 1, or 2 square roots in $\mathbb{Z}/(p)$ depending on the Legendre symbol $\left(\frac{a}{p}\right)$, much like a number $a \in \mathbb{R}$ has 0, 1, or 2 square roots depending on if it is positive, zero, or negative. Literally the same proof as for the reals produces the following result:

Theorem 7.7.9 (Mod-p Quadratic Formula)
Let p be an odd prime and $f(x) = ax^2 + bx + c \in \mathbb{Z}/(p)[x]$, with discriminant $\Delta = b^2 - 4ac$. Then

(i) If $\left(\frac{\Delta}{p}\right) = -1$, then f has no roots mod p.
(ii) If $\left(\frac{\Delta}{p}\right) = 0$, then f has the unique root $x = -\frac{b}{2a}$ (i.e., $x = -(2a)^{-1}b$).
(iii) If $\left(\frac{\Delta}{p}\right) = +1$, then f has two distinct mod-p roots, given by

$$x = (2a)^{-1}(-b \pm \sqrt{\Delta}),$$

where $\sqrt{\Delta}$ represents either of the two square roots of Δ in $\mathbb{Z}/(p)$.

▶ **Example 7.7.10** Let $f(x) = 2x^2 + x + 3$. Find all mod-p solutions to $f(x) = 0$ for $p = 19, 23, 29$.

Solution We compute $\Delta = 1^2 - 4(2)(3) = -23$. The for $p = 19$, we have

$$\left(\frac{-23}{19}\right) = \left(\frac{-4}{19}\right) = \left(\frac{-1}{19}\right)\left(\frac{2}{19}\right)^2 = -1,$$

so f has no roots mod 19. For $p = 23$, we have $\left(\frac{-23}{23}\right) = 0$, so f has a single root, and since $[4]^{-1} = [6]$ in $\mathbb{Z}/(23)$, this root occurs at

$$x = -(2a)^{-1}(b) = -(4)^{-1}(1) \equiv -6 \bmod 23.$$

Finally, in $\mathbb{Z}/(29)$, $\left(\frac{-23}{29}\right) = 1$, we compute $[4]^{-1} = [-7]$, and find by brute-force search the square root $[8]^2 = [-23]$. The quadratic formula gives

$$x = (2a)^{-1}(-b \pm \sqrt{\Delta}) \equiv -7(-1 \pm 8) \equiv -49 \text{ or } 63 \equiv 9 \text{ or } 5 \bmod 29,$$

so the only solutions are $x = [5]$ and $x = [9]$. □

Finally, we observe that the situation is only slightly more complicated working mod a composite n. There are more non-units mod n, and any given discriminant might have more than two square roots (e.g, in $\mathbb{Z}/(8)$, where all four units are square roots of 1). Theorem 7.7.9 continues to hold if we interpret $\sqrt{b^2 - 4ac}$ to mean that we run through *all* square roots of the discriminant.

7.8 Quadratic UFDs

Quadratic Reciprocity provides the final tool needed to generalize our understanding of Gaussian primes. Namely, whenever the ring of integers of $\mathbb{Q}[\sqrt{d}]$ is a UFD, quadratic reciprocity will reveal the behavior of primes in this world. We work out one more example in detail. Recall from Chapter 6 that we found $\mathbb{Z}[\sqrt{2}]$, the ring of integers of $\mathbb{Q}[\sqrt{2}]$, to possess unique factorization. Compare the following result to Theorem 7.1.1.

Theorem 7.8.1

The following are equivalent for all rational primes p:

(i) $p = \pm(a^2 - 2b^2)$
(ii) p factors in $\mathbb{Z}[\sqrt{2}]$, i.e., p is not prime in $\mathbb{Z}[\sqrt{2}]$

(iii) 2 is a square modulo p

(iv) $p \equiv \pm 1 \bmod 8$.

The proof of Theorem 7.1.1 occupied the entirety of Chapter 5, so it is noteworthy that we have now developed enough machinery to render problems of this type as procedural exercises.

Proof In fact, the list in the theorem is a rather truncated list of statements now known to us to be equivalent. Here is one cycle through a list of equivalent statements, which contains all of the ones in the statement of the theorem:

$$p \equiv \pm 1 \pmod 8 \;\Rightarrow\; \left(\frac{2}{p}\right) = 1$$
$$\Rightarrow\; m^2 \equiv 2 \pmod p \text{ for some } m \in \mathbb{Z}$$
$$\Rightarrow\; p \mid (m^2 - 2) \text{ for some } m \in \mathbb{Z}$$
$$\Rightarrow\; p \mid (m + \sqrt{2})(m - \sqrt{2}) \text{ in } \mathbb{Z}[\sqrt{2}]$$
$$\Rightarrow\; p \text{ is not prime in } \mathbb{Z}[\sqrt{2}]$$
$$\Rightarrow\; p \text{ is not irreducible in } \mathbb{Z}[\sqrt{2}]$$
$$\Rightarrow\; p = \pm(a^2 - 2b^2)$$

from which it follows that $p \equiv \pm 1 \bmod 8$ just by computing that the expression $a^2 - 2b^2$ is never congruent to ± 3 in $\mathbb{Z}/(8)$. This completes the cycle and so shows all of the conditions encountered along the way to be equivalent. $\square$

It is a valuable practice to justify each step of that series of implications as a consequence of a known (and typically rather significant!) theorem. That our workflow generalizes from $\mathbb{Z}[i]$ to $\mathbb{Z}[\sqrt{2}]$ (and to $\mathbb{Z}[\sqrt{3}]$ in the exercises) and beyond is testimony to how far we have come. But, as always, it is equally important to keep in mind how things differ. Unlike $\mathbb{Z}[i]$, for example, we saw in Corollary 6.3.9 that the ring $\mathbb{Z}[\sqrt{2}]$ has infinitely many units, like $7 + 5\sqrt{2}$. So we have in $\mathbb{Z}[\sqrt{2}]$ the prime $p = 3$ (since $p \equiv 3 \bmod 8$), we also have its infinitely many associates

$$3(7 + 5\sqrt{2}) = 21 + 15\sqrt{2}, \quad 3(7 + 5\sqrt{2})^2 = 297 + 210\sqrt{2}, \quad \text{etc.}$$

We also have to contend with the presence of units of norm -1, whose primary effect is the introduction of the $\pm$ term in part (i) of the Theorem.

Nevertheless, the phrase "up to associates" performs its usual job in suppressing unit difficulties, and we can continue to develop prime number theory in $\mathbb{Z}[\sqrt{2}]$ like

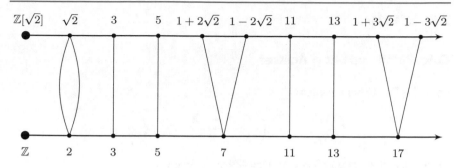

Fig. 7.5 The primes of $\mathbb{Z}$ and $\mathbb{Z}[\sqrt{2}]$

we did for $\mathbb{Z}[i]$. For example, from the previous result and some norm arithmetic we derive the analog of Theorem 5.6.6, classifying all of the primes of $\mathbb{Z}[\sqrt{2}]$.

Theorem 7.8.2
The primes of $\mathbb{Z}[\sqrt{2}]$ are, up to associates, precisely:

- The prime $\sqrt{2}$.
- The rational primes p that are congruent to 3 or 5 mod 8.
- The primes $a + b\sqrt{2}$ and $a - b\sqrt{2}$, where $a^2 - 2b^2 = \pm p \equiv \pm 1 \pmod{8}$.

Diagrammatically, we can see the splitting behavior of the primes of $\mathbb{Z}$ as they are imported to $\mathbb{Z}[\sqrt{2}]$ in Figure 7.5 just as we did for $\mathbb{Z}[i]$ in Figure 5.9.

The results of this section are in some sense the end of the story for factorization in quadratic UFDs, but like most "ends of stories" in mathematics, it also serves as the beginning of others. For example, there is still the question of the general phenomenon of units, suppressed in the consideration of primes but still of crucial importance in solving Diophantine equations (*all* infinitely many associates of $7 + 5\sqrt{2}$ present solutions to $x^2 - 2y^2 = \pm 1$, for example). The topic of units, specifically for real quadratic fields, is taken up in Section 9.4. Further, and more challenging, is the topic of what to do when we *don't* get unique factorization. We have ample evidence that the ring of integers of $\mathbb{Q}[\sqrt{d}]$ fails to be a UFD at least as frequently as not, and so number theory in these rings will require some new tools. These tools are at the forefront of modern algebraic number theory, and we give a brief introduction to these ideas in Section 9.5.

For the immediate future, however, we have a lingering but critical story left to complete—how to find *rational* solutions to conic equations—and it is to this story, and to the most exotic realms of numbers we've considered so far, that we turn next.

7.9 Exercises

Calculation and Short Answer

Exercise 7.1 Compute each of

$$\left(\frac{112}{659}\right) \qquad \left(\frac{5}{160465489}\right) \qquad \left(\frac{-2}{p}\right)$$

(You can assume that 659 and 160465489 are prime.)

Exercise 7.2 Compute each of the following:

$$\left(\frac{3}{97}\right) \ \left(\frac{5}{389}\right) \ \left(\frac{2033}{11}\right) \ \left(\frac{5!}{7}\right) \ \left(\frac{6!}{7}\right).$$

Generally for problems and discussions involving Legendre symbols, the phrase "find all primes..." should be interpreted as specifying a list of modular congruence conditions that such primes satisfy.

Exercise 7.3 Find all primes p for which $\left(\frac{13}{p}\right) = 1$. Repeat for $\left(\frac{19}{p}\right)$.

Exercise 7.4 Find all primes q for which $\left(\frac{-90}{q}\right) = 1$.

Exercise 7.5 Find primes p such that

$$\left(\frac{1}{p}\right) = \left(\frac{2}{p}\right) = \left(\frac{3}{p}\right) = \left(\frac{4}{p}\right) = \left(\frac{5}{p}\right) = +1.$$

What about

$$\left(\frac{1}{p}\right) = \left(\frac{2}{p}\right) = \left(\frac{3}{p}\right) = \left(\frac{4}{p}\right) = \left(\frac{5}{p}\right) = -1?$$

Exercise 7.6 Find all primes satisfying *both*

$$\left(\frac{5}{p}\right) = -1 \qquad \text{and} \qquad \left(\frac{-5}{p}\right) = +1.$$

Exercise 7.7 Explain why each of the following two identities are equivalent to the main Law of Quadratic Reciprocity:

$$\left(\frac{p}{q}\right)\left(\frac{q}{p}\right) = (-1)^{\frac{p-1}{2}\frac{q-1}{2}}.$$

and, with p^* is as in Corollary 7.7.5,

$$\left(\frac{p^*}{q}\right) = \left(\frac{q}{p}\right).$$

Exercise 7.8 Explore where things start to go awry with Legendre symbols modulo composite n. Draw the analog of the rings in Figure 7.1 for $\mathbb{Z}/(16)$. Find examples of results in the chapter that fail miserably if the modulus is not prime.

Exercise 7.9 Plot the line $y = 2x$ over $\mathbb{Z}/(7)$ (that is, in the grid $\mathbb{Z}/(7) \times \mathbb{Z}/(7)$). Describe how plotting lines in this context would work in general.

Exercise 7.10 Consider the sum

$$\left(\frac{1}{p}\right) + \left(\frac{2}{p}\right) + \left(\frac{3}{p}\right) + \left(\frac{4}{p}\right) + \left(\frac{5}{p}\right) + \left(\frac{6}{p}\right).$$

Find with justification all possible values of the sum, and one prime that produces each such value.

Exercise 7.11 For each $p \in \{5, 7, 13, 17, 19\}$, find all mod-$p$ roots of the polynomial

$$f(x) = 7x^2 + 15x + 5.$$

Exercise 7.12 For which primes p does the polynomial $x^2 + x + 3$ have 0, 1, and 2 roots mod p?

Exercise 7.13 The quadratic formula automatically fails for $p = 2$ since we cannot divide by $2a$. Fortunately, there are not many quadratic polynomials in $\mathbb{Z}/(2)[x]$. List them all along with their roots—is there a nice way to categorize them?

Formal Proofs

Exercise 7.14 Let p be prime. Prove that if $x^2 \equiv y^2 \pmod{p}$, then $x \equiv \pm y \pmod{p}$. Show this can fail for p composite.

Exercise 7.15 Prove that the squares of $\mathbb{Z}/(p)^\times$ form a subgroup of index 2, providing an alternate proof of Corollary 7.3.2.

Exercise 7.16 Let $\zeta = \zeta_p$ for p an odd prime. Suppose $a, b \in \mathbb{Z}$ are relatively prime to p. Prove that for some $c \in \mathbb{Z}$ we have

$$\frac{1 - \zeta^a}{1 - \zeta^b} = 1 + \zeta^b + \zeta^{2b} + \cdots + \zeta^{cb} \in \mathbb{Z}[\zeta],$$

and use this identity to conclude that $\frac{1-\zeta^a}{1-\zeta^b}$ is a unit in $\mathbb{Z}[\zeta]$. Then use the factorization of n in $\mathbb{Z}[\zeta_n]$ to conclude that $p = (1 - \zeta)^{p-1}u$ for some $u \in \mathbb{Z}[\zeta]^\times$.

Exercise 7.17 Let p be an odd prime, and let $f(x) = x^3 + sx^2 + tx + u \in \mathbb{Z}/(p)[x]$. Use a counting argument to show that there must be at least one value of $x \in \mathbb{Z}/(p)$ such that $f(x)$ is a quadratic residue mod p; i.e., that there is at least one solution over $\mathbb{Z}/(p)$ to the conic

$$y^2 = f(x).$$

Exercise 7.18 Consider the conic $y^2 = (x-1)(x-2)$ over $\mathbb{Z}/(p)$ for an arbitrary odd prime p, with the obvious point $(1, 0)$. Adapt the Diophantus Chord Method to show there are precisely $p-1$ points on this conic (as in Theorem 7.4.1) by intersecting the conic with the p lines $y = m(x-1)$ for $m \in \mathbb{Z}/(p)$ and then cleaning up special cases.

Exercise 7.19 . Revisit the proof of the second supplemental law given as Theorem 7.7.7. List and finish any details that were skipped along the way. Include the case $p \equiv \pm 3 \bmod 8$.

Exercise 7.20 Suppose that $\gcd(a, n) = 1$, and let f be a function on the integers mod n. Prove that $\displaystyle\sum_{k=1}^{n} f(k) = \sum_{k=1}^{n} f(ak)$.

Exercise 7.21 Prove that $p = \pm(a^2 - 3b^2)$ for non-zero integers a and b if and only if $p \equiv \pm 1 \pmod{12}$. (You may assume the true fact that $\mathbb{Z}[\sqrt{3}]$ is a UFD.)

Exercise 7.22 Find and prove congruence conditions on p equivalent to p being expressible in the form $\pm(a^2 - 7b^2)$ for some $a, b \in \mathbb{Z}$. (You may assume the true fact that $\mathbb{Z}[\sqrt{7}]$ is a UFD.)

Exercise 7.23 Use quadratic reciprocity to show that if $p, q \equiv 3 \bmod 4$ for distinct odd primes p and q, then there are no solutions to the Diophantine equation $x^2 - qy^2 = p$.

Exercise 7.24 Give a careful proof that the quadratic formula continues to work modulo an odd prime p, in that it provides the correct answers when the requisite square roots exist mod p, and has no solutions if they don't.

Computation and Experimentation

Exercise 7.25 Explore some sums of Legendre symbols. The Python worksheet "Conjectures on Sums of Legendre Symbols" provides an outline.

a) Start by computing

$$\left(\frac{0}{271}\right) + \left(\frac{1}{271}\right) + \left(\frac{2}{271}\right) + \cdots + \left(\frac{270}{271}\right).$$

Choose one or two more three-digit primes and calculate $\sum_{x=0}^{p-1} \left(\frac{x}{p}\right)$. Use your results to formulate a conjecture about $\sum_{x=0}^{p-1} \left(\frac{x}{p}\right)$. Give a brief argument as to why your conjecture is correct.

b) Choose a two- or three-digit prime p and a linear polynomial $f(x) = ax + b \in \mathbb{Z}[x]$ such that $p \nmid a$. Calculate

$$\sum_{x=0}^{p-1} \left(\frac{f(x)}{p} \right).$$

Repeat for another prime and another linear polynomial. Use your results to conjecture a value for $\sum_{x=0}^{p-1} \left(\frac{f(x)}{p} \right)$.

c) Repeat the previous experiment, but for a quadratic polynomial $f(x) = ax^2 + bx + c$ such that $p \nmid a$. Repeat for a couple of different values of p and $f(x)$. Compare the sum

$$\sum_{x=0}^{p-1} \left(\frac{f(x)}{p} \right)$$

to the value of $\left(\frac{a}{p} \right)$. Try to formulate a conjecture for $\sum_{x=0}^{p-1} \left(\frac{f(x)}{p} \right)$ given $f(x) = ax^2 + bx + c$ and $p \nmid a$. In formulating your conjecture, note whether p divides the discriminant $b^2 - 4ac$.

Exercise 7.26 The command `E=EllipticCurve(Integers(p),[s,t])` will construct in SageMath the elliptic curve $y^2 = x^3 + sx + t$ over the ring $\mathbb{Z}/(p)$. A slew of functions are available to call on E. For example `E.order()` will tell you the number of points on the curve. Experiment with the number of points as you:

a) Fix p and vary s and t.
b) Fix s and t and vary p.

Make some observations and compare them with the results of the section.

General Number Theory Awareness

Exercise 7.27 What are cyclotomic polynomials? What are the first few? What are their degrees? What do we know about their coefficients?

Exercise 7.28 Quadratic reciprocity decides the *existence* of square roots mod p, but not their *construction*. Research and implement the *Tonelli-Shanks algorithm* for doing just that.

Exercise 7.29 Theorem 7.6.3 shows that every element of $\mathbb{Z}[\zeta_n]$ is an algebraic integer, so that if R denotes the ring of algebraic integers contained in $\mathbb{Q}[\zeta_n]$, we have $\mathbb{Z}[\zeta_n] \subseteq R$. Are the two rings in fact equal? Find a reference with a proof or counter-example.

Exercise 7.30 Find a careful statement of the Hasse-Weil Theorem. Most statements have the estimate for $|E(\mathbb{F}_p)|$ as about $p + 1$ points, as opposed to our version which gave about p points. What gives?

Exercise 7.31 Application: What's a quadratic residue sound diffuser?

Exercise 7.32 Find an alternative proof of the 2nd Supplemental Law for quadratic reciprocity, one that mirrors the Wilson's Theorem approach to constructing an explicit square root of -1 mod p (when it exists).

Exercise 7.33 Explore alternative proofs of the main statement of the Quadratic Reciprocity Law. Is there a consensus on which is "best"?

Exercise 7.34 The Legendre symbol admits generalizations of various kinds, including *Jacobi symbols* and *Kronecker symbols*. What are these symbols and what do these extensions get us?

Number Theory Unleashed: Release $\mathbb{Z}_p$!

8

...wherein our cast of characters meets a new family.

8.1 The Analogy between Numbers and Polynomials

Consider the following all-too-common scene of frustration and despair.

NARRATOR: We enter mid-scene. Desperate for acceptance, the young $\mathbb{R}[x]$ is petitioning established aristocrat $\mathbb{Z}$ for membership in the upper ε's of ring society. It should all just be a matter of fitting in...

$\mathbb{R}[x]$: and so, you see, using the *degree* as the size of one of my elements serves to demonstrate the existence of a Division Algorithm...

$\mathbb{Z}$: [*aside*] Degree? How very proletariat. Size should speak for itself!

$\mathbb{R}[x]$: ...demonstrating the beloved property of Unique Factorization.

$\mathbb{Z}$: Naturally. I hope you haven't convened this meeting for the sake of *that* revelation.

$\mathbb{R}[x]$: [*taken aback*] I only mean to say that we have so much in common! We associate (and commute and distribute) within the same circles, we admit only the uniquest of factorizations, an infinitude of primes, and we have only a sliver's worth of inverses...

[$\mathbb{R}[x]$ *begins to pace with nervous energy*]

$\mathbb{R}[x]$: ...not to mention our fractions! Fields upon fields of rational numbers and rational functions, and how well we can approximate fractions! One quick *trip to the Taylor*, as they say, mends the rational function $\frac{1}{1-x}$ to the approximation $1 + x + x^2$.

$\mathbb{Z}$: [*warily*] Hm. Well, see, I don't really *do* approximations....

© Springer Nature Switzerland AG 2022
C. McLeman et al., *Explorations in Number Theory*, Undergraduate Texts in Mathematics,
https://doi.org/10.1007/978-3-030-98931-6_8

$\mathbb{R}[x]$: And it doesn't stop there! Even as unwieldy as a function like e^x can be tamed with the replacement $1 + x + \frac{x^2}{2} + \frac{x^3}{6}$. It's as good as new – take as many terms as you like!

$\mathbb{Z}$: [*silently dejected*] Well, of course, I know that e is between 2 and 3...

[*Exeunt.*]

No doubt we can all sympathize with $\mathbb{Z}$. Once the epitome of rings, the one ring to rule them all, our precious $\mathbb{Z}$ has rather fallen out of the limelight. Indeed, we started Chapter 3 with an eye toward determining the algebraic structures which are sufficiently "$\mathbb{Z}$-like" to do algebra in, but it is now common to find that other rings have their own equally enviable features. In particular, while unique factorization in $\mathbb{Z}$ is a hallmark of classical number theory, we've now established this same property in all kinds of rings. Examples include $\mathbb{Z}$, $\mathbb{Z}[i]$, $\mathbb{Z}[\sqrt{-2}]$, $\mathbb{Z}[\zeta_3]$, etc., and $\mathbb{R}[x]$ is yet another: given polynomials f and g, long division of polynomials guarantees the existence of quotient and remainder polynomials q and r such that $f = gq + r$ with $0 \leq \deg r < \deg g$. By Theorem 6.5.3, $\mathbb{R}[x]$ is a unique factorization domain whose primes are simply the irreducible (unfactorable) polynomials, which by the Fundamental Theorem of Algebra is just the linear polynomials and quadratic polynomials with no real roots. From an algebraic perspective, this puts $\mathbb{Z}$ and $\mathbb{R}[x]$ on a much more even footing than one might have expected by looking at their elements. But, as the scene above suggests, elements of the ring $\mathbb{R}[x]$ have auxiliary roles as *functions*, and thus the tools of calculus shed light on this ring in a way that *seeeeeems* (wink wink) to have no analogue in $\mathbb{Z}$. We have learned a lot about $\mathbb{Z}$ by working in other rings, e.g., modular rings and rings of integers. What does $\mathbb{R}[x]$ have to contribute to the cause?

Let's briefly review the relevant calculus. Suppose $f : \mathbb{R} \to \mathbb{R}$ is an everywhere infinitely differentiable function, so that for any real number x it is meaningful to talk about the nth derivative $f^{(n)}(x)$ at that point. Then at any point $a \in \mathbb{R}$, the *Taylor series of f at $x = a$* is the power series

$$\sum_{n=0}^{\infty} \frac{f^{(n)}(a)}{n!}(x - a)^n = f(a) + f'(a)(x - a) + \frac{f''(a)}{2!}(x - a)^2 + \cdots \quad (8.1)$$

Thorny questions concerning the convergence of such a series and where a function *equals* its Taylor series are resolved, like most questions, in books not equal to this one. Instead, we agree to restrict our attention to a *rational* function f, for which given any a in the domain of f, its Taylor series converges and equals the value of f for all x sufficiently close to a. For example, the Taylor series of $f(x) = \frac{1}{1-x}$ centered at $x = 0$ exists and converges for all $|x| < 1$ and so once we deduce the pattern of derivatives $f^{(n)}(0) = n!$, we obtain the equality

$$\frac{1}{1-x} = 1 + x + x^2 + x^3 + \cdots \qquad \text{for all } |x| < 1,$$

better known as the formula for an infinite geometric series.

Table 8.1 Several Taylor expansions for $f(x) = x^3 - 3x^2 + 4$

a	Taylor expansion of $f(x) = x^3 - 3x^2 + 4$ centered at a
0	$f(x) = 4 + 0(x - 0) - 3(x - 0)^2 + 1(x - 0)^3$
1	$f(x) = 2 - 3(x - 1) + 0(x - 1)^2 + 1(x - 1)^3$
-1	$f(x) = 0 + 9(x + 1) - 6(x + 1)^2 + 1(x + 1)^3$
2	$f(x) = 0 + 0(x - 2) + 3(x - 2)^2 + 1(x - 2)^3$

Underappreciated in the calculus version of this discussion is the power in choosing a variety of a's, and not just $a = 0$, producing a variety of Taylor series, even just for polynomials. Table 8.1 shows the Taylor series of $f(x) = x^3 - 3x^2 + 4$ centered at four different centers a, each of which can be verified either by checking the derivative calculations or simply by expanding each expression to see that they all resolve identically.

Note that we have written the terms of the polynomials in ascending order of degree, in contrast to the usual polynomial convention of writing the highest degree term first, to mirror the general expression for a power series. The organization also highlights some interesting data. For example, the value of f at each a can be easily read off of the corresponding Taylor series, since either by evaluation or by (8.1), $f(a)$ is always the degree-0 term of the expansion. We see that $f(0) = 4$, $f(1) = 2$, $f(-1) = 0$, and $f(2) = 0$. In particular, f has roots at $x = -1$ and $x = 2$. Further, in the expansion around $a = 2$, we are missing both the constant *and* linear terms, beginning only with the quadratic term $(x - 2)^2$. This shows that $(x - 2)^2$ divides *every* term in the expansion, and hence f itself:

$$f(x) = 3(x - 2)^2 + (x - 2)^3 = (x - 2)^2 (3 + (x - 2)) = (x - 2)^2(x + 1).$$

In general, if the first (lowest-degree) term of the Taylor expansion of f at a is $(x - a)^d$, then $(x - a)^d \mid f(x)$. This close tie between expansions and divisibility is the missing link on the integer side of the analogy between $\mathbb{R}[x]$ and $\mathbb{Z}$. We have studied divisibility questions throughout our consideration of $\mathbb{Z}$, but we lack a notion of a Taylor expansion of an integer. Fortunately, an appropriate analogue is available.

Table 8.2 Several base p expansions for $n = 75$

p	Base p expansion of $n = 75$
2	$75 = 1 + 1p + 0p^2 + 1p^3 + 0p^4 + 0p^5 + p^6$
3	$75 = 0 + 1p + 2p^2 + 2p^3$
5	$75 = 0 + 0p + 3p^2$
7	$75 = 5 + 3p + p^2$

Definition 8.1.1.

The **base p expansion** of a natural number n is the unique (Exercise 8.21) way of writing n in the form

$$n = a_0 + a_1 p + a_2 p^2 + \cdots + a_k p^k$$

for integer coefficients a_i with $0 \le a_i < p$. ◄

In the analogy between $\mathbb{R}[x]$ and $\mathbb{Z}$, irreducible polynomials correspond to primes, and so expanding a polynomial as a finite sum of powers of $(x - a)$ has as its natural analogue expanding a natural number as a sum of powers of a prime p. Just as with polynomials, varying p gives different shapes for the expansion, as in Table 8.2.

We will continue the practice of writing the expansion with a placeholder symbol p for the prime, even though we know its exact value (e.g., we write $5 + 3p + p^2$ instead of $5 + 3 \cdot 7 + 7^2$ in the last row of the table). In addition to reducing the horizontal space occupied by the expression, this convention has the benefits of providing a visual distinction between the prime p and the coefficients, mirroring the distinction between the variable x and the coefficients in the series expansion of a polynomial. Again the table reveals much. We note that in both the base 3 expansion and base 5 expansion the coefficient of p^0 is 0, and so each summand in those expansions is a multiple of p. We deduce that 3 and 5 are divisors of 75, and again we see the role of multiplicities of prime divisors in expansions: For $p = 5$, neither p^0 nor p^1 appears in the expansion, and so $p^2 \mid 75$. Generally, if the first (lowest power) term of the base p expansion of n is p^d, then $p^d \mid n$.

The number of times a given linear factor or prime divides a polynomial or natural number is an important metric to keep track of. Indeed we will soon use these numbers to quantify how big or small a number or polynomial is with respect to each of these factors. For polynomials, a long string of leading zero coefficients $(0 + 0x + 0x^2 + 5x^3 + 2x^4 + x^5)$ dictates the extent to which x plays a role in the factorization of the polynomial. Similarly, a long string of leading zero coefficients $(0 + 0p + 0p^2 + 5p^3 + 2p^4 + p^5)$ dictates the extent to which p plays a role in the factorization of a natural number. Fortunately, we already have encountered this idea in Chapter 3 (specifically, Definition 3.4.1) under the name of *p-adic valuations*.

Lemma 8.1.2

For a positive prime p, the p-**adic valuation** $v_p(n)$ of a natural number n is both:

- The largest power k of p such that $p^k \mid n$.
- The smallest power k of p that appears in the base p expansion of n. ◀

Proof The first of these bullets is the definition of $v_p(n)$, so we need to show that the two conditions are equivalent. Given p and n as in the definition, let k be the smallest power appearing in the base p expansion of n, so that

$$n = a_k p^k + a_{k+1} p^{k+1} + \cdots + a_m p^m = p^k \left(a_k + a_{k+1} p + \cdots + a_m p^{m-k} \right)$$

with $a_k \neq 0$. This immediately shows that $p^k \mid n$. Further, $p^{k+1} \nmid n$, as $\frac{n}{p^k}$ mod $p = a_k \neq 0$. We conclude that $k = v_p(n)$. □

Note that exactly the same equivalency can be made for polynomials f: the largest power k such that $(x - a)^k$ divides f is the smallest k such that $(x - a)^k$ appears in its Taylor expansion at a (and so, by construction of the Taylor series, implies the vanishing of the first k derivatives of f at $x = a$). For both numbers and polynomials, fluency in alternating one's thinking about valuations in these two contexts proves to be very valuable.

8.2 The p-adic World: An Analogy Extended

The analogy between polynomials and integers would be cute but not breathtaking if it were merely an observation that both types of rings enjoy similar structural properties. The real power of any mathematical analogy is the hope that the special properties of one spur us to investigate new ideas for the other. While using Taylor series for polynomials is of only middling interest in calculus, they are much more potent for studying rational functions. But on the integer side of analogy, it's not clear what we could even mean by the "base p expansion of a rational number." Let's give it a shot.

Recall our example from the previous section that for $|x| < 1$ we have

$$\frac{1}{1 - x} = 1 + x + x^2 + x^3 + \cdots .$$

Even without a careful study of convergence of Taylor series, it is easy to see that the hypothesis that $|x| < 1$ is crucial here. For example, naively evaluating the above at

$x = -5$ gives us the identity

$$\frac{1}{6} = \frac{1}{1-(-5)} = 1 + (-5) + (-5)^2 + (-5)^3 + \cdots = 1 - 5 + 25 - 125 + \cdots$$

But, as this latter series clearly diverges, this is simply nonsense, right?

> *...right?*

Are you ready for a magic trick? First, we begin with the distraction, the sleight-of-hand to distract your attention while we prepare to dazzle:

▶ **Example 8.2.1** Find 6^{-1} mod 25, 125, 625, and 3125.

Solution The first couple of these can be dispatched of quickly: $6^{-1} \equiv -4$ mod 25 since $6 \times (-4) = -24 \equiv 1$ mod 25 and $6^{-1} \equiv 21$ mod 125 since $6 \times 21 = 126 \equiv 1$ mod 25. In general, though, computing inverses requires an application of the extended Euclidean algorithm or Euler's Theorem, and since these get laborious by hand and we're here for the magic show, let us just report to you that $6^{-1} \equiv -104$ mod 625 and $6^{-1} \equiv 521$ mod 3125.   □

Now, observe—there is nothing up our sleeves—that these inverses are precisely the partial sums of the "impossible" representation for $\frac{1}{6} (= 6^{-1})$ as an infinite sum:

$$
\begin{aligned}
6^{-1} \bmod 5^1 &= 1 &= 1 \\
6^{-1} \bmod 5^2 &= -4 &= 1 - 5 \\
6^{-1} \bmod 5^3 &= 21 &= 1 - 5 + 25 \\
6^{-1} \bmod 5^4 &= -104 &= 1 - 5 + 25 - 125 \\
6^{-1} \bmod 5^5 &= 521 &= 1 - 5 + 25 - 125 + 625 \\
&\vdots \vdots &\ddots
\end{aligned}
$$

Was that your card? Show the audience. Miraculously, the single infinite divergent series

$$1 - 5 + 25 - 125 + 625 - \cdots$$

seems to know the inverse of 6 modulo *every* power of 5, simply by truncating the series after more and more terms. As with other magic tricks, the awe diminishes only slightly when you see how it is done, and it is indeed quite easy to prove that it continues to work indefinitely (and for any other prime p in place of 5), using the telescoping identity

$$(1 + 5)(1 - 5 + 5^2 - \cdots \pm 5^{n-1}) = 1 + (5 - 5) + \cdots + (5^{n-1} - 5^{n-1}) \pm 5^n$$
$$\equiv 1 \pmod{5^n}.$$

It is hard to believe that this prospect, an extension of the notion of a base p expansions to a comically divergent nonsense sum with infinitely many terms, is to be of any use to anyone. But these objects are not only non-nonsensical, they are of such pervasive interest in modern number theory that they earn the most prestigious moniker of them all – we dub them a new class of *number*.

Definition 8.2.2

For a prime $p \in \mathbb{N}$, a *p*-**adic (rational) number** is an expression of the form

$$\alpha = a_{-k}p^{-k} + a_{-k+1}p^{-k+1} + \cdots + a_{-1}p^{-1} + a_0 + a_1 p + a_2 p^2 + \cdots, \quad (*)$$

where $k \in \mathbb{Z}$ and each a_i is an integer with $0 \le a_i \le p - 1$. The collection of *p*-adic rational numbers is denoted $\mathbb{Q}_p$. The *p*-**adic integers**, $\mathbb{Z}_p$, consist of those *p*-adic numbers with $a_i = 0$ for all $i < 0$, that is, a *p*-adic number with no negative powers of *p* appearing. ◄

▶ **Example 8.2.3** The base *p* expansion of any natural number is a *p*-adic integer (e.g., $47 = 2 \cdot 1 + 4 \cdot 5 + 1 \cdot 5^2$), and conversely any *p*-adic integer with finitely many terms represents a natural number (e.g., $3 \cdot 1 + 5 \cdot 7 + 2 \cdot 7^2 = 136$).

▶ **Example 8.2.4** Writing down an element of $\mathbb{Z}_p$ is as simple as specifying a sequence of integers $0 \le a_i \le p-1$. For example, we could take the decimal digits of π and produce an element of $\mathbb{Z}_{11}$:

$$3 + p + 4p^2 + 1p^3 + 5p^4 + 9p^5 + 2p^6 + 6p^7 + \cdots \in \mathbb{Z}_{11}$$

Rational *p*-adics that aren't *p*-adic integers include finitely many negative powers of *p*; e.g.,

$$2p^{-2} + 7p^{-1} + 1 + 8p + 2p^2 + 8p^3 + 1p^4 + \cdots \in \mathbb{Z}_{11}$$

Again it is worth emphasizing the analogy. Though less studied in introductory Calculus courses, the analogue of *p*-adic rational numbers in the world of calculus is the notion of a *Laurent series*, a Taylor series in which we allow finitely many negative powers of $(x - a)$. For example, while we cannot take the Taylor series of $\frac{1}{x^2(1-x)}$ at $x = 0$ as the function is not defined there, it is perfectly sensible to write the Laurent series

$$\frac{1}{x^2(1 - x)} = \frac{1}{x^2} \cdot \frac{1}{1 - x} = \frac{1}{x^2}(1 + x + x^2 + x^3 + \cdots) = \frac{1}{x^2} + \frac{1}{x} + 1 + x + x^2 + \cdots$$

We will be occupied for some time with viewing more and more rational numbers as *p*-adic expansions. So far we have only applied base *p* expansions to natural numbers, but the Laurent series above shows how we might vastly expand our scope. As a first example, to find the 5-adic expansion of the rational number $\frac{697}{125}$, we can write (using $p = 5$)

$$\frac{697}{125} = \frac{2 + 4p + 2p^2 + p^4}{p^3} = 2p^{-3} + 4p^{-2} + 2p^{-1} + p.$$

The arithmetic of p-adic numbers will clearly take some getting used to, but as we do so, it is worth remembering the hallmark of these expansions in the first place: they are almost as simple to work with as the Taylor series that they analogize. Taylor series are a phenomenal substitute for rational and even transcendental functions because they behave algebraically just like polynomials. Likewise, doing arithmetic in the p-adic world will not be so different from doing arithmetic with integers. In fact, many of our basic notions of arithmetic generalize almost trivially.

Definition 8.2.5

Given a p-adic integer

$$\alpha = a_0 + a_1 p + a_2 p^2 + \cdots \in \mathbb{Z}_p,$$

we define **mod-p reduction of** α by $\alpha \bmod p = a_0$ and more generally

$$\alpha \bmod p^k = a_0 + a_1 p + a_2 p^2 + \cdots + a_{k-1} p^{k-1}.$$

We will often abbreviate this, the truncation of α after k terms, as simply α_k. (Note that α_k does not have a p^k term.) As we will see, it is not unreasonable to think of α as a limit, as in $\alpha = \lim_{k \to \infty} \alpha_k$. ◄

Notice that this generalizes the usual notion of mod-p reduction in the sense that if the $\alpha \in \mathbb{Z}_p$ in question is actually the finite base-p expansion of a natural number n, then the notions of $\alpha \bmod p$ and $n \bmod p$ coincide. Since reducing a p-adic integer modulo p^k produces a regular integer, we can feel free to make use of our previous understanding of congruence, e.g., that

$$\alpha \equiv \beta \bmod p^k$$

if and only if $\alpha_k = \beta_k$. Finally, recall from Lemma 8.1.2 that the p-adic valuation $v_p(n)$ of a natural number can be read off of its base p expansion as simply the least power of p appearing.

Definition 8.2.6

For non-zero $\alpha \in \mathbb{Q}_p$ as in (∗), the p-**adic valuation** $v_p(\alpha) = -k$ is the least power of p appearing in its p-adic expansion. As in Lemma 8.1.2, we can alternatively view it as the largest power of p dividing every term in the p-adic expansion of α. By factoring it out, we can write any $\alpha \in \mathbb{Q}_p$ uniquely in the form

$$\alpha = p^{v_p(\alpha)}\alpha' \tag{8.2}$$

where $\alpha' \in \mathbb{Z}_p$ and $v_p(\alpha') = 0$ (i.e., α' mod $p \neq 0$). As we did in $\mathbb{Z}$, define $v_p(0) = \infty$. Note the *p*-adic integers $\mathbb{Z}_p$ are characterized precisely as those *p*-adic numbers α with $v_p(\alpha) \geq 0$. ◄

► **Example 8.2.7** For any prime p (> 8), the element

$$\begin{aligned}
\alpha &= 2p^{-2} + 7p^{-1} + 1 + 8p + 2p^2 + 8p^3 + 1p^4 + \cdots \\
&= p^{-2}(2 + 7p + 1p^2 + 8p^3 + 2p^4 + 8p^5 + 1p^6 + \cdots)
\end{aligned}$$

has $v_p(\alpha) = -2$.

Exploration O

p-adic Dot-Dot-Dots ◄

O.1 Suppose $p = 7$. Find the base p expansion of the sum

$$(3 + 4p + p^4) + (5 + 2p + 6p^2).$$

More generally, describe an algorithm for finding the base p expansion of the sum of two-base p expansions of natural numbers.

Addition of p-adic numbers will be similar to the above but require an infinite chain of computations.

O.2 Compute $68 + \alpha$ in $\mathbb{Q}_7$, where

$$\alpha = 2 + 4 \cdot 7 + 5 \cdot 7^2 + 6 \cdot 7^3 + 6 \cdot 7^4 + 6 \cdot 7^5 + 6 \cdot 7^6 + 6 \cdot 7^7 + 6 \cdot 7^8 + \cdots,$$

with the pattern of coefficients continuing as all 6's. What does your result tell you about how to interpret α?

O.3 Propose a p-adic expansion for $-1 \in \mathbb{Z}_p$. For -2?

We will develop formal rules for multiplication but for now let's see what mileage we can get out of assuming that the normal rules of arithmetic (in particular, the distributive law) apply. As a first test, note that our early observation that $\frac{1}{6} \in \mathbb{Z}_5$ via the identity

$$\frac{1}{6} = 1 - 5 + 5^2 - 5^3 + \cdots$$

violates our definition of 5-adic integers by including coefficients of -1. We claim the following version works just as well, using 4 in replace of -1:

$$\frac{1}{6} = 1 + 4 \cdot 5 + 5^2 + 4 \cdot 5^3 + 5^4 + 4 \cdot 5^5 + \cdots,$$

with coefficients alternating between 1 and 4.

O.4 Attempt to verify this claim by multiplying 6 by the expansion above. That is, see if you can compute in $\mathbb{Z}_5$ that the product

$$(1 + p)(1 + 4p + p^2 + 4p^3 + p^4 + \cdots) = 1.$$

Be mindful of any assumptions you make.

O.5 Starting with the 7^0 term and working upwards, compute terms of this 7-adic square until you feel ready to make a conjecture about what it is:

$$(3 + 7 + 2 \cdot 7^2 + 6 \cdot 7^3 + 7^4 + 2 \cdot 7^5 + 7^6 + 2 \cdot 7^7 + 4 \cdot 7^8 + 6 \cdot 7^9 + \cdots)^2$$

8.3   *p*-adic Arithmetic: Making a Ring

If we enjoy the set of p-adic numbers, then we really ought to endow it with an algebraic structure by equipping it with notions of addition and multiplication[1].

> **Theorem 8.3.1**
> The set $\mathbb{Z}_p$ is a ring.

   As practiced in the previous exploration, this is not much more substantive than the (admittedly unusual) process of adding and multiplying natural numbers by their base p expansion. The point of this section will be to more formally establish the well-definedness of these p-adic operations, and we begin with an example:

▶ **Example 8.3.2** Evaluate $21 + 33$ in $\mathbb{Q}_5$ and $\mathbb{Q}_7$.

Solution The answer should of course be 54 in either case, but the question for now is what this addition looks like when done in terms of p-adic expansions. When $p = 5$, we have $21 = 1 + 4p$ and $33 = 3 + p + p^2$, so

$$21 + 33 = (4p + 1) + (p^2 + p + 3) = 4 + p + 4p + p^2 = 4 + 5p + p^2 = 4 + 2p^2,$$

which upon substituting $p = 5$ gives 54. Similarly, when $p = 7$, we have

$$21 + 33 = 3p + (4p + 5) = 7p + 5 = p^2 + 5 = 54. \qquad \square$$

This problem displays the entire knack to p-adic addition, viewing p simultaneously as a placeholder for the expansion *and* a specific fixed prime value. This is in stark contrast with polynomial addition, where the powers of x serve only as placeholders for their coefficients. Of course, this more complicated process for doing arithmetic with expansions is nothing new to elementary school students—addition of natural numbers in decimal notation necessitates precisely the same "carrying the 1" as we just saw above. What about multiplication?

▶ **Example 8.3.3** Evaluate $12 \cdot 32 \in \mathbb{Q}_5$:

---

[1] Beyoncé. *Single Ladies.* Columbia Records, 2008. (paraphrased)

Solution We take the 5-adic expansions $12 = 2 + 2p$ and $32 = 2 + p + p^2$ and then multiply the results as if polynomials in p:

$$
\begin{aligned}
12 \cdot 32 &= (2 + 2p) \cdot (2 + p + p^2) \\
&= (4 + 2p + 2p^2) + (4p + 2p^2 + 2p^3) \\
&= 4 + 6p + 4p^2 + 2p^3 \\
&= 4 + p + 5p + 4p^2 + 2p^3 \\
&= 4 + p + p^2 + 4p^2 + 2p^3 \\
&= 4 + p + 5p^2 + 2p^3 \\
&= 4 + p + p^3 + 2p^3 \\
&= 4 + p + 3p^3.
\end{aligned}
$$

The result is 384, achieved by both standard multiplications in $\mathbb{Z}$ and by evaluating this last p-adic expansion by substituting $p = 5$.   □

That these last couple of examples gave the correct result is not surprising—it is just regular integer arithmetic intentionally obfuscated by re-writing numbers in terms of their p-adic expansions. Nevertheless, we encourage the reader to work through every step of that last solution carefully, as it contains the critical manipulations needed to make progress. The validity of the corresponding additions and multiplications will be less trivial when done with infinite expansions.

Exposure to a little bit of practice doing arithmetic in this new world is pretty compelling evidence that everything can be made to work, provided that we're willing to sweep some $\cdots$'s under the rug. The p-adic world necessitates an infinite amount of calculation to multiply two elements via their p-adic expansions, and while this isn't inherently scandalous (after all, we could in principal evaluate $\pi + e$ from their infinite decimal expansions), it does require some care. Addition and multiplication of p-adic integers should look a lot like addition and multiplication of formal power series:

$$
\begin{aligned}
(a_0 + a_1 p + a_2 p^2 + \cdots) &+ (b_0 + b_1 p + b_2 p^2 + \cdots) \\
&= (a_0 + b_0) + (a_1 + b_1)p + (a_2 + b_2)p^2 + \cdots \\
(a_0 + a_1 p + a_2 p^2 + \cdots) &\cdot (b_0 + b_1 p + b_2 p^2 + \cdots) \\
&= (a_0 b_0) + (a_1 b_0 + a_0 b_1)p + (a_2 b_0 + a_1 b_1 + a_0 b_2)p^2 + \cdots
\end{aligned}
$$

Indeed, with x's in place of p's, this is precisely power series arithmetic, and it continues to hold in the p-adic world with the caveat that one must "carry" powers of p in order to re-write the result in standard form (e.g., as seen repeatedly in Example 8.3.3). The need to be careful about this is probably best exemplified by instances where an infinite number of such carries are required. Consider for example the following sum of two 5-adic integers:

▶ **Example 8.3.4** Compute in $\mathbb{Z}_5$ the sum of 1 and

$$\alpha = 4 + 4p + 4p^2 + \cdots,$$

the 4-adic integer with all coefficients equal to 4.

Solution As with the standard algorithm for long addition of natural numbers, we compute the sum by starting from the "ones digit," (here the p^0 term) resolving any carries, then the linear term (the p^1 term), etc. The current sum will involve infinitely many carries, repeatedly using the relationship $5p^k = p^{k+1}$ in $\mathbb{Z}_5$:

$$
\begin{aligned}
1 + \alpha &= 1 + (4 + 4p + 4p^2 + 4p^3 + 4p^4 + \cdots) \\
&= 5 + 4p + 4p^2 + 4p^3 + 4p^4 + \cdots \\
&= p + 4p + 4p^2 + 4p^3 + 4p^4 + \cdots \\
&= 0 + 5p + 4p^2 + 4p^3 + 4p^4 + \cdots \\
&= 0 + p^2 + 4p^2 + 4p^3 + 4p^4 + \cdots \\
&= 0 + 0p + 5p^2 + 4p^3 + 4p^4 + \cdots \\
&= 0 + 0p + 0p^2 + 5p^3 + 4p^4 + 4p^5 + \cdots \\
&= 0 + 0p + 0p^2 + 0p^3 + 5p^4 + 4p^5 + 4p^6 + \cdots \\
&= \vdots \\
&= 0 + 0p + 0p^2 + 0p^3 + 0p^4 + 0p^5 + 0p^6 + \cdots \\
&= 0.
\end{aligned}
$$
□

Whoa. Whoa, there. Did those $\vdots$ stand as a placeholder for the "argument"

> *well, as we keep going all of the powers eventually go away, so after infinitely many steps there's nothing left, so the final answer is zero[2] ? Or something?*

Well, in short, yes. Indeed, there is a sense in which fluent *p*-adic reasoning looks exactly like this. But in long, undoubtedly that egregious lack of precision raises your mathematical hackles, and a formalization of this approach is to introduce notions of convergence completely analogous to the way that we measure convergence of sequences and series in calculus. Re-writing basic notions in the language of limits is not difficult: if $\alpha, \beta \in \mathbb{Z}_p$, recalling the notation of α_n for the n-th truncation of α, we have

$$\alpha = \beta \quad \Longleftrightarrow \quad \alpha_n = \beta_n \text{ for all } n \geq 1 \quad \Longleftrightarrow \quad \lim_{n \to \infty} v_p(\alpha_n - \beta_n) = \infty$$

---

[2] Of course, we also try not to reason in run-on sentences, but if we're being cavalier about the mathematics anyway...

(see Exercise 8.22 for the argument that p-adic expansions, and hence their truncations, are unique). This suggests that α and β are *close* to equal if $v_p(\alpha_n - \beta_n)$ is large, and generally that numbers with large p-adic valuation should be considered to be p-adically small (close to zero). The analogous statement in $\mathbb{R}$ is that two decimal expansions are close if their difference is small, i.e., begins with a long string of 0's. The way we encode "size" in rings where our usual notions either do not apply or seem to give the wrong answer is to define an absolute value on the ring. The above provides us a perfect idea for how to measure size: the more leading 0s in your p-adic expansion (i.e., the higher your p-adic valuation), the smaller you should be considered. The following definition accomplished precisely this.

Definition 8.3.5

Define the *p-adic absolute value* $|\cdot|_p$ for $\alpha \in \mathbb{Q}_p$ by

$$|\alpha|_p = p^{-v_p(\alpha)}.$$

(Interpret $p^{-\infty} = 0$, so $|0|_p = 0$ for all primes p). ◀

This definition starts to make the formal process of reasoning through "infinite dot-dot-dot calculations" feel both a little more rigorous, and also more analogous to the arguments one sees in a real analysis course. Continuing the above chain of logical equivalences for testing equality, for example, we have

$$\alpha = \beta \iff \lim_{n \to \infty} v_p(\alpha_n - \beta_n) = \infty \iff \lim_{n \to \infty} |\alpha_n - \beta_n|_p = 0.$$

That is, much like convergence of infinite real sums, two p-adic numbers are equal if and only if their sequence of partial sums approach each other p-adically. As a second example, the off-handed comment in Definition 8.2.5 that $\alpha = \lim_{n \to \infty} \alpha_n$ is now seen to be a precise and accurate statement, and we can also resolve the uncertainty of convergence of calculations having infinitely many "carries." For example, to see that the α of Example 8.3.4 is indeed the additive inverse of 1 in $\mathbb{Z}_5$, we need only verify that $1 + \alpha = 0$ by showing that $v_p(1 + \alpha_n)$ tends to infinity with n. This is a quick exercise in reasoning with finite geometric series (note $p = 5$, so $p - 1 = 4$):

$$v_p(1 + (4 + 4p + 4p^2 + \cdots + 4p^{n-1})) = v_p\left(1 + 4(1 + p + p^2 + \cdots + p^{n-1})\right)$$

$$= v_p\left(1 + 4 \cdot \frac{p^n - 1}{p - 1}\right)$$

$$= v_p\left(p^n\right) = n \to \infty.$$

Again, this is more formal that we typically intend to be with calculations of this type, choosing instead to view this as the implicit rigorous justification underlying

the somewhat more free-wheeling use of $\cdots$'s. Nevertheless, we can now close the
story and formally define *p*-adic arithmetic.

Definition 8.3.6

For $\alpha, \beta \in \mathbb{Z}_p$, we define *p*-**adic addition and multiplication** by

$$\alpha + \beta = \lim_{n \to \infty} \alpha_n + \beta_n \qquad \text{and} \qquad \alpha \cdot \beta = \lim_{n \to \infty} \alpha_n \cdot \beta_n.$$

For *p*-adic rationals $\alpha, \beta \in \mathbb{Q}_p$, we can bootstrap the definition of *p*-adic integer
addition by factoring out appropriate powers of *p*: If $\alpha = p^k \alpha'$ and $\beta = p^\ell \beta'$
with $\alpha', \beta' \in \mathbb{Z}_p$ and without loss of generality $k \geq \ell$, then

$$\alpha\beta = p^{k+\ell}(\alpha'\beta') \qquad \text{and} \qquad \alpha + \beta = p^k \alpha' + p^\ell \beta' = p^\ell(p^{k-\ell}\alpha' + \beta'),$$

where both concluding terms in parentheses are operations in $\mathbb{Z}_p$ and hence
defined above. ◄

Finally, having established addition and multiplication as operations that do, in
fact, totally exist, it is worth recording for future use their interaction with the valuation function.

Lemma 8.3.7

For $\alpha, \beta \in \mathbb{Q}_p$, and $n \in \mathbb{N}$, we have

   (i) $v_p(\alpha\beta) = v_p(\alpha) + v_p(\beta)$.
   (ii) $v_p(\alpha^n) = n v_p(\alpha)$.
   (iii) $v_p(\alpha + \beta) \geq \min\{v_p(\alpha), v_p(\beta)\}$.

                                                                                    ◄

Proof First assume that $\alpha, \beta \neq 0$. Write $\alpha = p^k \alpha'$ and $b = p^\ell \beta'$ as in (8.2). Then
$\alpha'\beta' \bmod p \neq 0$ since both $\alpha', \beta' \not\equiv 0 \bmod p$, and so

$$v_p(\alpha\beta) = v_p(p^{k+l}\alpha'\beta') = k + \ell = v_p(a) + v_p(b).$$

The second bullet follows from the first by induction. For the final part, let's assume
without loss of generality that $k \geq \ell$. Then $\ell = \min\{v_p(\alpha), v_p(\beta)\}$ and

$$v_p(\alpha + \beta) = v_p(p^k \alpha' + p^\ell \beta') = v_p(p^\ell(p^{k-\ell}\alpha' + \beta')) \geq \ell.$$

When $\alpha = 0$, we can see that the convention from Definition 8.2.6 that $v_p(0) = \infty$
was hand-picked to make these three properties hold. We leave the details for Exercise
8.30.   □

We wish to assure the reader that the p-adic absolute value absolutely has value beyond making things look more difficult than they seem. To the contrary, this notion of absolute value is one of the most intriguing aspects of the p-adic world, as it provides the segue to analogies with numerous other branches of mathematics. Absolute values beget distances, distances beget lengths, and lengths beget—dunh dunh dunh!—*geometry*.

Exploration P

p-adic Geometry ◄

The first few questions are intended to explore the behavior of this new (and counterintuitive) measurement of size. Train your mind to reinterpret all measurement words as dependent on the absolute value in question: α being "5-adically small" means that the real number $|\alpha|_5$ is small in the usual sense, and α and β being "3-adically close" means that $|\alpha - \beta|_3$ is small in the usual sense.

P.1 What elements of $\mathbb{Q}_5$ are 5-adically big? Small?

P.2 How far apart can two elements of $\mathbb{Z}_5$ be? How close? What about $\mathbb{Q}_5$?

P.3 Find examples of integers that are 5-adically close but 7-adically far apart, and vice versa.

P.4 Find pairs of numbers that are p-adically close for lots of p. All p?

P.5 Pick a prime p, and then choose three elements a, b, c in $\mathbb{Z}_p$ (take natural numbers to begin with) and compute $|a - b|_p$, $|a - c|_p$, and $|b - c|_p$. Repeat for several choices of a, b, and c, recording your results.

a	b	c	$|a-b|_p$	$|a-c|_p$	$|b-c|_p$

P.6 In $\mathbb{R}^2$, one makes triangles by a similar process: a triangle in $\mathbb{R}^2$ is formed from three points A, B, and C, and its side lengths are given by $|A - B|$, $|A - C|$, and $|B - C|$. The triangle inequality says that any of these three lengths is at most the sum of the other two. Is there something analogous to the triangle inequality in $\mathbb{Z}_p$? What else is remarkable about p-adic "triangles"?

8.4 Which numbers are p-adic?

It is clear that every *natural* number has an interpretation as an element of $\mathbb{Z}_p$ since base-p expansions are just finite p-adic expansions. Further, the previous sections show how to interpret -1 as an element of $\mathbb{Z}_p$:

$$-1 = (p-1) + (p-1)p + (p-1)p^2 + (p-1)p^3 + \cdots,$$

and the process is not much different for any negative integer. What about rational numbers? We have already discovered (Exploration O.4) the identity

$$\frac{1}{6} = 1 + 4p + p^2 + 4p^3 + p^4 + \cdots \in \mathbb{Z}_5,$$

but the source of this was a bit of a one-off idea coming from Taylor series. The most naive of techniques for coming up for other such expansions turns out to be one of the most effective. We employ the *method of undetermined coefficients*, a fancy name for the mathematically pervasive technique of assuming that a solution exists and then determining what its coefficients must have been, one at a time.

▶ **Example 8.4.1** Decide if $\frac{3}{8} \in \mathbb{Z}_7$ and if so find the 7-adic expansion of $\frac{3}{8}$.

Note that by the statement $\frac{3}{8} \in \mathbb{Z}_7$ we mean the existence of an element $\alpha \in \mathbb{Z}_7$ such that $8\alpha = 3$.

Solution Since $8 = 1 + p \in \mathbb{Z}_7$, we are attempting to solve for the sequence of a_i's in the equation

$$(1 + p)(a_0 + a_1 p + a_2 p^2 + a_3 p^3 + \cdots) = 3.$$

We distribute the product on the left, equate coefficients, and then solve for each a_i recursively to find them all. Equivalently, we work modulo p^n for increasingly large n. Equating

$$a_0 + (a_0 + a_1)p + (a_1 + a_2)p^2 + (a_2 + a_3)p^3 + \cdots = 3 + 0p + 0p^2 + 0p^3 + \cdots$$

shows that we must have $a_0 = 3$ for this equality to hold (the mod-p condition). With this value in place, we have the equality

$$3 + (3 + a_1)p + (a_1 + a_2)p^3 + (a_2 + a_3)p^3 + \cdots = 3 + 0p + 0p^2 + 0p^3 + \cdots$$

For this to hold mod p^2, we need for $3 + (3 + a_1)p \equiv 3 + 0p$ mod p^2, so $(3 + a_1)p \equiv 0$ mod p^2, so $(3 + a_1) \equiv 0$ mod p. This forces $a_1 = 4$. Substituting this choice in we get

$$3 + (3 + 4)p + (4 + a_2)p^2 + (a_2 + a_3)p^3 + \cdots =$$
$$3 + (1 + 4 + a_2)p^2 + (a_2 + a_3)p^3 + \cdots = 3 + 0p + 0p^2 + 0p^3 + \cdots$$

Reasoning identically mod p^3 forces $a_2 = 2$ (to give the p^2 a coefficient of zero), and then reasoning mod p^4 gives the coefficient of p^3 as $(1+2+a_3)$, forcing $a_3 = 4$, etc. With some pattern-deduction (more formally, an induction proof), we conclude that

$$\frac{3}{8} = 3 + 4p + 2p^2 + 4p^3 + 2p^4 + 4p^5 + 2p^6 + 4p^7 + \cdots \in \mathbb{Z}_7,$$

with eventually periodic coefficients alternating between 2 and 4. $\qquad\square$

This process generalizes nicely to finding p-adic representations of all sorts of rational numbers. It is worth stressing that the somewhat involved notation of the following proof is simply the formalization via induction of the pattern recognition in the example just completed.

Lemma 8.4.2

The units $\mathbb{Z}_p^\times$ of $\mathbb{Z}_p$ are precisely those elements $\alpha \in \mathbb{Z}_p$ with $v_p(\alpha) = 0$. That is, for $\alpha \in \mathbb{Z}_p$, we have the equivalent conditions

$$\alpha \text{ is a unit} \iff v_p(\alpha) = 0 \iff |\alpha|_p = 1 \iff \alpha \bmod p \neq 0. \qquad \blacktriangleleft$$

Proof The last three conditions are easily seen to be equivalent, so we focus our attention on the first and last conditions, that α is a unit if and only if $\alpha \bmod p \neq 0$. If α is a unit, then $\alpha\beta = 1$ for some $\beta \in \mathbb{Z}_p$, and so $\alpha\beta \bmod p = 1$, which would be impossible if $\alpha \bmod p = 0$. For the reverse direction, suppose $\alpha \bmod p = a_0 \neq 0$. Given the coefficients a_i of α's p-adic expansion, we wish to solve, for unknown coefficients b_i $(0 \le b_i \le p - 1)$, the equation

$$(a_0 + a_1 p + a_2 p^2 + a_3 p^3 + \cdots)(b_0 + b_1 p + b_2 p^2 + b_3 p^3 + \cdots) = 1 + 0p + 0p^2 + 0p^3 + \cdots.$$

We proceed inductively. Looking at the constant term, we need $a_0 b_0 \equiv 1 \bmod p$, so choose b_0 to be the integer from 0 to $p - 1$ that is congruent to $a_0^{-1} \bmod p$ (which exists since $a_0 \not\equiv 0 \bmod p$). Now suppose that we have computed $b_0, \ldots, b_n$ so that

$$(a_0 + a_1 p + \cdots + a_n p^n)(b_0 + b_1 + \cdots + b_n p^n) \equiv 1 \bmod p^{n+1}.$$

Consequently, this product on the left can be written as $c_{n+1} p^{n+1} + 1$ for some $c_{n+1} \in \mathbb{Z}$. We want to find the next term in the expansion, i.e., work modulo p^{n+2} and solve

$$\begin{aligned}
1 &\equiv (a_0 + a_1 p + \cdots + a_n p^n + a_{n+1} p^{n+1})(b_0 + b_1 p + \cdots + b_n p^n + b_{n+1} p^{n+1}) \\
&\equiv (a_0 + \cdots + a_n p^n)(b_0 + \cdots + b_n p^n) + a_{n+1} b_0 p^{n+1} + a_0 b_{n+1} p^{n+1} \\
&\equiv 1 + p^{n+1}(c_{n+1} + a_{n+1} b_0 + a_0 b_{n+1}) \pmod{p^{n+2}}.
\end{aligned}$$

The last condition is equivalent to $c_{n+1} + a_{n+1} b_0 + a_0 b_{n+1} \equiv 0 \bmod p$, so we can take b_{n+1} to be the unique integer from 0 to $p - 1$ satisfying

$$b_{n+1} \equiv a_0^{-1}(-c_{n+1} - a_{n+1} b_0) \bmod p.$$

Having constructed its sequence of coefficients (b_n) by induction, we now have the p-adic number β such that $\alpha\beta = 1$.   □

▶ **Remark 8.4.3** This again mirrors the situation in calculus. If f is a rational function with $f(a) \neq 0$ (that is, the constant term of f's Taylor expansion at a is not zero), then $\frac{1}{f}$ is also a rational function with a Taylor expansion at a. But if $f(a) = 0$, then $\frac{1}{f}$ is not defined at a and so does not have a Taylor expansion there, and we must instead use a Laurent expansion.

Corollary 8.4.4

Every non-zero $\alpha \in \mathbb{Q}_p$ can be written uniquely as a power of p times a p-adic unit, i.e., in the form
$$\alpha = p^{v_p(\alpha)}u \quad (u \in \mathbb{Z}_p^\times) \qquad ◀$$

Proof This is direct from Definition 8.2.6 and the lemma.   □

Corollary 8.4.5

The ring $\mathbb{Z}_p$ is a unique factorization domain with p its unique irreducible element (up to associates). Further, $\mathbb{Q}_p$ is a field, and equals the set of fractions of p-adic integers. ◀

In algebraic jargon, the last claim is that $\mathbb{Q}_p$ is the *field of fractions* of $\mathbb{Z}_p$.

Proof First, since we have not made the point explicitly yet, note that $\mathbb{Z}_p$ and $\mathbb{Q}_p$ are both integral domains (neither the product of powers of p nor of two units can be zero) and $\mathbb{Q}_p$ is further a field ($\alpha^{-1} = p^{-v_p(\alpha)}u^{-1}$). The irreducibility of p in $\mathbb{Z}_p$ follows from a valuation calculation: if $p = \alpha\beta$, then
$$1 = v_p(p) = v_p(\alpha\beta) = v_p(\alpha) + v_p(\beta),$$
and since both $v_p(\alpha)$ and $v_p(\beta)$ are non-negative integers, one of them must be 0, making that factor a unit by Lemma 8.4.2. A similar argument shows p is also prime. This also gives unique factorization: if $\alpha \in \mathbb{Z}_p$, then since $\alpha = p^{v_p(\alpha)}u$ for some unit u, α can be written as a product of irreducibles in $\mathbb{Z}_p$, and uniquely so since the valuation $v_p(\alpha)$ is well-defined. Finally,
$$\mathbb{Q}_p = \left\{ \frac{\alpha}{\beta} : \alpha, \beta \in \mathbb{Z}_p, \beta \neq 0 \right\}$$
by mutual inclusion (Exercise 8.36).   □

▶ **Remark 8.4.6** Instead of directly invoking the factorization to show that $\mathbb{Z}_p$ is a UFD, we could invoke once more *The Path*: one checks (Exercise 8.31) that the

p-adic valuation on $\mathbb{Z}_p$ defines a Euclidean norm in the sense of Definition 6.4.1, from which unique factorization follows by Theorem 6.5.3.

Now we return to $\mathbb{Q}$. By prime factorization in $\mathbb{Z}$, every rational number $r \in \mathbb{Q}$ can be written in the form

$$r = p^{v_p(r)}\frac{a}{b},$$

where $\gcd(a, p) = \gcd(b, p) = 1$. Since a and b are integers relatively prime to p, by Lemma 8.4.2, they are units in $\mathbb{Z}_p$, and thus so is the product ab^{-1}. That is, every rational number can be written in the form

$$r = p^{v_p(r)}u,$$

where u is a unit of $\mathbb{Z}_p$. Since $\mathbb{Z}_p$ is characterized as the elements of $\mathbb{Q}_p$ with $v_p(r) \geq 0$, rational numbers with $v_p(r) \geq 0$ have a p-adic expansion that actually lives in $\mathbb{Z}_p$.

Corollary 8.4.7

For any prime p, "$\mathbb{Q} \subseteq \mathbb{Q}_p$." ◀

The scare quotes on "$\mathbb{Q} \subseteq \mathbb{Q}_p$" are there to emphasize that what we *really* mean is that every rational number admits a p-adic expansion and so can be viewed as an element of $\mathbb{Q}_p$. We will maintain this slight abuse of notation, writing, e.g., $\frac{3}{8} \in \mathbb{Q}_7$ instead of any of the more pedantic options. (Note one does not typically stress about writing $\mathbb{Q} \subseteq \mathbb{R}$ despite a similar setup.)

It may not be clear at this point that $\mathbb{Q}$ and each $\mathbb{Q}_p$ aren't essentially the same ring (or to reuse the fancy word from Remark 4.7.7, *isomorphic*). After all, for generic elements $r \in \mathbb{Q}$ and $\alpha \in \mathbb{Q}_p$ we can compare the general forms

$$r = p^{v_p(r)}\frac{a}{b} \in \mathbb{Q} \qquad \text{versus} \qquad \alpha = p^{v_p(\alpha)}u \in \mathbb{Q}_p$$

with $u \in \mathbb{Z}_p^\times$. Having learned that $\frac{a}{b} \in \mathbb{Z}_p^\times$ as well, it is plausible to believe that we could reverse the construction, and to each $\alpha \in \mathbb{Q}_p$ determine the rational number that gives rise to it. As it turns out, this impression is quite false, as the vast majority of elements of $\mathbb{Q}_p$ are not the p-adic expansion of any rational number. We sketch two arguments, both of which can be seen as direct analogues of the same arguments for $\mathbb{R}$. First is by explicit description of the elements: if we view a real number as an infinite decimal expansion, then the rational numbers correspond to those decimal expansions that are eventually periodic. Though not intuitively clear, the set of eventually periodic expansions forms a rather thin subset of $\mathbb{R}$. By a near-verbatim argument replacing decimal expansions with p-adic ones, the p-adic expansions that represent rational numbers are those with eventually periodic coefficients. A second argument proceeds set-theoretically: The same diagonalization argument that shows $\mathbb{R}$ uncountable also shows that the collection of p-adic expansions $\mathbb{Z}_p$, $\mathbb{Q}_p$, and even $\mathbb{Z}_p^\times$, are uncountable.

The next reasonable-but-false thing to believe is that all of the $\mathbb{Q}_p$ are essentially the same. Of course, all the $\mathbb{Q}_p$'s are structurally very similar, and each contains all of the rational numbers, so it is somewhere in those uncountably many "irrational" elements of $\mathbb{Q}_p$ that we will have to search for differences. The next section will cover our principal tool for deducing the explicit presence of such elements.

8.5   Hensel's Lemma

A natural starting place for exploring the massive gap between the countably many rational elements of $\mathbb{Q}$ and the uncountably many elements of $\mathbb{Q}_p$ is to explore questions of "algebraic" elements of $\mathbb{Q}_p$. Does $\mathbb{Q}_p$ contains roots of integer-coefficient polynomials that are not themselves rational? Exploring this question will also resolve the concluding question of the previous section, that despite their structural similarities, the various $\mathbb{Q}_p$ are genuinely quite different. The answer is generically yes, but the more specific answer is that *which* algebraic numbers are present depends heavily on the choice of p! We only have to go as far as quadratic irrationals to see our first example:

Claim 8.5.1

There is a square root of 7 in $\mathbb{Q}_3$ but not in $\mathbb{Q}_5$ or $\mathbb{Q}_7$. ◀

▶ **Remark 8.5.2** As with real numbers, we might encode the first claim as the expression $\sqrt{7} \in \mathbb{Q}_3$, though in this case only while grimacing slightly internally. In $\mathbb{R}$, the symbol $\sqrt{7}$ references "the unique positive real number whose square is 7." In $\mathbb{Q}_3$, without an obvious notion of positiveness, the $\sqrt{7}$ then somewhat ambiguously refers to one of the two such elements. That will not stop us from occasionally employing this abuse of notation for visual or dramatic effect. Just don't tell anyone.

Solution Consider the valuation of a purported solution of $\alpha^2 = 7$ in $\mathbb{Q}_p$ for some prime p. Taking the p-adic valuation of both sides gives

$$v_p(\alpha) = \frac{1}{2}\,v_p(\alpha^2) = \frac{1}{2}v_p(7) = \begin{cases} 0 & \text{if } p \neq 7 \\ \frac{1}{2} & \text{if } p = 7. \end{cases}$$

This is already a contradiction if $p = 7$, so one of our three claims is disposed of immediately. For $p \neq 7$, we learn that $v_p(\alpha) = 0$, so can write

$$\alpha = a_0 + a_1 p + a_2 p^2 + a_3 p^3 + \cdots.$$

Determining the existence of such an α for a given p is equivalent to solving (or demonstrating the inability to solve) via the method of undetermined coefficients the equation

$$(a_0 + a_1 p + a_2 p^2 + a_3 p^3 + \cdots)^2 = 7.$$

When $p = 5$, we write $7 = 2 + p$ and so need to solve the system of coefficients

$$a_0^2 + (2a_0 a_1)p + (a_1^2 + 2a_0 a_2)p^2 + \cdots = 2 + 1p + 0p^2 + 0p^3 + \cdots$$

Beginning with the constant coefficient, we need $a_0^2 \equiv 2 \bmod 5$, and from explorations with squares mod p, we see that there is no such value for a_0 (the Legendre symbol $\left(\frac{2}{5}\right) = -1$). Thus there is no square root of 7 in $\mathbb{Q}_5$!

When $p = 3$, we write $7 = 1 + 2p$ and find that this time it appears we can solve the system. Setting

$$a_0^2 + (2a_0 a_1)p + (a_1^2 + 2a_0 a_2)p^2 + \cdots = 1 + 2p + 0p^2 + 0p^3 + \cdots,$$

we can choose $a_0 = 1$, which then after mild arithmetic forces $a_1 = 1$, then $a_2 = 1$, then $a_3 = 0$, then $a_4 = 2$, etc. We can verify our work

$$\begin{aligned}
\alpha^2 &= (1 + p + p^2 + 2p^4 + \cdots)^2 \\
&= 1 + 2p + 3p^2 + 2p^3 + 5p^4 + 4p^5 + \cdots \\
&= 1 + 2p + 0p^2 + 3p^3 + 5p^4 + 4p^5 + \cdots \\
&= 1 + 2p + 0p^2 + 0p^3 + 6p^4 + 4p^5 + \cdots \\
&= \quad \vdots \\
&= 1 + 2p \\
&= 7,
\end{aligned}$$

as long as we're willing to believe that we can indefinitely continue the process of finding the next coefficient to extend the streak of 0 coefficients. □

Thus the elephant in the example is, yet again, the tantalizingly oblique "etc" or "$\cdots$" that we slide in so cavalierly as we go. Having established the first few coefficients of the p-adic expansion of an ostensible $\sqrt{7} \in \mathbb{Q}_3$, can we be sure that the process can be continued indefinitely? After all, the search for $\sqrt{7} \in \mathbb{Q}_5$ fell flat on its face at the very first coefficient!

The answer is a resounding yes, both for finding square roots and the vastly more general context of solving polynomial equations (recall that finding $\sqrt{7} \in \mathbb{Q}_p$ is more precisely encoded as finding a root of the polynomial $x^2 - 7$ in $\mathbb{Q}_p$). Roughly, the mantra provided to us by the upcoming Hensel's Lemma, a generalization of the argument for producing multiplicative inverses, is that if you can *start* the method of undetermined coefficients (find the a_0 term), then you can finish it (find all a_i).

Theorem 8.5.3 (Hensel's Lemma)

Let $f \in \mathbb{Z}_p[x]$, and suppose we have an integer a_0, with $0 \le a_0 \le p - 1$, such that $f(a_0) \equiv 0 \pmod{p}$ and $f'(a_0) \not\equiv 0 \pmod{p}$. Then there is a unique $\alpha \in \mathbb{Z}_p$ such that $f(\alpha) = 0$ and $\alpha \bmod p = a_0$.

The punchline of this remarkable result is that once we satisfy a mild condition on the *derivative*[3] of f, having a solution in $\mathbb{Z}_p$ is *equivalent* to having one in $\mathbb{Z}/(p)$. So if nothing else, Hensel's Lemma is remarkable for replacing an uncountably infinite solution space with a finite solution space which, at least in principal, could be searched by systematically checking every element.

Proof Suppose $f \in \mathbb{Z}_p[x]$ and $a_0 \in \mathbb{Z}$ satisfy the conditions of the theorem. We aim to find $\alpha \in \mathbb{Z}_p$ such that $f(\alpha) = 0$ and $\alpha \bmod p = a_0$. Write

$$\alpha = a_0 + a_1 p + a_2 p^2 + \cdots$$

and let $\alpha_n = a_0 + \cdots + a_{n-1} p^{n-1}$ be its n-th truncation, so $\alpha_1 = a_0$ and $\alpha_{n+1} = \alpha_n + a_n p^n$. We show that we can inductively solve for a_n, with the hypothesis $f(\alpha_1) = f(a_0) \equiv 0 \bmod p^1$ being the base case. Now suppose that we have found the unique α_n so that $f(\alpha_n) \equiv 0 \bmod p^n$. We must find a_n so that $\alpha_{n+1} = \alpha_n + a_n p^n$ satisfies $f(\alpha_{n+1}) \equiv 0 \bmod p^{n+1}$.

Write $f(x) = \sum \beta_k x^k$, so $f'(x) = \sum k \beta_k x^{k-1}$. We compute

$$
\begin{aligned}
f(\alpha_{n+1}) &= f(\alpha_n + a_n p^n) \\
&= \sum \beta_k (\alpha_n + a_n p^n)^k \\
&\equiv \sum \beta_k (\alpha_n^k + k p^n a_n \alpha_n^{k-1}) \pmod{p^{n+1}},
\end{aligned}
$$

where the last step employs the binomial theorem—in the expansion of an arbitrary $(a + b p^n)^k$, all but two terms contain a factor of p^{2n} and so vanish mod p^{n+1} for $n \ge 1$. Continuing, we note

$$
\begin{aligned}
\sum \beta_k (\alpha_n^k + k p^n a_n \alpha_n^{k-1}) &= \sum \beta_k \alpha_n^k + a_n p^n \sum k \beta_k \alpha_n^{k-1} \\
&= f(\alpha_n) + a_n p^n f'(\alpha_n).
\end{aligned}
$$

For this expression above to reduce to 0 modulo p^{n+1} we need

$$f(\alpha_n) + a_n p^n f'(\alpha_n) \equiv 0 \bmod p^{n+1}.$$

---

[3] Calculus has been pervasive in this chapter as an analogy, but in this instance appears as the literal derivative of a polynomial. We the authors are wowed by this and would understand if you needed to take a break to marvel – we'll still be here when you get back.

Dividing through by p^n (and recalling $f(\alpha_n) \equiv 0 \pmod{p^n}$ by the induction hypothesis), we can equivalently write this as

$$\frac{f(\alpha_n)}{p^n} + a_n f'(\alpha_n) \equiv 0 \bmod p$$

Since $f'(\alpha_n) \equiv f'(a_0) \not\equiv 0 \bmod p$, we can divide and then solve for a_n: Let a_n be the unique integer $0 \le a_n \le p - 1$ such that

$$a_n \equiv -\frac{1}{f'(\alpha_n)} \frac{f(\alpha_n)}{p^n} \bmod p,$$

completing the construction and demonstrating uniqueness. □

▶ **Remark 8.5.4 Lifting** is a convenient verb for the act of taking an element of $\mathbb{Z}/(p)$ and finding an appropriate element of $\mathbb{Z}_p$ (or sometimes $\mathbb{Z}$) that reduces to it mod p. In this language, Hensel's Lemma provides conditions under which we can *lift* a $\mathbb{Z}/(p)$-root of a polynomial f to a $\mathbb{Z}_p$-root.

▶ **Remark 8.5.5** Tracing through the proof, we find that the inductive procedure for lifting a $\mathbb{Z}/(p)$-root to a $\mathbb{Z}_p$-root is to first find α_1 by working mod p, and then to recursively use the formula

$$\alpha_{n+1} \equiv \alpha_n - \frac{f(\alpha_n)}{f'(\alpha_n)} \pmod{p^{n+1}}.$$

Especially in light of our motivating analogy with polynomials, it is striking that this is exactly the same formula as Newton's method for finding roots of a differentiable function! Newton's method turns out not to always work, but Hensel's Lemma does.[4]

The question of which p-adic numbers have p-adic square roots is almost completely resolved by the lemma, and the resolution happily rests on something we are quite familiar with, the completely analogous question about modular square roots. With quadratic reciprocity at our disposal, the problem is essentially solved. The bulk of the work is figuring out square roots of units, so we begin with this case.

Corollary 8.5.6

Suppose p is an odd prime, and let $\beta \in \mathbb{Z}_p^\times$. Then β has a square root in $\mathbb{Z}_p$ if and only if $\beta \bmod p$ has a square root in $\mathbb{Z}/(p)$. ◀

Proof One direction is clear by looking at constant terms. If $\alpha^2 = \beta$ in $\mathbb{Z}_p$, then their leading coefficients satisfy $a_0^2 \equiv b_0 \pmod p$. The converse is Hensel's Lemma

---

[4] p-adic analysis: 1, real analysis: 0. Take that, calculus!

applied to the polynomial $f(x) = x^2 - \beta$: assume that $b_0 \equiv a_0^2 \pmod{p}$ for some integer $0 \le a_0 \le p - 1$. Then since $b_0 \not\equiv 0 \bmod p$ (since β is a unit), it is also true that $a_0 \not\equiv 0 \bmod p$, and so $f'(a_0) = 2a_0 \not\equiv 0 \bmod p$ (recall p is odd). By Hensel's Lemma, there exists a unique $\alpha \in \mathbb{Z}_p$ such that $f(\alpha) = 0$ and $\alpha \equiv a_0 \bmod p$. Such an α is a p-adic square root of β.   □

When $p = 2$, the situation is, again, slightly more complicated. *But why?* Why must 2 *always* be *such* a nuisance? Loosely, the culprit is that the dual usages of 2 in this setting—the explicit 2 in working with the 2-adics and the implicit 2 when talking about *square* roots—interfere with each other. For example, we can see the coefficient of 2 and the placeholder $p = 2$ interact differently than for p odd in any squaring computation:

$$(1 + a_1 p)^2 = \begin{cases} 1 + 2a_1 p + a_1^2 p^2 & p \ne 2 \\ 1 + (a_1 + a_1^2) p^2 & p = 2 \end{cases}$$

This shows, for example, that every odd square must have a zero coefficient in front of 2^1, and in fact 2^2 as well, since $a_1 + a_1^2$ is always even. This already shows the impossibility of a direct analogue of Corollary 8.5.6. Taking $\beta = 5 = 1 + 2^2 \in \mathbb{Z}_2$, for example, we see that $\beta \bmod 2$ is a square in $\mathbb{Z}/(2)$, but β cannot be a square in $\mathbb{Z}_2$. We can also see where the hypotheses of Hensel's Lemma fail: if $f(x) = x^2 - \beta \in \mathbb{Z}_2[x]$, then for *every* $\alpha \in \mathbb{Z}_p$ we have $f'(\alpha) = 2\alpha \equiv 0 \bmod 2$, violating the hypotheses of Hensel's Lemma. Sheesh! That pesky 2! Stronger forms of Hensel's exist for this purpose, but we can handle the case of square roots directly.

Lemma 8.5.7

Let $\beta \in \mathbb{Z}_2^\times$. Then β has a square root in $\mathbb{Z}_2$ iff $\beta \equiv 1 \bmod 8$. ◄

Proof First, suppose $\alpha^2 = \beta$ in $\mathbb{Z}_2$. Then since $\beta \in \mathbb{Z}_2^\times$, we know $\beta \bmod 2 = 1$, and so $\alpha \bmod 2 = 1$ as well. This forces $\alpha \bmod 8 \in \{1, 3, 5, 7\}$, and for any of these choices we get $\beta \equiv \alpha^2 \equiv 1 \bmod 8$.

Conversely, suppose $\beta \equiv 1 \bmod 8$. We proceed as in the proof of Hensel's Lemma, inductively computing more and more coefficients of the 2-adic expansion of a square root. We will show that we can construct the sequence of coefficients $a_n \in \{0, 1\}$ so that $\alpha_n = a_0 + \cdots + a_{n-1} p^{n-1}$ satisfies $\alpha_n^2 \equiv \beta \bmod 2^{n+1}$ for[5] all $n \ge 1$. The base case of $n = 1$ is handled by $\alpha_1 = 1$, since then both α^2 and β are congruent to 1 mod 2^{1+1}. Suppose for the sake of induction that we have found α_n so that $\alpha_n^2 \equiv \beta \bmod 2^{n+1}$ and so $\alpha_n^2 - \beta = c_n 2^{n+1}$ for some $c_n \in \mathbb{Z}_2$. We wish to choose a_n so that $\alpha_{n+1} = \alpha_n + a_n 2^n$ gives the next correct coefficient of β when squared:

---

[5] The exponent of $n + 1$ here means that we check that the new a_n makes the next *two* coefficients (the n-th and $(n + 1)$-st) of α_n^2 work out correctly. This, as opposed to the n in the proof of Hensel's Lemma, is essentially the solution to the problem that $p = 2$ causes.

$\alpha_{n+1}^2 \equiv \beta \bmod 2^{n+2}$. We compute

$$\begin{aligned} \alpha_{n+1}^2 - \beta &= \alpha_n^2 + 2\alpha_n a_n 2^n + a_n^2 2^{2n} - \beta \\ &\equiv \alpha_n^2 - \beta + a_n \alpha_n 2^{n+1} \\ &\equiv 2^{n+1}(c_n + a_n \alpha_n) \pmod{2^{n+2}} \end{aligned}$$

For this to be zero, it suffices for the parenthetical to be even, which is equivalent to simply choosing $a_n = c_n \bmod 2$ since $\alpha_n \equiv \alpha_1 \equiv 1 \bmod 2$. $\qquad\square$

▶ **Example 8.5.8** Find all p for which $\sqrt{7} \in \mathbb{Z}_p$.

Solution Since $7 \not\equiv 1 \bmod 8$, there is no $\sqrt{7} \in \mathbb{Z}_2$. For odd p, $\sqrt{7} \in \mathbb{Z}_p$ if and only if $\left(\frac{7}{p}\right) = 1$, a problem which is solvable using quadratic reciprocity. Namely, we found in Example 7.5.7 that this occurs if and only if $p \equiv \pm 1, \pm 3$, or $\pm 9 \pmod{28}$. $\qquad\square$

Broadening our scope of square root inquiry from $\mathbb{Z}_p^\times$ to $\mathbb{Z}_p$, and even $\mathbb{Q}_p$, does not increase the difficulty, as the difference between these sets is just the allowable p-adic valuations, and the meat of the square criterion comes instead from the unit part.

Corollary 8.5.9 (Squares in $\mathbb{Q}_p$)

Suppose $\alpha \in \mathbb{Q}_p$ is written in the form

$$\alpha = p^{v_p(\alpha)} u \qquad (u \in \mathbb{Z}_p^\times).$$

- For p odd, α is a square in $\mathbb{Q}_p$ if and only if $v_p(\alpha)$ is even and $\left(\frac{u \bmod p}{p}\right) = 1$.
- For $p = 2$, α is a square in $\mathbb{Q}_p$ if and only if $v_p(\alpha)$ is even and $u \equiv 1 \bmod 8$.

◀

Proof If $v_p(\alpha)$ is odd, α has no square roots by valuation considerations. If $v_p(\alpha)$ is even, then $p^{v_p(\alpha)}$ has the obvious square root $p^{v_p(\alpha)/2}$ so α has a square root if and only if u does. $\qquad\square$

Generally speaking, Hensel's Lemma reduces polynomial root-finding in $\mathbb{Z}_p$ to that of $\mathbb{Z}/(p)$, an uncountafold reduction in our search space. Quadratic reciprocity picks up the torch for quadratic such equations, allowing us to quickly solve equations for all p. While we have not discussed versions of quadratic reciprocity for higher degree equations (which is not to say they don't exist!), we still have a slew of modular arithmetic tools that can be brought to bear.

▶ **Example 8.5.10** Decide which elements of $\mathbb{Q}_{11}$ are cubes.

Solution Let $p = 11$ and take $\alpha = p^{v_p(\alpha)}u \in \mathbb{Q}_p$. Then as in the proof of the corollary, it is clearly necessary that $3 \mid v_p(\alpha)$, and if $3 \mid v_p(\alpha)$ then it is sufficient to check that u is a cube. By Hensel's Lemma, this will happen as long as u mod 11 is a cube, but since $11 \equiv 2 \pmod 3$, *every* element of $\mathbb{Z}/(11)$ is a cube (Corollary 4.6.24). Thus for α to be a cube in $\mathbb{Q}_{11}$ it is both necessary and sufficient for $v_p(\alpha)$ to be a multiple of 3. □

▶ **Remark 8.5.11** The argument can be repeated nearly verbatim for any prime $p \equiv 2$ mod 3 in place of 11. If $p \equiv 1 \pmod 3$, then there are three cube roots of unity mod p and only a third of the elements of $\mathbb{Z}/(p)$ will have cube roots. We would need a "cubic reciprocity law" to streamline results in the same way as we did in Example 8.5.8. Such laws exist but are beyond the scope of this book.

As a final example of copying rational algebra to p-adic algebra, in the last chapter we fawned over another family of algebraic numbers, the roots of unity. Suppose we were to ask the analogous p-adic question—can we find roots of unity in $\mathbb{Z}_p$? Since such elements are precisely the roots of the polynomial $x^n - 1$, our approach is natural.

Pickle 8.5.12

Which roots of unity are in $\mathbb{Z}_p$? ◀

Solution Suppose p is odd and α is a primitive n-th root of unity. Then α reduces mod p to an element of order n in $\mathbb{Z}/(p)$, and Lagrange's Theorem tells us the order of any non-zero $a \in \mathbb{Z}/(p)$ divides $|\mathbb{Z}/(p)^{\times}| = p - 1$. So primitive n-th roots of unity can only exist for n dividing $p - 1$, and since an n-th root of unity is also a $(p-1)$-st root of unity when $n \mid p - 1$, it suffices to restrict our attention to those. But *every* non-zero element $a \in \mathbb{Z}/(p)^{\times}$ satisfies $a^{p-1} = 1$, and the derivative condition

$$f'(a) = (p-1)a^{p-2} \not\equiv 0 \pmod p$$

is also clearly satisfied, so all $(p - 1)$ of the $(p - 1)$-st roots of unity in $\mathbb{Z}/(p)^{\times}$ lift to roots of unity in $\mathbb{Z}_p$. We conclude that $\mathbb{Z}_p$ contains the primitive n-th roots of unity if and only if $n \mid p - 1$. For $p = 2$, it's clear that the 2nd roots of unity ± 1 are in $\mathbb{Z}_2$, and Exercise 8.27 will show that there are no others. □

To close the motivating line of inquiry, the various fields $\mathbb{Q}_p$, despite being similar in description and structure, differ in arithmetically interesting ways. Half of the $\mathbb{Q}_p$'s have a square root of 7 and half don't, half have a square root of 11 and half don't, etc., and no two $\mathbb{Q}_p$'s contain precisely the same set of square roots (Exercise 8.35).

And, at the risk of waxing too metaphysical, it is this diversity of $\mathbb{Q}_p$'s that provides their greatest strength. Much like the particle physicist's process of studying new particles by smashing all known particles into them and seeing what happens, we will revisit classical number-theoretic questions and see what happens if we smash all known primes into them, *localizing* our study on that prime in particular. By seeing how the problem reacts to each prime, we gain insight into the overall (*global*) solution to the problem.

8.6 The Local-Global Philosophy and the Infinite Prime

Chapter 2 introduced us to the idea that we can view a polynomial equation with integer coefficients, like

$$x^2 + y^2 = z^2, \tag{$*$}$$

as an interesting mathematical object of study independent of the ring in which x, y, z reside. If C denotes such an equation, then we can ask about the set $C(R)$ of solutions in any of the rings R we've encountered[6]. Diophantine analysis often focuses on the set $C(\mathbb{Z})$, but we have also had cause to explore sets like $C(\mathbb{Z}/(n))$, $C(\mathbb{Q})$, $C(\mathbb{R})$, $C(\mathbb{Z}[i])$, $C(\mathbb{C})$, $C(\mathbb{Z}/(p^k))$, $C(\mathbb{Z}_p)$, $C(\mathbb{Q}_p)$, etc. It seems fair to say that the principal benefit to the algebraic approach to Diophantine equations is the observation that these sets are often intimately related. For example, if (x, y, z) is a point in $C(\mathbb{Z})$, then (by Lemma 4.3.3), we have the point $([x], [y], [z]) \in C(\mathbb{Z}/(n))$.

We have made potent use of these relationships, even when we did not call them out by name. For one, a consequence of the relationship above is that if $C(\mathbb{Z})$ is non-empty, then $C(\mathbb{Z}/(n))$ must be non-empty as well. The contrapositive of this implication allowed us to deduce, for example, that there are no integer points on the curve

$$y^2 = 5x^3 + 7$$

as there are no solutions in $\mathbb{Z}/(5)$ (the equation reduces mod 5 to $y^2 = 2$, but $\left(\frac{2}{5}\right) = -1$). This is a powerful tool – we have infinitely many primes modulo which to reduce, and if *any one* of these reductions results in a congruence with no solution, then there is no integer solution to the original equation either.

---

[6] A particularly fancy viewpoint, which this notation encourages, is to view the equation as a function that inputs a ring and outputs the corresponding set of solutions!

But thus far, showing the *lack* of solutions to a Diophantine equation is the limit of this technique, as the existence of a solution mod n does *not* guarantee the existence of an integer solution. Indeed, an equation as simple as $x^2 = -1$ clearly has no integer solutions, but half of all primes p admit a solution to this equation mod p. Worse, it is possible for a Diophantine equation to have a solution modulo *every* prime p and yet still have no integer solutions:

▶ **Example 8.6.1** Consider the polynomial equation $f(x) = 0$, where

$$f(x) = (x^2 - 3)(x^2 - 13)(x^2 - 39).$$

Then the equation $f(x) = 0$ clearly has no integer solutions (its only real solutions are $\pm\sqrt{3}, \pm\sqrt{13}$, and $\pm\sqrt{39}$), but we claim it has a solution modulo *any* prime p. Modulo 3 or 13, we can take $x = 0$, and modulo any other prime p, multiplicativity of the Legendre symbols tells us that

$$\left(\frac{3}{p}\right)\left(\frac{13}{p}\right) = \left(\frac{39}{p}\right).$$

Thus not all of 3, 13, and 39 can be non-squares modulo p, and so we can find a solution x as a square root of whichever one is a square.

What a depressing situation! So imagine what a miracle it would be, then, to have a method for reversing this process, of deducing the conclusive existence of integer solutions to a Diophantine equation by working in a collection of simpler worlds. As we've come to understand the word "miraculous" and "p-adic" as being essentially synonymous, it will come as little shock that the p-adic worlds are precisely the worlds to save the day, at least for some quadratic Diophantine equations.

First, let us fact-check the claim that $\mathbb{Z}_p$ is "simple," despite being a still-unfamiliar place to work. We have commented before that *fields* are in some sense arithmetically uninteresting because there are no primes at all. The other extreme is a ring like $\mathbb{Z}$ with so many primes that we write full textbooks[7] trying to understand them. The rings $\mathbb{Z}_p$ form a remarkable intermediate realm along this spectrum, with only one solitary prime each (Corollary 8.4.5) on which to rest all of their arithmetic. Better yet, the primes of the various $\mathbb{Z}_p$ together comprise precisely all of the primes of $\mathbb{Z}$, and the perspective that naturally arises is that of a division of labor: to understand a problem of arithmetic involving primes in $\mathbb{Z}$, we try to understand that same problem in each $\mathbb{Z}_p$ in turn, where the scarcity of primes may be of great benefit.

Figure 8.1 represents this approach to studying classical number theory from a p-adic viewpoint (we'll be back to that ∞ in a minute!), that the überfield $\mathbb{Q}$ is built up of many primes, and we can simplify many questions by studying them one at a time. To focus on that one prime, we move to the corresponding p-adic world, which removes all other primes from arithmetic significance. The common

---

[7] ...well, certain poor souls do, anyway.

Fig. 8.1 Studying $\mathbb{Q}$ through its various completions

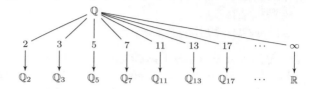

metaphor, enshrined in the terminology, is to think of $\mathbb{Q}$ as a *global field*, from which we can zoom in to any of the *local fields* $\mathbb{Q}_p$.

The cleanest statement of this phenomenon needs one more addendum, one more "locality" in the global map of $\mathbb{Q}$, shown at the far right of Figure 8.1. In terms of their construction, both $\mathbb{R}$ and the various $\mathbb{Q}_p$ can be thought of as being formed from $\mathbb{Q}$ by the analytical process of *completion*: in both cases you begin with $\mathbb{Q}$ and observe that certain limits that you'd like to exist simply do not. For example, while the decimal truncations

$$1, \quad 1.4, \quad 1.41, \quad 1.414, \quad 1.4142, \quad \ldots$$

of $\sqrt{2}$ form a sequence of rational numbers that converges ($|a_i - a_j| \to 0$ as $i, j \to \infty$), its limit is not rational. We expand from $\mathbb{Q}$ to $\mathbb{R}$ by including all limits of convergent sequences of rational numbers. Likewise, while the sequence of p-adic truncations

$$3, \quad 3+p, \quad 3+p+2p^2, \quad 3+p+2p^2+6p^3, \quad 3+p+2p^2+6p^3+p^4, \quad \ldots$$

of $\sqrt{2} \in \mathbb{Z}_7$ form a sequence of rational numbers that p-adically converges ($|a_i - a_j|_p \to 0$ as $i, j \to \infty$), its limit is not rational. We expand from $\mathbb{Q}$ to $\mathbb{Q}_p$ by including limits of p-adically convergent sequences of rational numbers.

The strength of this analogy turns out to be rather staggering. The fields $\mathbb{Q}_p$ and $\mathbb{R}$, though miles apart algebraically, are analytically very similar in construction, and this similarity suggests that we should include $\mathbb{R}$ alongside all of the various $\mathbb{Q}_p$ in Figure 8.1. Lest the reader be concerned that we could continue coming up with bizarre metrics, topologies, completions, etc., giving rise to more and more bizarre worlds in which to do arithmetic, rest assured that this is not the case. A remarkable result called *Ostrowski's Theorem* states, roughly, that the fields in the figure ($\mathbb{R}$ and the various $\mathbb{Q}_p$) are the *only* completions of $\mathbb{Q}$ (see Exercise 8.46).

To package this encompassing idea in compact notation, we adopt the convention that $\mathbb{Q}_\infty = \mathbb{R}$ (and $|\cdot|_\infty = |\cdot|$, the usual absolute value), completing the lastmost column in Figure 8.1. Informally, then, a *Local-Global Principle* is a statement that a certain problem submits to the type of resolution described above, that resolving it in $\mathbb{Q}$ (the global solution) is equivalent to solving a related one in each $\mathbb{Q}_p$ (a local solution) for each "$p \le \infty$" (a shorthand for "for all regular primes p and the prime at infinity"). Exploration Q takes us through some examples.

Local-Global Principles ◀

Q.1 Here we explore the *Local-Global Product Formula* relating the p-adic absolute values $|\cdot|_p$ for all $p \le \infty$. Choose some rational numbers r and compute $|r|_p$ for each prime $p \le \infty$. Finally, for each such r compute

$$\prod_{p \le \infty} |r|_p.$$

What happens and why?

Q.2 Prove the *Local-Global Principle for Units*:
   $n \in \mathbb{Z}$ is a unit if and only if n is a unit in $\mathbb{Z}_p$ for all primes $p \in \mathbb{N}$.

Q.3 Prove the *Local-Global Principle for Divisibility*:
   For $a, b \in \mathbb{Z}$, we have $a \mid b$ in $\mathbb{Z}$ if and only if $a \mid b$ in $\mathbb{Z}_p$ for all $p < \infty$.
Note that divisibility in $\mathbb{Z}_p$ is very peculiar: For example, $7 \mid 13$ in $\mathbb{Z}_5$!

Q.4 Prove the *Local-Global Principle for Cubeness*:
   $a \in \mathbb{Q}$ is a cube if and only if a is a cube in $\mathbb{Q}_p$ for each $p \le \infty$.

Q.5 Explain why the following claim doesn't makes sense:
   $\alpha \in \mathbb{Q}_p$ is a cube if and only if it is a cube in $\mathbb{Q}_q$ for all $q \ne p$.

Q.6 Explain why the *Local-to-Global Principle for Squareness* is false if we neglect to include the prime at infinity:
   $a \in \mathbb{Q}$ is a square if and only if a is a square in $\mathbb{Q}_p$ for all $p < \infty$, but *is* true if we change it to "for all primes $p \le \infty$."
Why is this change important for squares but unnecessary for cubes?

Q.7 The symbol ∞ does not, as one has to meticulously and repeatedly explain to young children trying desperately to weaken your resolve with pleas of "I love you times infinity," represent an actual number. On the other hand, the defining condition for being prime is that $p \mid ab \Rightarrow p \mid a$ or $p \mid b$. Come up with reasonable conventions for a divisibility notion in $\mathbb{Z} \cup \{\infty\}$ that would make ∞ a prime in this world. Discuss: is this a reasonable thing to do, or has mathematics finally jumped the shark?

8.7 The Local-Global Principle for Quadratic Equations

While local-global principles arise in a variety of algebraic contexts, many of them can be crisply encoded as statements about solutions of equations. The questions "is a a square?", "is a a cube?", and "is a a unit?" from Exploration Q can all be rephrased as the existence of solutions to the respective equations

$$x^2 = a, \qquad x^3 = a, \qquad \text{and} \qquad xa = 1.$$

That we have the prospect of resolving such questions by invoking local-global principles is certainly a heart-warming thought, but any fledgling romantic is well advised to accept both the good and the bad in their partner, and finding oneself with a newfound love for local-global principles is no different. We note two potential red flags. First is that local-global principles do not hold for every problem: Example 8.6.1 combines with Hensel's Lemma to provide a polynomial equation with no solution in $\mathbb{Q}$ but solutions in $\mathbb{Q}_p$ for each $p \leq \infty$ (Exercise 8.33). Nor is it true that local-global principles arc necessarily helpful! The examples on the Exploration also highlight the second concern—even when a local-global principle exists, it is not clear that it is of any practical use. For example, in most contexts it is certainly much easier to say directly whether a given $n \in \mathbb{Z}$ is a unit than it is to export it into infinitely many p-adic worlds and do work there.

The point of the current section is to discuss one of the more remarkable local-global principles, that of solutions to *quadratic* Diophantine equations. The quadratic world emerges as a perfect intermediate between the occasional mundanity of linear affairs and the impenetrability of higher-degree analogues. Namely, it is not hard to prove a local-global principle for linear Diophantine equations, that for $a, b, c \in \mathbb{Z}$ the equation $ax + by = c$ has a solution in $\mathbb{Z}$ if and only if it has one in each $\mathbb{Z}_p$ (Exercise 8.43). The quadratic analog, however, is much a more exhilarating version of the local-global philosophy. Much like the case of linear Diophantine equations, where a solution is constructed by the explicit Extended Euclidean Algorithm, the proof of the following principle is algorithmic in nature. The algorithm itself is just marginally outside the realm of plausibility for doing by hand, so we do not detail all the steps here. The procedure has been implemented in various mathematical software packages, and we refer the reader to Exercise 8.51 for details. We hope to instead illustrate the significance of the corresponding principle and present a perspective on how it unifies much of the work done in this course. Here is the statement:

Theorem 8.7.1 (Hasse's Local-Global Principle)
For fixed non-zero rational numbers a and b, there exists a rational solution to the equation

$$ax^2 + by^2 = 1$$

if and only if there exists a p-adic solution to this equation for all $p \leq \infty$.

The form of the equation may at first glance appear rather restrictive, but we will see that by employing change of variables, it is quite broad in scope. It is also true that Hasse's principle extends to quadratic equations in many variables, but the two-variable case has sufficient meat to keep us entertained for the present.

Taking the Hasse principle as given, what remains is to address the analogous question in each $\mathbb{Q}_p$: given a, b fixed, for which p does there exist a $\mathbb{Q}_p$ solution to $ax^2 + by^2 = 1$? We will proceed analogously to the one-variable case. There, we determined the existence of solutions to the equation $ax^2 = 1$ in $\mathbb{Z}/(p)$ (and hence in $\mathbb{Q}_p$ by Hensel's Lemma for p odd) by calculations with Legendre symbols. There exists a solution if and only if $\left(\frac{a}{p}\right) = 1$ (since $x^2 = 1/a$ is solvable if and only if $x^2 = a$ is solvable). The major theorems in Chapter 7 provide us with an algorithm to compute these symbols. Here for simplicity of exposition, we will reverse the process somewhat: the next definition is a calculational definition for a new symbol $(a, b)_p$ that evaluates to either ± 1, and we will soon show that it encodes the existence of a p-adic solution to $ax^2 + by^2 = 1$ just as the Legendre symbol did for mod-p solutions to $x^2 = a$. As with many other such definitions, we will have to pull out $p = 2$ and $p = \infty$ as special cases.

Definition 8.7.2

We define the p-**adic Hilbert symbol** $(a, b)_p$ for $a, b \in \mathbb{Q}^\times$ (or $\mathbb{Q}_p^\times$), and $p \leq \infty$ a prime, as follows:

- For $2 < p < \infty$, write $i = v_p(a)$ and $j = v_p(b)$, so $a = p^i u$ and $b = p^j v$ with $u, v \in \mathbb{Z}_p^\times$. Then set

$$(a, b)_p = \left(\frac{(-1)^{ij} u^j v^i}{p}\right) = \left(\frac{-1}{p}\right)^{ij} \left(\frac{u}{p}\right)^j \left(\frac{v}{p}\right)^i = (-1)^{ij(p-1)/2} \left(\frac{u}{p}\right)^j \left(\frac{v}{p}\right)^i$$

- For $p = 2$, take i, j, u, v as above, and set

$$(a, b)_2 = (-1)^e \quad \text{where} \quad e = \frac{u-1}{2} \cdot \frac{v-1}{2} + \frac{j(u^2-1)}{8} + \frac{i(v^2-1)}{8}$$

- For $p = \infty$, define

$$(a, b)_\infty = \begin{cases} +1 & \text{if } a > 0 \text{ or } b > 0 \\ -1 & \text{if } a < 0 \text{ and } b < 0. \end{cases}$$

◄

▶ **Remark 8.7.3** Notational conventions: For $u \in \mathbb{Z}_p^{\times}$ and p odd, the expression $\left(\frac{u}{p}\right)$ means the usual Legendre symbol $\left(\frac{u \bmod p}{p}\right)$. By Lemma 4.3.3, this is equivalent to asking if u is a square in $\mathbb{Z}_p$. In the second bullet we encounter a power of (-1) to a 2-adic integer, and here for $\alpha \in \mathbb{Z}_2$ we take $(-1)^{\alpha}$ to mean $(-1)^{\alpha \bmod 2}$, evaluating to $+1$ if $2 \mid \alpha$ and -1 otherwise.

First, a word on the two special cases: The relative intricacy of the formula for $p = 2$ is inherited from the difference between the two cases in Corollary 8.5.9 and the fact that we have the convenient Legendre symbol to simplify the expression for odd p. The $p = \infty$ case is a bit of a triviality, with the symbol $(a, b)_{\infty}$ succinctly encoding the test for whether or not $ax^2 + by^2 = 1$ has any real solutions: If a is positive, we can take the solution $(\frac{1}{\sqrt{a}}, 0)$, and likewise for $(0, \frac{1}{\sqrt{b}})$ when b is positive. If a and b are both negative, then there can be no solution to $ax^2 + by^2 = 1$ since the left-hand side is negative.

▶ **Example 8.7.4** Compute $(9, 15)_p$ for various p.

Solution Let's start with $p = 3$. We have $a = 9 = 3^2 \cdot 1$, so in the notation of the definition, we have $i = 2$ and $u = 1$; and $b = 15 = 3^1 \cdot 5$, so $j = 1$ and $v = 5$. Then

$$(9, 15)_3 = \left(\frac{(-1)^{ij} u^j v^i}{3}\right) = \left(\frac{25}{3}\right) = \left(\frac{1}{3}\right) = +1.$$

Similarly, to compute $(9, 15)_5$, we write $9 = 5^0 \cdot 9$ and $15 = 5^1 \cdot 3$, so

$$(9, 15)_5 = \left(\frac{(-1)^0 \cdot 9^1 \cdot 3^0}{5}\right) = \left(\frac{9}{5}\right) = \left(\frac{4}{5}\right) = +1.$$

Next for $p = 7$, we have $9 = 7^0 \cdot 9$, so $i = 0$ and $u = 9$, and $15 = 7^0 \cdot 15$, so $j = 0$ and $v = 15$. Now

$$(9, 15)_7 = \left(\frac{(-1)^0 \cdot 9^0 \cdot 15^0}{7}\right) = \left(\frac{1}{7}\right) = +1.$$

The simplicity of this last example will in fact repeat for any odd prime p other than 3 and 5, as we will have $i = j = 0$ so $(9, 15)_p = +1$. That leaves only the two special cases: for $p = \infty$, the definition immediately provides $(9, 15)_{\infty} = 1$, and for $p = 2$ we have $u = 9$, $v = 15$, and $i = j = 0$, so

$$(9, 15)_2 = (-1)^{(u-1)/2 \cdot (v-1)/2} = (-1)^{4 \cdot 7} = +1.$$

We conclude that $(9, 15)_p = +1$ for all $p \leq \infty$. Intriguing... $\qquad \square$

We saw in the example that the only primes that can possibly produce a -1 from a Hilbert symbol calculation of $(a, b)_p$ are $p = 2$, $p = \infty$, and primes dividing either a or b, as for any other prime we get $i = j = 0$, so $(a, b)_p = +1$. So much like Legendre symbols, the calculation of Hilbert symbols is made remarkably more efficient by use of symbolic manipulation and general deductions. We collect this and other key properties below.

Lemma 8.7.5

Let $a, b, c \in \mathbb{Q}^\times$. Then for any prime p, we have the following properties of the Hilbert Symbol:

(1) Irrelevant primes: If $p \neq 2, \infty$ and $p \nmid a, b$, then $(a, b)_p = +1$.
(2) Symmetry: $(a, b)_p = (b, a)_p$.
(3) Multiplicativity: $(a, bc)_p = (a, b)_p(a, c)_p$ and $(ab, c)_p = (a, c)_p(b, c)_p$.
(4) Irrelevance of Squares: $(a, b)_p = (ac^2, b)_p$
(5) $(a, -a)_p = 1$.
(6) For all $a \neq 1$, $(a, 1 - a)_p = 1$.   ◄

Proof Symmetry is clear from the definition for all three cases, and the rest of the claims each require up to 3 separate proofs for the even, odd, and infinite cases. We present here the arguments for p odd and leave $p = 2$ and $p = \infty$ to Exercise 8.37.

For multiplicativity, we piggyback on multiplicativity of the Legendre symbol: write $a = p^i u$, $b = p^j v$ and $c = p^k w$, so that $bc = p^{j+k} vw$ and

$$(a, bc)_p = \left(\frac{(-1)^{i(j+k)} u^{j+k}(vw)^i}{p}\right) = \left(\frac{(-1)^{ij} u^j v^i}{p}\right)\left(\frac{(-1)^{ik} u^k w^i}{p}\right)$$
$$= (a, b)_p(a, c)_p.$$

Multiplicativity in the first slot then follows by symmetry (or repeating this argument), and (4) follows immediately as a consequence, as

$$(ac^2, b) = (a, b)(c^2, b) = (a, b)(c, b)^2 = (a, b)$$

since $(c, b)^2 = (\pm 1)^2 = +1$. For (5), write $a = p^i u$. Then

$$(a, -a)_p = \left(\frac{(-1)^{i \cdot i} u^i (-u)^i}{p}\right) = \left(\frac{(-1)^{i(i+1)} u^{2i}}{p}\right) = +1$$

since both $2i$ and $i(i + 1)$ are necessarily even.

Finally, for (6), we have $a = p^i u$ and $1 - a = 1 - p^i u$, so at least one of i or j is zero, and if both are zero the Hilbert symbol is automatically $+1$. Otherwise, if $i = 0$, then $j > 0$ and $p \mid v = 1 - u$. Thus $u \equiv 1 \bmod p$. If $i > 0$ and $j = 0$, then $v = 1 - p^i u \equiv 1 \bmod p$. Thus each case evaluates to one of

$$(a, 1 - a)_p = \left(\frac{(-1)^{i \cdot j} u^j v^i}{p}\right) = \left(\frac{1^j}{p}\right) = 1 \text{ or } \left(\frac{(1 - p^i u)^i}{p}\right) = \left(\frac{1}{p}\right) = 1. \quad \square$$

We can now make the connection to quadratic Diophantine equations, after which we will move to solving specific equations. We reiterate the punchline of the section, that the symbols $(a, b)_p$ serve a function completely analogous to the symbols $\left(\frac{a}{p}\right)$, providing a calculable quantity that returns a yes/no answer to a question "Does this equation have a solution?" By Theorem 8.7.1, to answer this question in $\mathbb{Q}$ is equivalent to being able to answer it in each $\mathbb{Q}_p$, and so the following theorem neatly decides the whole story.

Theorem 8.7.6

Fix $a, b \in \mathbb{Q}_p^\times$. Then there is a solution $x, y \in \mathbb{Q}_p$ to

$$ax^2 + by^2 = 1$$

if and only if $(a, b)_p = +1$.

The proof is surprisingly explicit, though requires breaking into several cases depending on whether p is two, odd, or infinity, and then further depending on the p-adic valuations of a and b. The case $p = \infty$ is immediate (there are no solutions in $\mathbb{R}$ if and only if $a, b < 0$), and we leave the $p = 2$ case as Exercise 8.42, leaving the case of an odd prime p. The difficulty before us is transitioning from abstract knowledge of positive Hilbert symbols to an explicit solution to an equation.

Pickle 8.7.7

In Example 8.7.4 we found that $(9, 15)_p = +1$ for all p. How do we go from this to, say, a $\mathbb{Q}_7$-solution to
$$9x^2 + 15y^2 = 1?$$
◄

Solution We find a solution (x_0, y_0) mod 7 and then apply Hensel's Lemma. A brute force search finds the solution $(x_0, y_0) = (3, 2)$:

$$9(3)^2 + 15(2)^2 \equiv 1 \bmod 7.$$

We will take $y = 2 \in \mathbb{Z}_7$. To find x, re-writing the above expression $3^2 \equiv (1 - 15(2)^2)9^{-1} \bmod 7$ shows that the 7-adic integer $\alpha = \frac{1 - 15(2^2)}{9}$ is congruent to a square (namely, 3^2) modulo 7. By Corollary 8.5.6, $\alpha = x^2$ is thus itself a square in $\mathbb{Z}_7$. Using the techniques of the previous section, we can work with explicit expansions whenever desired, and such calculations would produce[8]

$$\alpha = \frac{1 - 15(2^2)}{9} = 2 + 5 \cdot 7 + 3 \cdot 7^2 + 7^3 + 6 \cdot 7^4 + 3 \cdot 7^5 + \cdots$$
$$= (3 + 3 \cdot 7 + 7^2 + 3 \cdot 7^4 + 3 \cdot 7^5 + \cdots)^2 = x^2 \in \mathbb{Z}_7$$

We have now constructed our solution, as

$$9x^2 + 15y^2 = 9 \cdot \left(\frac{1 - 15(2^2)}{9}\right) + 15 \cdot (2)^2 = 1.$$

Note that since 7 divides neither 9 nor 15, the Hilbert symbol hypothesis $(9, 15)_7 = +1$ presented us with no extra information to use. For the finitely many but more interesting examples where p divides a or b, the Hilbert symbol evaluating to $+1$ will be crucial.   □

One remark on the solution above before we begin the proof in earnest: we managed to brute force the solution $(3, 2)$ to $9x^2 + 15y^2 \equiv 1 \bmod 7$, but brute force is less effective mid-proof. How do we know we are guaranteed to be able to find a mod-p solution? One option is a slight modification of Theorem 7.4.1, which guarantees in fact quite a number of points on the curve modulo p. Alternatively, a simpler argument can guarantee the existence of *at least one*: as we run through squaring all p values for each of $x, y \in \mathbb{Z}/(p)$, there are $\frac{p+1}{2}$ different values obtained by ax^2 and $\frac{p+1}{2}$ obtained by $1 - by^2$, and since $\frac{p+1}{2} + \frac{p+1}{2} > p = |\mathbb{Z}/(p)|$, at least one value is hit by both expressions. That is, there must exist $(x_0, y_0) \in \mathbb{Z}/(p)$ such that $ax_0^2 \equiv 1 - by_0^2$.

Proof (of Theorem 8.7.6) We deal here with the case of p an odd prime (again, the case $p = \infty$ is trivial and for $p = 2$, see Exercise 8.42). For the first direction, assume that there exist $x, y \in \mathbb{Q}_p$ such that $ax^2 + by^2 = 1$. Then as long as $x, y \neq 0$, properties (4) and (6) in Lemma 8.7.5 give

$$(a, b)_p = (ax^2, b)_p = (ax^2, by^2)_p = (ax^2, 1 - ax^2)_p = +1.$$

And if one of x or y is zero (let's say x), then $by^2 = 1$, so $b = (y^{-1})^2$ and

$$(a, b)_p = (a, y^{-1})^2 = +1.$$

For the converse, we break into cases by $v_p(a)$ and $v_p(b)$. Note that neither condition—either that $(a, b)_p = 1$ or that $ax^2 + by^2 = 1$ has a solution—is affected by multiplying nor dividing a or b by a square. This follows from the irrelevance of squares property for the former and by absorbing squares into x or y for the latter. So

---

[8] We do not claim either of these coefficient patterns are obvious, only that we have developed enough machinery to know that we could continue to compute them indefinitely.

by multiplying by a suitable power of p^2 we can without loss of generality assume that $v_p(a)$ and $v_p(b)$ are both either 0 or 1.

The case $v_p(a) = v_p(b) = 0$ is the case where both a and b are p-adic units, and we can proceed as in Pickle 8.7.7: find a mod-p solution (x_0, y_0) to $ax_0^2 + by_0^2 \equiv 1$ mod p. First, if $x_0 \equiv 0$ mod p, then $1 - by_0^2 \equiv 0$, so b^{-1} is a square mod p, and hence in $\mathbb{Z}_p$. Let y be a square root of b^{-1} in $\mathbb{Z}_p$, giving us the solution $a(0) + by^2 = a(0) + b \cdot b^{-1} = 1$. Next, if $x_0 \neq 0$, let y be any lift of y_0 to $\mathbb{Z}_p$, so that the p-adic integer $(1 - by^2)a^{-1} \equiv (1 - by_0^2)a^{-1} \equiv x_0^2$ mod p is a non-zero square mod p. By Corollary 8.5.6, we can choose an $x \in \mathbb{Z}_p$ to be a square root of $a^{-1}(1 - by^2)$ congruent to x_0 mod p. Now (x, y) is a solution to $ax^2 + by^2 = 1$.

If one of $v_p(a)$ or $v_p(b)$ equals 1 (say $v_p(b)$) and the other is 0, then the hypothesis that $(a, b)_p = +1$ gives us that

$$1 = (a, b)_p = \left(\frac{(-1)^{0 \cdot 1} a^1}{p} \right) = \left(\frac{a}{p} \right),$$

so a is a square mod p. Hensel's Lemma implies that a, and hence $\frac{1}{a}$, is a square in $\mathbb{Q}_p$. Let $x \in \mathbb{Q}_p$ be a square root of $\frac{1}{a}$. Then

$$ax^2 + b(0)^2 = a \left(\frac{1}{a} \right) + b(0^2) = 1$$

provides the solution $(x, 0)$. The final case where $v_p(a) = v_p(b) = 1$ reduces to the previous cases by a substitution. See Exercise 8.41.   □

Exploration R

Hilbert Symbols ◄

The upcoming section will have us do a lot of Hilbert symbol calculations rather quickly, so here's our chance to build some fluency.

R.1 Compute $(18, 3)_p$ for all $p \leq \infty$.

R.2 Explore/simplify the expressions $(a, p)_p$, $(a, 1)_p$, and $(a, -1)_p$.

R.3 Deduce the identity
$$(a, a)_p = (-1, a)_p$$
in two different ways: First using the explicit definition of the Hilbert symbol, and second, using properties of the Hilbert symbol.

R.4 Use properties of the Hilbert symbol to verify each step below:

$$\left(12, -\frac{5}{3}\right)_3 = (3, -15)_3 = (3, -3)_3(3, 5)_3 = (3, 5)_3 = \left(\frac{5}{3}\right) = -1.$$

R.5 Use properties of the Hilbert symbol to verify each step below:

$$\left(\frac{3}{7}, -\frac{5}{9}\right)_p \left(\frac{3}{7}, -5\right)_p = (3 \cdot 7, -5)_p = (21, -20)_p = +1.$$

R.6 Using the prior problems as motivation (or not), come up with your own simplifying property of Hilbert symbols. It need not be particularly deep, only a convenient shorthand for an identity you could see yourself finding handy on occasion.

Our recent result, Theorem 8.7.6, shows it would have been logically equivalent to make the following definition for the p-adic Hilbert symbol:

$$(a, b)_p = \begin{cases} +1 & \text{if the equation } ax^2 + by^2 = 1 \text{ has a solution } (x, y) \in \mathbb{Q}_p \\ -1 & \text{if the equation } ax^2 + by^2 = 1 \text{ has no solution } (x, y) \in \mathbb{Q}_p \end{cases}$$

R.7 Using the above definition of $(a, b)_p$, verify properties (2), (3), (5), and (6) of Lemma 8.7.5.

R.8 We have thus far constantly insisted that the symbol $(a, b)_p$ requires $a, b \in \mathbb{Q}^\times$ or $\mathbb{Q}_p^\times$. Using the quadratic equation definition of Hilbert symbols, what if we allow a or b to be zero? Are the resulting equations solvable? Do the properties of the symbol continue to hold?

8.8   Computations: Quadratic Equations Made Easy

The synthesis of the results of the previous section provides us a pleasantly easy-to-use procedure for resolving quadratic equations:

Theorem 8.8.1

Fix $a, b \in \mathbb{Q}^\times$. There is a rational solution to

$$ax^2 + by^2 = 1$$

if and only if $(a, b)_p = +1$ for $p = 2$, $p = \infty$, and for each prime p for which either $v_p(a)$ or $v_p(b)$ is nonzero.

It is important to emphasize that this is a *finite* calculation. With prior local-global principles, we occasionally bemoaned that passing to the myriad p-adic worlds was of little benefit if we had to do computations in infinitely many of them. But Theorem 8.8.1 shows that for quadratic equations only the divisors of the coefficients and the two special cases need checking. Even simpler, to check *non-existence* requires only one lucky hit.

▶ **Example 8.8.2**   Show that there are no rational solutions to

$$7x^2 + 11y^2 = 1.$$

Solution We need only compute $(a, b)_p$ for $p \in \{2, 7, 11, \infty\}$, and can immediately check off that $(a, b)_\infty = 1$. Direct computation gives $(7, 11)_{11} = \left(\frac{7}{11}\right) = -1$, which proves the non-existence of a $\mathbb{Q}_{11}$ solution, and hence the non-existence of a rational solution.   ☐

Note we do not claim that Hilbert symbol calculations are the *only* way to make these deductions, especially for non-existence answers. The previous example could be replaced with a mod-11 argument if we were working with integer solutions, and a similar argument applies to rational solutions if we clear denominators. Instead, we claim they serve as a unifying system to tackle many seemingly disparate such equations all at once.

▶ **Example 8.8.3**   Show there exist rational solutions to

$$15x^2 - 126y^2 = 1.$$

Solution For any prime p, we have

$$(15, -126)_p = (15, 9)_p (15, -14)_p = (15, 3)_p^2 (15, -14)_p = 1 \cdot (15, -14)_p = +1$$

by property (6) of Lemma 8.7.5. Note how much faster we can proceed via the abstract calculus of Hilbert symbols, compared to, say, our more explicit computations in Example 8.7.4.   □

Of course, demonstrating the existence of a solution can also be done by finding one (for example, $15(1^2) - 126(\frac{1}{3})^2 = 1$), but there is no guarantee that explicit solutions will be forthcoming, nor is it clear how to search for them efficiently. Next, to fulfill an earlier promise, let us begin to document how much broader in scope the technique is then merely tackling equations of the form $ax^2 + by^2 = 1$.

▶ **Example 8.8.4** Decide if there are any rational points on the hyperbola

$$u^2 + 12uv - v^2 = 3.$$

Solution By completing the square, we can re-write the equation as

$$(u + 6v)^2 - 37v^2 = 3,$$

so the substitution $x = u + 6v$ and $y = v$ puts the equation in the form addressed by Theorem 8.8.1:

$$\tfrac{1}{3}x^2 - \tfrac{37}{3}y^2 = 1$$

We can resolve the solution-existence problem with judicious use of irrelevance of squares and other properties of Hilbert symbols:

$$\left(\frac{1}{3}, -\frac{37}{3}\right)_p = (3, -3 \cdot 37)_p = (3, -3)_p(3, 37)_p = (3, 37)_p.$$

These symbols are trivially $+1$ for $p \notin \{2, 3, 37, \infty\}$, and the Hilbert symbols for these four special cases can be individually computed to be $+1$, so there is indeed a rational solution. Two with small denominator are $(x, y) = (\frac{7}{2}, \frac{1}{2})$ and $(x, y) = (\frac{8}{3}, \frac{1}{3})$, corresponding to $(u, v) = (\frac{1}{2}, \frac{1}{2})$ and $(u, v) = (\frac{2}{3}, \frac{1}{3})$.   □

Completing the square can eliminate both cross-terms and linear terms, rendering even the most generic quadratic equation

$$ax^2 + bxy + cy^2 + dx + ey = f$$

susceptible to this line of attack.

Another avenue of generalization is in application to Diophantine equations. Recall that finding rational points on the unit circle is essentially equivalent to the Diophantine problem of finding Pythagorean Triples. Namely, by clearing denominators, any rational solution to $x^2 + y^2 = 1$ gives an integer solution to $x^2 + y^2 = z^2$.

▶ **Example 8.8.5** There are no solutions to the Diophantine equation

$$7a^2 + 11b^2 = c^2$$

since if there were, dividing through by c^2 (and checking $c = 0$ separately) would give the solution $(x, y) = (\frac{a}{c}, \frac{b}{c})$ to the rational equation $7x^2 + 11y^2 = 1$, contradicting our work in Example 8.8.2.

▶ **Example 8.8.6** Diophantus, on a quest for right triangles whose area is a square greater than one of its side lengths[9], comes across the equation

$$15x^2 - 36 = y^2$$

and argues that it has no integer solutions. In fact, there are no rational solutions either, as we can see by re-writing the expression as

$$\frac{15}{36}x^2 - \frac{1}{36}y^2 = 1,$$

and then computing (again repeatedly using Lemma 8.7.5)

$$\left(\frac{15}{36}, -\frac{1}{36}\right)_3 = (15, -1)_3 = (3, -1)_3 = \left(\frac{-1}{3}\right) = -1. \qquad \square$$

Before continuing, we have one more remarkable property of Hilbert symbols to present, which will help simplify some of the calculations.

Theorem 8.8.7 (Hilbert Symbols Product Formula)
For all $a, b \in \mathbb{Q}$, we have

$$\prod_{p \leq \infty} (a, b)_p = +1.$$

Proof By multiplicativity, it suffices to check the formula when a and b are each either 2, -1, or an odd prime. Let's evaluate three of these cases, in each case noting that the vast majority of terms in the product evaluate to be 1 (and switching to r for our indexing prime for aesthetic reasons):

- Case I: If $a = p$ and $b = q$ are two odd primes: Then

$$\prod_{r \leq \infty} (a, b)_r = \prod_{r \leq \infty} (p, q)_r = (p, q)_p (p, q)_q (p, q)_2 = \left(\frac{q}{p}\right)\left(\frac{p}{q}\right)(-1)^{\frac{p-1}{2} \frac{q-1}{2}}$$

---

[9] Look, not all of his problems could be winners. You do you, Diophantus.

- Case II: If $a = p$ and $b = -1$, then

$$\prod_{r \leq \infty} (a, b)_r = \prod_{r \leq \infty} (p, -1)_r = (p, -1)_p (p, -1)_2 = \left(\frac{-1}{p}\right) \cdot (-1)^{(p-1)/2}$$

- Case III: If $a = p$ and $b = 2$, then

$$\prod_{r \leq \infty} (a, b)_r = \prod_{r \leq \infty} (p, 2)_r = (p, 2)_p (p, 2)_2 = \left(\frac{2}{p}\right) \cdot (-1)^{(p^2-1)/8}$$

The statement that these three products evaluate to $+1$ is precisely the statement of quadratic reciprocity and its two supplemental laws. The remaining cases, where both $a, b \in \{-1, 2\}$ are much simpler by comparison and left to Exercise 8.40.   □

The product formula could just as well be called Hilbert's Reciprocity Law due to its equivalence to the quadratic reciprocity law (including the two supplementary laws). It is intriguing to think through a (pedagogically unwise) course on number theory that begins with p-adic numbers, arrives at the product formula, and deduces quadratic reciprocity as a consequence. In any case, while the product formula is principally of import for its deep structural significance, it more practically has computational utility:

Corollary 8.8.8

For all $a, b \in \mathbb{Q}^\times$, the number of primes $p \leq \infty$ such that $(a, b)_p = -1$ is even.

◀

Proof Note that the number of such primes is finite by Property (1) of 8.7.5. Further, this number is even by the product formula, as an odd number of primes with $(a, b)_p = -1$ would give a global product of -1.   □

This provides a minor albeit pleasant simplification to Theorem 8.8.1 as it shortens the list of primes to check by one. It cannot be the case that exactly one prime p gives $(a, b)_p = -1$, so if all but one of them gives $+1$, we can conclude that they are all $+1$. Moreover, the prime to be omitted can be chosen at will. Perhaps you never want to run the $p = 2$ check ever again, or perhaps in a specific problem there's one prime that promises to be cumbersome to work with. The corollary allows you to skip it! Let's continue working on examples, occasionally making use of this revelation to unearth new hidden mysteries of the universe.

▶ **Example 8.8.9** There is at least one prime $p < \infty$ such the equation

$$x^2 + y^2 = -1$$

has no solutions in $\mathbb{Q}_p$.

Proof The equivalent equation $-x^2 - y^2 = 1$ obviously has no real solutions, so $(-1, -1)_\infty = -1$. By Corollary 8.8.8, there must exist at least one other prime p such that $(-1, -1)_p = -1$. Done! In fact, since no prime divides $a = -1$ or $b = -1$, the only remaining possibility is $p = 2$, and so it must be that $(-1, -1)_2 = -1$. So while -1 is only a square in half of the $\mathbb{Q}_p$'s, it is a sum of two squares in all but two of them, $\mathbb{Q}_2$ and $\mathbb{Q}_\infty = \mathbb{R}$. □

▶ **Example 8.8.10** Is there an element of $\mathbb{Q}(\sqrt{-13})$ of norm 30?

Solution Since it's been a while, remember that an element of $\mathbb{Q}(\sqrt{-13})$ takes the form $x + y\sqrt{-13}$, and its norm is given by

$$(x + y\sqrt{-13})(x - y\sqrt{-13}) = x^2 + 13y^2.$$

We thus seek a rational solution to $x^2 + 13y^2 = 30$ and so are left to evaluate the Hilbert symbols

$$\left(\tfrac{1}{30}, \tfrac{13}{30}\right)_p = (30, 13 \cdot 30)_p = (30, -30)_p (30, -13)_p = (30, -13)_p.$$

This can only be -1 for $p \in \{2, 3, 5, 13, \infty\}$, and we find, for example, that $(30, -13)_3 = -1$. Thus there are no elements of $\mathbb{Q}(\sqrt{-13})$ of norm 30. □

The previous example hints at the tie-in from Hilbert symbols to the study of algebraic number fields, and indeed the use of the p-adic worlds in this study. We will not formally study extensions of p-adic fields, but analogous to the construction of $\mathbb{Q}(\sqrt{-13})$ from $\mathbb{Q}$, we can introduce fields like

$$\mathbb{Q}_p(\sqrt{-13}) = \{a + b\sqrt{-13} : a, b \in \mathbb{Q}_p\}.$$

An interpretation of Example 8.8.10 is that $30 \in \mathbb{Q}$ is not a norm from $\mathbb{Q}(\sqrt{-13})$ because $30 \in \mathbb{Q}_3$ is not a norm from $\mathbb{Q}_3(\sqrt{-13})$ (where we again define $N(a + b\sqrt{-13}) = a^2 + 13b^2$). It is a theorem of Hilbert that a local-global principle holds for this situation—roughly, a rational number $r \in \mathbb{Q}$ is a norm from $\mathbb{Q}(\sqrt{d})$ if and only if for all $p \le \infty$, the corresponding $r \in \mathbb{Q}_p$ is a norm from $\mathbb{Q}_p(\sqrt{d})$. The typical benefits of the p-adic worlds thus come into play as well. For example, since d will already *have* a square root in many of the $\mathbb{Q}_p$, we have $\mathbb{Q}_p(\sqrt{d}) = \mathbb{Q}_p$.

Since our much-investigated example of writing numbers as sums of two squares has already been seen as a question about norms (since $N(x + yi) = x^2 + y^2$), we can use this as yet another perspective on this question. We can also introduce here another source of interesting problems: quadratic equations in which one or both of the coefficients contains a parameter, and we ask how changing the parameter influences the existence of solutions. Here is one we tackled before in Chapter 5, seen in a new light: the existence of *rational* solutions to $x^2 + y^2 = p$.

▶ **Example 8.8.11** For which primes p can we write $p = x^2 + y^2$ for $x, y \in \mathbb{Q}$?

Solution Certainly, we have solutions for $p = 2$, so we assume that p is an odd prime. Re-writing the equation as $\frac{1}{p}x^2 + \frac{1}{p}y^2 = 1$, we see that it is necessary and sufficient to have for all primes $q \leq \infty$ that

$$1 = \left(\frac{1}{p}, \frac{1}{p}\right)_q = (p, p)_q$$

The only relevant primes are $q \in \{2, p, \infty\}$, and since $(p, p)_\infty = +1$ trivially and we can skip $q = 2$ by the product formula[10], this only leaves $q = p$ itself to check. Directly from the definition we get

$$(p, p)_p = \left(\frac{-1}{p}\right) = \begin{cases} +1 & \text{if } p \equiv 1 \pmod 4 \\ -1 & \text{if } p \equiv 3 \pmod 4. \end{cases}$$

So by Hasse's Local-Global Principle, $p = x^2 + y^2$ if and only if $p \equiv 1 \pmod 4$. The condition for rational solutions is identical to that for integer solutions! □

Exercise 8.28 has you repeat this exercise moving from prime p to arbitrary integers n, repeating quickly what took us a deep understanding of arithmetic in $\mathbb{Z}[i]$ to complete in Chapter 5 (Corollary 5.7.2).

8.9 Synthesis and Beyond: Moving Between Worlds

The problem of writing primes in the form $x^2 + y^2$, which consumed much of our energy for much of the book, highlights well the amount of progress we've made in understanding how to move between various mathematical worlds in order to solve problems. We initially made progress on the equation $x^2 + y^2 = p$ using modular arithmetic. When $p \equiv 3 \bmod 4$, the lack of solutions to the reduced equation $x^2 + y^2 \equiv 3 \pmod 4$ implies the non-existence of integer solutions as well. Modular arithmetic is typically not sufficient, however, to guarantee the existence of a solution, and we had to move to the world $\mathbb{Z}[i]$ of Gaussian integers to finish the problem: primes that are congruent to 1 mod 4 factor as a product of two conjugate Gaussian primes in $\mathbb{Z}[i]$, and writing such a prime as the norm of either Gaussian prime is equivalent to writing it as a sum of two squares. The p-adic worlds shed further light on the modular arithmetic approach: Hensel's Lemma shows that the ability to find solutions to modular equations can in many cases be lifted to construct corresponding integer solutions, and Hasse's principle allows us to glue all the conditions together, providing a classification of which primes p can be written as a sum of rational

---

[10] How awesome is that?!?

squares. In this case, the conditions turn out to be the same as those for the integral squares.

Much of the simplicity of this story, however, comes from the simplicity of the form $x^2 + y^2$. In an attempt to illustrate both how much progress we have made and how many mysteries still remain, let us undertake a case study of a similar question, that of representability of primes p by the expression $x^2 + 11y^2$. Being freshest in our memory, rational solutions will come first. Omitting $p = 11$ for now due to an obvious such representation, the equation

$$x^2 + 11y^2 = p$$

has a rational solution if and only if

$$\left(\frac{1}{p}, \frac{11}{p}\right)_q = (p, 11p)_q = (p, -p)_q(p, -11)_q = (p, -11)_q = +1$$

for all $q \in \{2, 11, p, \infty\}$. These amount to the conditions

$$(p, -11)_{11} = \left(\frac{p}{11}\right) = +1 \qquad \text{and} \qquad (p, -11)_p = \left(\frac{-11}{p}\right) = +1,$$

two conditions that are equivalent by quadratic reciprocity. Thus

$$x^2 + 11y^2 = p \text{ has a rational solution} \iff p = 11 \text{ or } \left(\frac{p}{11}\right) = 1$$

$$\iff p \equiv 0, 1, 3, 4, 5, 9 \bmod 11.$$

Picking a prime satisfying one of these congruences, say $p = 23$, provides us with the conic

$$x^2 + 11y^2 = 23$$

now guaranteed to have at least one rational solution. But, as you no doubt recall(!), this was the starting point for applying the Diophantus chord method from Chapter 3 (2.2.6) to deduce there are infinitely many solutions. Indeed, a principal point made at the time was "It's often hard to tell if there are solutions, but *if* there is one solution, then there are infinitely many." But this particular dilemma has now been resolved, and an explicit example can be found by the algorithmic form of the Local-Global Principle. For this example, a brute force search gives a first rational solution of $(x, y) = \left(-\frac{9}{2}, \frac{1}{2}\right)$, and then taking a line of any rational slope through this point connects us to another. For example, taking $m = -\frac{1}{2}$ gives the second solution $\left(-\frac{19}{30}, -\frac{43}{30}\right)$.

As with Pythagorean triples, "clearing the denominator" shows that having rational solutions to $x^2 + 11y^2 = 23$ provides us with integer solutions to $x^2 + 11y^2 = 23z^2$, e.g.,

$$9^2 + 11 \cdot 1^2 = 23 \cdot 2^2 \qquad \text{and} \qquad 19^2 + 11 \cdot 43^2 = 23 \cdot 30^2$$

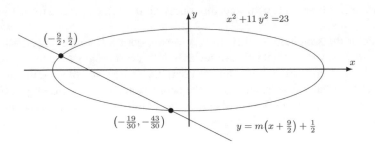

Fig. 8.2 Remember this? From back in the day? Good times!

from the two displayed points in Figure 8.2. Running over all rational slopes $m = \frac{u}{v}$ parameterizes for us all solutions to this Diophantine equation.

Despite the positive progress, some tantalizing mysteries still remain. For example, moving to *integer* solutions, we find that unlike $x^2 + y^2 = p$, there is a difference between primes for which there exist integer solutions to $x^2 + 11y^2 = p$ and ones for which there exist rational solutions. For an integer solution to $x^2 + 11y^2 = p$ to exist, it is *necessary* that $\left(\frac{p}{11}\right) = +1$, as seen by reducing the equation mod 11. This condition is the same as the one just derived above for rational solutions to this equation, but it is no longer sufficient. For example, we can see by brute force testing of small values of y that there are no integer solutions to

$$x^2 + 11y^2 = 23,$$

even though there are no mod-11 or mod-23 obstructions. This is true despite the fact that there are not only $\mathbb{Q}_p$-solutions for each $p \leq \infty$, but also $\mathbb{Z}_p$-solutions to this equation for each finite p: the solution $(\frac{9}{2}, \frac{1}{2})$ is a p-adic integer solution for all $p > 2$, and for $p = 2$ we can find the solution $(6\sqrt{-7}, 5)$ (noting that $\sqrt{-7} \in \mathbb{Z}_2$ by Corollary 8.5.7). (You should verify that these are solutions.) This shows that there is no analogous "integral" version of the Local-Global Principle.

Turning back to the algebraic number theory side of things reveals the surprising difficulty: it is our perpetual frenemy, the question of unique factorization. The "sum of two squares" form $x^2 + y^2$ is precisely the norm of a generic element $x + yi \in \mathbb{Z}(i)$. Our characterization of primes expressible in the form $x^2 + y^2$ hinged crucially on the fact that $\mathbb{Z}[i]$ is a unique factorization domain. On the other hand, the form $x^2 + 11y^2$ is the norm of an element from the quadratic field $\mathbb{Q}(\sqrt{-11})$, whose ring of integers is *not* $\mathbb{Z}[\sqrt{-11}]$ but rather (Theorem 6.3.5) the ring $\mathbb{Z}\left[\frac{1+\sqrt{-11}}{2}\right]$. While this larger ring is a unique factorization domain, its subring $\mathbb{Z}[\sqrt{-11}]$ is *not*: We have, for example,

$$2^2 \cdot 3 = (1 + \sqrt{-11})(1 - \sqrt{-11})$$

with all of these factors being non-associate irreducibles. And while 23 was seen above *not* the norm of an element of $\mathbb{Z}[\sqrt{-11}]$, it *is* the norm of an integral element of $\mathbb{Q}(\sqrt{-11})$:

$$23 = N\left(4 + \tfrac{1+\sqrt{-11}}{2}\right).$$

We will have to part ways before studying what happens when, like for $n = 11$, the ring of integers of $\mathbb{Q}[\sqrt{-n}]$ does not have unique factorization, except to tantalize you with the answer that it is not that we have gone too far with our inclusion of new numbers, but that we have not yet gone far enough! The study of quadratic phenomena has scratched the surface of the vast array of higher-degree number theory left to explore. Indeed, the solution to our $x^2 + 11y^2$ integer-representability problem turns out to be a *cubic* one. If we let α be the unique real root of $x^3 - x^2 - x - 1$ and consider the field

$$K = \mathbb{Q}(\alpha) = \{a + b\alpha + c\alpha^2 : a, b, c, \in \mathbb{Q}\}$$

and its ring of integers R, then for any prime p other than 11 we have

$$p = x^2 + 11y^2 \quad (x, y \in \mathbb{Z}) \iff p = \pi_1\pi_2\pi_3 \quad (\pi_i \text{ a prime in } R)$$
$$\iff x^3 - x^2 - x - 1 \text{ has three roots modulo } p.$$

We leave as a mystery the question of whence came the polynomial $x^3 - x^2 - x - 1$ other than to note that the resolution to such questions involves precisely the topics we have begun to study in this book: number fields and their rings of integers, algebraic structures and unique factorization, p-adic worlds and their generalizations, and even the study of elliptic curves and their points in various solution rings. Present-day mathematicians constantly tackle problems that arise via this type of thinking, and call upon arguments from this breathtakingly broad and deep pool of mathematics. We hope the reader will consider diving in.

8.10 Exercises

Calculation & Short Answer

Exercise 8.1 Compute the Taylor series of $f(x) = x^3 - 11x^2 + 35x - 25$ centered at $x = a$ for each $a \in \{1, 2, 3, 4, 5\}$. Use your results to factor $f(x)$.

Exercise 8.2 Find the base p expansion of $n = 70$ for each $p \in \{2, 3, 5, 7\}$.

Exercise 8.3 Find the base p expansion of $n = 12$ for every prime p.

Exercise 8.4 Find the 2-adic expansion of -3.

Exercise 8.5 Find the 11-adic expansion of -200.

Exercise 8.6 Find the 11-adic expansion of $\frac{1}{3}$.

Exercise 8.7 Show that the equation $x^3 = 2$ has no solutions in $\mathbb{Z}_7$.

Exercise 8.8 Show that there are no rational solutions to

$$7x^2 + 10y^2 = 1.$$

Exercise 8.9 Show that there are no integer solutions to the Diophantine equation

$$7x^2 + 6y^2 = z^2.$$

Exercise 8.10 For $\alpha \in \mathbb{Z}_p$ and r a positive real number, define

$$B_p(\alpha, r) = \{x \in \mathbb{Z}_p : |x - \alpha|_p < r\},$$

the *open ball* of radius r in $\mathbb{Z}_p$ centered at α. Use congruence conditions to describe the sets

$$B_{11}(50, 1) \qquad B_{11}(50, 0.1) \qquad B_{11}(50, .01) \qquad B_{11}(50, .001)$$

Exercise 8.11 For $\alpha \in \mathbb{Z}_p$ and r a positive real number, define

$$\overline{B}_p(\alpha, r) = \{x \in \mathbb{Z}_p : |x - \alpha|_p \le r\},$$

the *closed ball* of radius r in $\mathbb{Z}_p$ centered at α. Use congruence conditions to describe the sets

$$\overline{B}_7(5, 0.1) \qquad \overline{B}_7(47, 0.1) \qquad \overline{B}_7(-21, 0.1)$$

Comment on anything of interest.

Exercise 8.12 Find all primes p for which there exists a solution in $\mathbb{Z}_p$ to

$$x^2 - 5x + 3 = 0.$$

Exercise 8.13 Use properties of the Hilbert symbol to deduce the identity

$$(-ac, -bc)_p = (-1, -1)_p(-1, a)_p(-1, b)_p(-1, c)_p(a, b)_p(a, c)_p(b, c)_p.$$

Exercise 8.14 For which n and p do we have $(n, n)_p = +1$?

Exercise 8.15 Provide simple formulas for $(1, n)_p$ and $(-1, n)_p$.

Exercise 8.16 As with $\mathbb{Q}$, we often adjoin missing elements to $\mathbb{Q}_p$, e.g., considering $\mathbb{Q}_p[\sqrt{p}] = \{a + b\sqrt{p} : a, b \in \mathbb{Q}_p\}$. Assuming that it is possible to extend the p-adic absolute value to this ring, what would $v_p(\sqrt{p})$ have to be? What about $v_5(x)$, where x satisfies the polynomial equation

$$x^3 + 50x^2 + 25x + 35 = 0?$$

Exercise 8.17 Silently direct consternation toward someone who uses the notation $\mathbb{Z}_p$ to mean $\mathbb{Z}/(p)$. (Optional: Repeat this exercise without the "silently" hypothesis.)

Exercise 8.18 The ring $\mathbb{Z}/(p) \times \mathbb{Z}/(p^2) \times \mathbb{Z}/(p^3) \times \cdots$ consists of ∞-tuples

$$(x_1, x_2, x_3, x_4, \ldots)$$

where $x_i \in \mathbb{Z}/(p^i)$. Call an element of this ring *coherent* if x_n mod $p^m = x_m$ whenever $m < n$. Explain how specifying a coherent sequence is equivalent to specifying a p-adic integer. What do natural numbers look like as coherent sequences?

Exercise 8.19 What is the probability that a randomly selected p-adic rational number will be a p-adic integer? (Caution: This question is nonsense. Can you come up with justifications for different answers?)

Exercise 8.20 For p odd, suppose informally that you were to pick a "random" element of $\mathbb{Q}_p$. What's the probability that this element is a square in $\mathbb{Q}_p$? How does this change if $p = \infty$? If $p = 2$?

Formal Proofs

Exercise 8.21 Let n be a natural number and p a (positive, rational) prime. Prove by induction on n that n has a unique base-p expansion as described in Definition 8.1.1.

Exercise 8.22 Distinct infinite decimal expansions can converge to the same number in $\mathbb{R}$ (e.g., $1.000\ldots = 0.999\ldots$). Prove that this is not the case for p-adic expansions. (Hint: Suppose two p-adic expansions first differ in the coefficient of p^k. Use the p-adic absolute value to show the corresponding p-adic numbers are not equal.)

Exercise 8.23 The so-called *Harmonic numbers*

$$H_n = 1 + \frac{1}{2} + \frac{1}{3} + \frac{1}{4} + \cdots + \frac{1}{n}$$

tend to infinity with n, as the partial sums of the divergent harmonic series. Prove that H_n is never an integer by considering $v_2(H_n)$.

Exercise 8.24 Prove any of the following three local-global principles from Exploration Q that you didn't get to at the time[11]:

(a) Units   (b) Divisibility   (c) Cubes

Exercise 8.25 Verify that the set

$$\mathbb{Z}_{(p)} = \left\{ \frac{a}{b} : a, b \in \mathbb{Z}, p \nmid b \right\}$$

---

[11] or really, whichever ones your instructor tells you to

defines a subring of $\mathbb{Q}$ (called the *localization* of $\mathbb{Z}$ at p). Explain/prove the relationship $\mathbb{Z}_p \cap \mathbb{Q} = \mathbb{Z}_{(p)}$.

Exercise 8.26 Determine the units and primes of the ring $\mathbb{Z}_{(p)}$ from the previous problem.

Exercise 8.27 Finish up the $p = 2$ case for p-adic roots of unity begun in Pickle 8.5.12. Which roots of unity are in $\mathbb{Z}_2$? Consider working modulo 4 instead of 2.

Exercise 8.28 Use the Local-Global Principle to determine which (not necessarily prime) integers n can be written as a sum of two rational squares. Your answer should match the results of Chapter 5.

Exercise 8.29 Prove that a p-adic expansion represents a rational number if and only if it is eventually periodic.

Exercise 8.30 Verify the properties of the p-adic valuation in Lemma 8.3.7 in the case that $\alpha = 0$. What arithmetic conventions about ∞ do you have to adopt to make everything work?

Exercise 8.31 Prove that the valuation v_p is a Euclidean norm on $\mathbb{Z}_p$.

Exercise 8.32 Describe divisibility, gcds, and the Euclidean algorithm in $\mathbb{Z}_p$.

Exercise 8.33 Expanding on Example 8.6.1, use Hensel's Lemma to show that the Local-Global Principle fails for the polynomial equation $f(x) = 0$ with

$$f(x) = (x^2 - 3)(x^2 - 13)(x^2 - 39).$$

That is, show that the equations have roots in $\mathbb{Q}_p$ for each $p \leq \infty$ despite having no rational solutions.

Exercise 8.34 See Exercise 8.10 for a definition of the p-adic open ball. Prove the (unsettling) observation that every element of a p-adic ball is a center of that ball. That is, if $x \in B(\alpha, r)$, then $B(\alpha, r) = B(x, r)$.

Exercise 8.35 Show that given two natural primes p and q, there is a natural number n that has a square root in both $\mathbb{Q}_p$ and $\mathbb{Q}_q$.

Exercise 8.36 Complete the proof of Corollary 8.4.5 by verifying that

$$\mathbb{Q}_p = \left\{ \frac{\alpha}{\beta} : \alpha, \beta \in \mathbb{Z}_p, \beta \neq 0 \right\}.$$

Exercise 8.37 Prove the properties in Lemma 8.7.5 for $p = 2$ and $p = \infty$.

Exercise 8.38 For $a, b, c \in \mathbb{Q}_p^\times$, prove that there is a $\mathbb{Q}_p$-solution to

$$ax^2 + by^2 + cz^2 = 0$$

if and only if $(-1, -abc)_p = (a, b)_p (a, c)_p (b, c)_p$.

Exercise 8.39 Present with proof a classification of which elements of $\mathbb{Q}_p$ can be written as a sum of two squares.

Exercise 8.40 Finish the remaining cases in the proof of Theorem 8.8.7, those where both $a, b \in \{-1, 2\}$.

Exercise 8.41 Show that the substitution $a \to -ab^{-1}$ reduces the case of Theorem 8.7.6 where $v_p(a) = v_p(b) = 1$ to one we have already solved ($v_p(a)=0$, $v_p(b)=1$) by showing:

- $(a, b)_p = +1 \iff (-ab^{-1}, b)_p = +1$
- $ax^2 + by^2 = 1$ has a solution in $\mathbb{Q}_p$ if and only if $-ab^{-1}x^2 + by^2 = 1$ does.

Exercise 8.42 Mirror the proof of the odd case of Theorem 8.7.6 to finish the theorem for $p = 2$. The principal task is to unpack the condition $(a, b)_2 = 1$ in terms of mod-8 congruence, and then use our knowledge of 2-adic squares.

Exercise 8.43 A Local-Global Principle for linear Diophantine equations might say that for $a, b, c \in \mathbb{Z}$, the equation $ax + by = c$ has a solution in $\mathbb{Z}$ if and only if it has one in each $\mathbb{Z}_p$, for each prime $p \in \mathbb{N}$. Fill in the details of the following proof of that result:

$$\exists x, y \in \mathbb{Z} : ax + by = c \iff \gcd(a, b) \mid c$$
$$\iff v_p(c) \geq \min\{v_p(a), v_p(b)\} \text{ for all } p$$
$$\iff \exists x, y \in \mathbb{Z}_p : ax + by = c \text{ for all } p$$

Computation and Experimentation

SageMath can compute Hilbert symbols with the syntax

$$(a, b)_p = \texttt{hilbert\_symbol(a,b,p)}$$

using $p = -1$ for the prime at infinity.

Exercise 8.44 Verify experimentally that SageMath's implementation of this symbol agrees with the definition in the section. Repeat for some of the section's principal theorems, e.g., the Local-Global product formula.

Exercise 8.45 We have primarily dealt with binary quadratic forms $ax^2 + by^2$, but most of the theory of the section can be developed for arbitrary quadratic forms. The *Hasse-Minkowski invariant* of the quadratic form

$$f = a_1 x_1^2 + a_2 x_2^2 + \cdots + a_n x_n^2$$

is the constant

$$c_f = \prod_{1 \leq i < j \leq n} (a_i, a_j)_p.$$

Write a function in SageMath to compute Hasse-Minkowski invariants and experiment with them. Are they more frequently 1 or -1? Do some research into its
significance. Where does c_f appear in the section?

General Number Theory Awareness

Exercise 8.46 Research and give a precise statement of Ostrowski's Theorem (and
its ingredients). How does the theorem justify our claim that "the only completions
of $\mathbb{Q}$ are our $\mathbb{Q}_p$ for $p \leq \infty$"?

Exercise 8.47 In principle, there's nothing that inherently goes wrong with defining
n-adics for an arbitrary (that is, not necessarily prime) integer n. Why aren't they
as prevalent as p-adics in modern number theory? Possible starting points include
units, zero-divisors, or a p-adic Sunzi's Theorem.

Exercise 8.48 Much like 0, negative numbers, imaginary numbers, etc., p-adic numbers had a rocky path to full acceptance by the mathematics community. Hensel himself inadvertently contributed to suspicion around their use as valid mathematical
objects via an erroneous proof of a famous mathematical theorem. What happened?

Exercise 8.49 A p-adic interpretation of Sunzi's Theorem is that we can find integers
satisfying any finite collection of local absolute value conditions. Make sense of this
claim, and then look up the "weak approximation theorem." Find an understandable
version of this result, and explain how it expands from Sunzi's Theorem to incorporate
the prime at infinity.

Exercise 8.50 Predating Hilbert symbols for finding points on rational conics were
results of Gauss and Legendre phrased in terms of Legendre symbols. Explore the
history of these results and prove the equivalence of Legendre's approach to Hilbert's.

Exercise 8.51 Describe the current status of the algorithmic approach to finding a
point on rational conics. Searchable terms include Holzer's Theorem, or efficient
solution of rational conics. How much of this process has been incorporated into
current computer algebra systems?

The Adventure Continues

9

...wherein you embark on a mighty quest.

After conquering the *Fires of Euclidea* (Figure 9.1), your party disbanded and left you without a guide. Having explored worlds heretofore unknown and having met some of the most outlandish creatures the Numberverse has to offer, you had intended to return to the safety of your home town[1] of $\mathbb{Z}_0^{rk}$. Easily distracted, however, you can't help but feel the pull of several unexplored regions lying before you. It's dangerous to strike out on your own, but all the more rewarding to choose your own course...

```
You see paths heading North, East, South, and West.
```

```
 > Your arch-nemesis blocks your way North. To charge
   forward and take a stab at Fermat's Last Theorem,
   turn to Section 9.1.
```

Chapter 2 began with a historical discussion of Fermat's Last Theorem. We now have the tools (e.g., unique factorization in the Eisenstein integers) to mirror the early days of the theory, when leading mathematicians would knock off special cases of this theorem one *n* at a time.

```
 > To head South down a square-covered path, turn to
   Section 9.2.
```

Chapter 5 explored the question of which primes were expressible as a sum of *two* squares. But more questions remain. The prime 31, for example, cannot be written as a sum of two squares, nor can it be written as a sum of *three* squares.

Supplementary Information The online version contains supplementary material available at https://doi.org/10.1007/978-3-030-98931-6_9.

---

[1] Section 9.6 welcomes you home with a moment to reflect on your hero's journey. You could head there now, but we warn you—it is not the paths taken in life that haunt you, but the paths not taken.

© Springer Nature Switzerland AG 2022
C. McLeman et al., *Explorations in Number Theory*, Undergraduate Texts in Mathematics,
https://doi.org/10.1007/978-3-030-98931-6_9

Fig. 9.1 The road stretches out before you.

The best we can do is 4, as in $31 = 5^2 + 2^2 + 1^2 + 1^2$. Addressing the general phenomenon leads us to introduce yet another new type of number, namely the quaternions and their *Hurwitz Integers*.

> If it is secrecy you seek, brace yourself for some wild curves and turn East via Section 9.3.

Beneath the electronic communications and commerce that permeate our modern life lie many of the beautiful algebraic structures we have touched on in this text. Not only do the modular fields of Chapter 4 play an important role, elliptic curves (such as the Mordell Curves $y^2 = x^3 + d$ for various d) have paramount significance in securing information. Peek behind the veil of modern secrecy …if you dare.

> To turn West, where things get real, see Section 9.4.

Chapter 6 highlighted briefly the distinction between real and complex quadratic fields, with a tantalizing sneak preview of one of the most significant differences—their *units*. Characterizing the units of real quadratic fields finishes off the basics of doing number theory in such fields and turns out to have interesting applications to the study of rational numbers.

> Also tempting is to just dig into the field. To burrow
> down into the depths of abstraction, turn to Section 9.5.

While we celebrate unique factorization in quadratic fields whose rings of integers are UFDs, we also shed tears for those fields for which there is no analog of the Fundamental Theorem of Arithmetic. But despair not, for the failure of unique factorization is not the end of the story but rather the beginning of another. The notion of an *ideal* within the ring of integers turns out to save the day, and becomes a central object of study.

9.1 Exploration: Fermat's Last Theorem for Small n

The second chapter of this book recounts the origin and resolution of Fermat's Last Theorem, from Diophantus to Fermat to Wiles. But this cursory overview leaves out a tremendous amount of mathematical history in the intervening years, an elaborate story in which many of the mathematical superstars of history played a part. We first recall the statement:

Theorem 9.1.1 (Fermat's Last Theorem (1637))
Let $n > 2$ be a natural number. Then there are no positive integer solutions to the equation
$$x^n + y^n = z^n.$$

Fermat's contributions to the proof of this theorem certainly fell shy of the full result, but this is not to say that these contributions were negligible. In this section you will prove the $n = 3$ and $n = 4$ cases of this theorem using two principal ideas that Fermat applied to the problem. The first is the transition from an additive problem to a multiplicative one by factorizations of the type seen in Section 7.6, and in particular the identity

$$z^n = x^n + y^n = (x + y)(x + \zeta y)(x + \zeta^2 y) \cdots (x + \zeta^{n-1} y)$$

for $\zeta = \zeta_n$ a primitive n-th root of unity. That is, our approach[2] will be to pass to the ring $\mathbb{Z}[\zeta_n]$ of cyclotomic integers and hope that arithmetic in that ring is sufficiently

---

[2] Of course, Fermat would not have thought in these terms, nor do we know for sure what his imagined proof of Fermat's Last Theorem was.

accommodating to rule out the existence of solutions. The prospect of applying the Power Lemma is particularly enticing here, as it would give us a path toward proving that each $x + \zeta^i y$ is itself an n-th power, a condition often strong enough to deduce a contradiction. But the Power Lemma does not come for free in a general ring, and indeed hinges crucially on unique factorization. So for each n, the study of the ring $\mathbb{Z}[\zeta_n]$ is of increasing significance.

The second idea is Fermat's proof technique of *infinite descent*, a mere stone's throw away from the principle of mathematical induction:

Lemma 9.1.2 (Principle of Infinite Descent)

If the existence of a natural number n that satisfies some set of properties automatically implies the existence of a smaller natural number $n' < n$ that also satisfies the properties, then there cannot exist any natural number satisfying the given properties. ◄

You have likely seen this principle in use even if not by name. For example, an infinite descent argument for the irrationality of $\sqrt{2}$ argues that if $\sqrt{2} = \frac{a}{b}$, then the re-writing $2b^2 = a^2$ eventually shows that a and b must both be even, and so dividing each by two gives a solution with smaller a and b. Since the (positive) integers a and b could not possibly be made smaller and smaller indefinitely, the supposition that there existed *any* such a and b must have been incorrect.

We begin with a proof of the $n = 4$ case, one of the few complete proofs we have provided by Fermat himself. The proof will be somewhat historically authentic, touching upon several problems that Fermat studied over his life. Afterward, we will move to the case $n = 3$, where it fell to Euler to write up an essentially complete proof. Following his argument will feel much more modern by comparison, making arguments based on unique factorization in the ring of Eisenstein integers.

The case $n = 4$

A notion dating back to Diophantus is that of a *congruum*, a common difference between a three-term arithmetic progression of integer squares. For example, 24 is a congruum because we have the three-term arithmetic progression of squares 1, 25, 49 with common difference 24. If we expand our scope to arithmetic progressions of *rational* squares, then we get the notion of a *congruent number*. For example, 6 is a congruent number since it is the common difference of the arithmetic sequence $\left(\frac{1}{2}\right)^2, \left(\frac{5}{2}\right)^2, \left(\frac{7}{2}\right)^2$. At first glance, congruent numbers seem hard to come by, but as we will prove momentarily, they are precisely those numbers equal to the areas of right triangles with rational sides. For several different values of n, Table 9.1 gives both a Pythagorean triple (a, b, c) corresponding to a triangle of area n and a triple (x, y, z) of rational numbers whose squares form an arithmetic progression with common difference n.

Table 9.1 Congruent Numbers

n	(a, b, c)	(x, y, z)
6	$(3, 4, 5)$	$\left(\frac{1}{2}, \frac{5}{2}, \frac{7}{2}\right)$
24	$(6, 8, 10)$	$(1, 5, 7).$
30	$(5, 12, 13)$	$\left(\frac{7}{2}, \frac{13}{2}, \frac{17}{2}\right).$
84	$(7, 24, 25)$	$\left(\frac{17}{2}, \frac{25}{2}, \frac{31}{2}\right).$

The proof of the following prompt is largely tantamount to pattern recognition, deducing the relationship between the last two columns of the table. Given an (a, b, c), how does one construct an (x, y, z), and vice versa?

▶ **Prompt 9.1.3** Prove that $n \in \mathbb{Q}$ is a congruent number if and only if there exists a right triangle with positive rational sides and area n.

Modern authors typically use the interpretation in terms of right triangles as the *definition* of congruent numbers, and we shall feel free to move between the two as needed.

▶ **Prompt 9.1.4** Prove using either interpretation that if $n \in \mathbb{Q}$ is a congruent number, then so are nd^2 and $\frac{n}{d^2}$ for all non-zero $d \in \mathbb{Q}$.

The combination of these two prompts provides us with a simple way to generate lots of congruent numbers by leveraging our classification of Pythagorean triples (see Corollary 3.4.11).

▶ **Prompt 9.1.5** Find all primitive Pythagorean triples with $u \leq 5$ in our standard (u, v)-parameterization and, by computing the areas of the corresponding triangles, give some examples of congruent numbers. Then use the previous prompt to conclude that, less obviously, 5 and 21 are also congruent.

Fermat's proof of His Last Theorem for exponent 4 comes down to the remarkably simple fact that $n = 1$ is not a congruent number.

Theorem 9.1.6 (Fermat)
There does not exist a right triangle with positive rational side lengths and area 1.

▶ **Prompt 9.1.7** Argue by clearing denominators that it suffices to prove that no integer square is the area of a primitive Pythagorean triple[3].

Moving from rationals to integers allows us to (a) unambiguously use "square" to mean "integer square," but more importantly to (b) employ infinite descent: we suppose for the sake of contradiction that we have a perfect square d^2 that is the area of a primitive Pythagorean triple, and subsequently argue that we can construct a smaller such perfect square. The following few prompts achieve this, using yet again our standard (u, v)-parameterization of all primitive Pythagorean triples.

▶ **Prompt 9.1.8** Show that if d^2 (for $d \in \mathbb{N}$) is the area of a primitive (u, v)-Pythagorean triple, then u, v, $u + v$, and $u - v$ must all be perfect squares.

Thus, to continue, we may write $u = j^2$, $v = k^2$, $u + v = \ell^2$, and $u - v = m^2$.

▶ **Prompt 9.1.9** Show that $\gcd(\ell - m, \ell + m) = 2$ and that $2k^2 = (\ell + m)(\ell - m)$ to deduce that $\ell + m$ and $\ell - m$ must take the form $2s^2$ and $4t^2$ (in some order) with s odd.

▶ **Prompt 9.1.10** Prove that the triple $(s^2, 2t^2, j)$ is a primitive Pythagorean triple with a perfect square area less than d^2, establishing the descent. Conclude that d^2, and hence 1, is not congruent.

Finally, back to Fermat's Theorem:

▶ **Prompt 9.1.11** Suppose we have a positive integer solution to $x^4 + y^4 = z^4$. Show that the Pythagorean triple parameterized by $(u, v) = (z^2, y^2)$ has area a perfect square, providing a contradiction.

▶ **Prompt 9.1.12** In fact, the proof above also proves that there are no non-trivial solutions to the Diophantine equation $x^4 + y^4 = z^2$ as well. How?

Note that Fermat's argument, as roughly paraphrased here, does not make explicit use of the factorization

$$z^4 = x^4 + y^4 = (x + y)(x - y)(x + iy)(x - iy),$$

but we can see all of the ingredients lurking in our Gaussian integer interpretation of the classification of Pythagorean triples. Moving on to $n = 3$, we will not be able to be so circumspect about working in other rings.

---

[3] that is, the area of a triangle corresponding to a primitive Pythagorean triple

The case $n = 3$

Euler's 1770 proof of Fermat's Last Theorem for $n = 3$ begins, like so many of our problems, by embedding the problem in a different ring. We recall the ring

$$\mathbb{Z}[\zeta_3] = \{a + b\zeta_3 : a, b \in \mathbb{Z}\} \qquad \zeta_3 = \frac{-1 + \sqrt{-3}}{2}$$

of Eisenstein integers, and in particular that we have shown that $\mathbb{Z}[\zeta_3]$ is a unique factorization domain. Thus, arithmetic in this ring allows us to make sense of gcds and relative primeness, to reason about factorization and valuations, and to apply milestone results like the Power Lemma. Our goal, therefore, is to move to this ring, use the properties above, and prove the following expanded result:

> **Theorem 9.1.13 (Euler, 1770)**
> There are no solutions to the equation
>
> $$\alpha^3 + \beta^3 + u\omega^3 = 0 \qquad\qquad (*)$$
>
> with non-zero $\alpha, \beta, \omega \in \mathbb{Z}[\zeta_3]$ and $u \in \mathbb{Z}[\zeta_3]^\times$.

This is certainly sufficient, as taking $u = -1$ implies Fermat's Last Theorem for $n = 3$ in the ring $\mathbb{Z}[\zeta_3]$, which contains $\mathbb{Z}$. Thus, in particular, this result must hold for integers $\alpha, \beta, \omega \in \mathbb{Z}$. Note however that the generalization carries a little risk since in principal by working in a larger ring and with more general coefficients, we might have introduced solutions that weren't in $\mathbb{Z}$. You will show below that this does not in fact happen. The remainder of the section should be considered a proof of this theorem, with notation being kept consistent from one prompt to the next.

▶ **Prompt 9.1.14** Argue that without loss of generality, in $(*)$ we can take α, β, and ω to be pairwise relatively prime.

Our approach to infinite descent will be to show that given a solution (α, β, ω) to $(*)$, we can factor out $\sqrt{-3}$ from ω to get a new, smaller, solution. Since there can't be infinitely many $\sqrt{-3}$'s to factor out of ω, this leads to a contradiction by descent. We begin by showing that $\sqrt{-3}$ must divide *one* of α, β, ω, and that we can assume without loss of generality that $\sqrt{-3}$ divides ω. (We will exploit this "no loss of generality" several times—this is the point of allowing the unit u in Euler's extension of the problem).

▶ **Prompt 9.1.15** Explore the congruence classes modulo $\sqrt{-3}$ in $\mathbb{Z}[\zeta_3]$: show that $\zeta_3 \equiv 1 \pmod{\sqrt{-3}}$ and that the only congruence classes modulo $\sqrt{-3}$ are $\{0, \pm 1\}$. What are the possible congruence classes of units?

▶ **Prompt 9.1.16** Prove that $\sqrt{-3}$ is prime in $\mathbb{Z}[\zeta_3]$. Deduce that if $\sqrt{-3} \mid \sigma^3$ for some $\sigma \in \mathbb{Z}[\zeta_3]$, then $\sqrt{-3} \mid \sigma$ and thus $3\sqrt{-3} \mid \sigma^3$.

▶ **Prompt 9.1.17** Determine the congruence classes of cubes mod 9: if $\sigma \in \mathbb{Z}[\zeta_3]$ is not congruent to 0 mod $\sqrt{-3}$, what can you deduce about σ^3 mod 9? (Hint: Work with congruence classes mod $\sqrt{-3}$ and use the previous prompts. For example, what if $\sigma \equiv 1 \pmod{\sqrt{-3}}$?)

▶ **Prompt 9.1.18** Conclude from the previous prompt that in every solution to (∗), exactly one of α, β, ω is divisible by $\sqrt{-3}$. Argue that without loss of generality we can assume $\sqrt{-3} \mid \omega$.

Now we begin ... the descent! Suppose a solution to (∗) exists. Then out of *all* such solutions, choose one with the smallest possible $\sqrt{-3}$-adic valuation $v_{\sqrt{-3}}(\omega)$, i.e., such that $\omega = \sqrt{-3}^k \delta$ for the smallest possible k. We will establish the existence of a solution with a smaller such k, concluding the descent.

▶ **Prompt 9.1.19** The previous prompt argues that $v_{\sqrt{-3}}(\omega) \geq 1$. In fact, show we must have $v_{\sqrt{-3}}(\omega) \geq 2$. (Hint: Take (∗) mod 9.)

Though we have already been playing with $\sqrt{-3}$ for some time now, it is the next prompt, the ability to factor $\alpha^3 + \beta^3$ in $\mathbb{Z}[\zeta_3]$, that retroactively justifies the insight to work in $\mathbb{Z}[\zeta_3]$ in the first place.

▶ **Prompt 9.1.20** Since $\sqrt{-3}^3 \mid u\omega^3 = \alpha^3 + \beta^3$, the factorization at the start of the section suggests factoring $\alpha^3 + \beta^3$ as a product of three factors τ_1, τ_2, τ_3 such that $\sqrt{-3} \mid \tau_i$ for each i. Do so[4].

Next, the construction phase of the descent:

▶ **Prompt 9.1.21** Show that $\frac{\tau_1}{\sqrt{-3}}, \frac{\tau_2}{\sqrt{-3}}$, and $\frac{\tau_3}{\sqrt{-3}}$ are relatively prime[5] and thus (up to units) are cubes in $\mathbb{Z}[\zeta_3]$, say, of elements α', β', ω'. Argue without loss of generality that we can choose ω' to be the only one divisible by $\sqrt{-3}$, and so have $0 < v_{\sqrt{-3}}(\omega') < v_{\sqrt{-3}}(\omega)$.

To summarize, we began with the equation $\alpha^3 + \beta^3 - u\omega^3 = 0$, factored $\alpha^3 + \beta^3 = \tau_1\tau_2\tau_3$, and realized that each τ_i must be divisible by $\sqrt{-3}$ and that the quotient $\frac{\tau_i}{\sqrt{-3}}$

---

[4] We're just calling them τ_i so as not to spoil the game for you. You'll need to figure out what elements of $\mathbb{Z}[\zeta_3]$ they actually are.

[5] This technically requires three checks. We have used the "the other cases are similar" card sufficiently many times on you that we feel you can use one back on us here.

itself must be a cube (up to units). The three such cubes will be our smaller solution for the purposes of descent. That is, we will conclude by showing that

$$(\alpha')^3 + (\beta')^3 + u'(\omega')^3 = 0 \qquad (**)$$

for some unit $u' \in \mathbb{Z}[\zeta_3]$.

▶ **Prompt 9.1.22** Find a unit linear combination $v_1(\alpha')^3 + v_2(\beta')^3 + v_3(\omega')^3$ equal to 0, taking for coefficients the units 1, ζ_3, ζ_3^2 in some order (recall $1 + \zeta_3 + \zeta_3^2 = 0$). It may help to do this for τ_1, τ_2, τ_3 in place of the cubes first.

▶ **Prompt 9.1.23** The linear combination $v_1(\alpha')^3 + v_2(\beta')^3 + v_3(\omega')^3 = 0$ of the previous prompt nearly provides $(**)$. Note we can dispose of v_1 by division. Finally, argue we can dispose of the remaining unit coefficient $v_1^{-1}v_2$ of $(\beta')^3$ using our previous work about congruence classes mod 9 and mod $3\sqrt{-3}$.

▶ **Prompt 9.1.24** Synthesize your argument. Why are we now done? Finally, bask in the glow of a job well done.

...and beyond

Proving Fermat's Last Theorem for $n = 3$ and $n = 4$ could be glibly dismissed as having merely proven the first two of infinitely many cases. Of course, there is some punch to this criticism, as it certainly was not clear how to generalize these results to other n. But, first note that some cases come along for free:

▶ **Prompt 9.1.25** Argue that as a consequence of our work so far, we can quickly deduce Fermat's Last Theorem for $n = 6, 8$, and 9. Generalize.

But most importantly, there was meat enough in these two proofs of Fermat and Euler to keep mathematicians busy for centuries with their generalizations.

▶ **Prompt 9.1.26** Research: The previous prompt shows that only $n = 5$ and $n = 7$ remain as single-digit potential counterexamples to Fermat's Last Theorem. When were these two cases resolved? How? By whom?

▶ **Prompt 9.1.27** Explore the development of Fermat's Last Theorem between Fermat and Wiles. When were early cases proved? Who were the main players? Pay particular attention to false proofs and where they went wrong.

▶ **Prompt 9.1.28** Research: For which n does $\mathbb{Z}[\zeta_n]$ have unique factorization? Most n? Just a few? Chronicle the historical progress towards the answers to these questions, and their current status.

Finally, we note that questions surrounding this material remain at the forefront of mathematical research even today. For example, we saw Fermat's proof for $n = 4$ depended crucially on the argument that 1 is not a congruent number. To this day, we still do not have a provably efficient algorithm for deciding if a given n is congruent. As testimony to its difficulty, note that $n = 23$ is congruent thanks to the rational Pythagorean triple

$$(a, b, c) = \left(\frac{80155}{20748}, \frac{41496}{3485}, \frac{905141617}{72306780} \right).$$

Our most promising lead on this problem is to relate the question to one about (surprise!?) elliptic curves.

▶ **Prompt 9.1.29** Show that if n is a congruent integer, say from Pythagorean triple (a, b, c), then the point $(x, y) = \left(\frac{c}{2}, \frac{c(a-b)(a+b)}{8} \right)$ defines a rational point on the elliptic curve $y^2 = x^3 - n^2 x$.

In fact, by repeatedly using the chord/tangent method developed in Chapter 2, we can construct *infinitely many* rational points on the elliptic curve corresponding to a congruent n. Furthermore, the result is reversible—if the curve $y^2 = x^3 - n^2 x$ has infinitely many rational points, then n is congruent. Questions surrounding the infinitude of rational solutions to an elliptic curve form a large research subdiscipline within number theory.

▶ **Prompt 9.1.30** Research challenge: What is known about the "fraction" of elliptic curves that have infinitely many rational solutions? Look up Tunnell's Theorem: How does it imply the existence of an algorithmic check to decide whether a number is congruent?

9.2   Exploration: Lagrange's Four-Square Theorem

We have countless times expanded our notion of number to accommodate new ideas arising in our attempts to solve problems. One of the first such instances was using the Gaussian integers for problems involving sums of two squares, thanks to the norm-factorization $a^2 + b^2 = (a + bi)(a - bi) = N(a + bi)$. As a consequence, we were able to classify in Theorem 5.7.1 precisely those integers that can be written as a sum of two squares. For those that *can't*, some require at least three squares ($6 = 2^2 + 1^2 + 1^2$), while still others require four ($7 = 2^2 + 1^2 + 1^2 + 1^2$). A natural question is whether there is some maximum number of squares ever required or if there are natural numbers that require arbitrarily many. Rather that maintain excessive suspense, we answer the question here, with the goal of getting more quickly to the new class of numbers we will introduce to study the problem.

> **Theorem 9.2.1 (Lagrange's Four-Square Theorem)**
> Every natural number can be written as a sum of at most four squares of natural numbers.

That is, alleviating our worry of three sentences ago, there is no natural number that requires more than four squares. Let's outline our proof strategy, much of which will seem familiar when compared to the analogous argument for sums of two squares. We will introduce a new number system, the quaternions, and their ring of integers, the Hurwitz integers. What norms of Gaussian integers did for sums of two squares, norms of Hurwitz integers will do for sums of four squares. But unlike the Gaussian integers, the Hurwitz integers form a *non-commutative* ring, and so the principal theoretical track of the section revolves around replicating a version of *The Path* (Figure 6.1) in this non-commutative setting. Namely, we'll show that the Hurwitz integers have a perfectly lovely Division Algorithm that gets them "close enough" to being a Euclidean Domain, in that we'll be able to show that for integer primes p the notions of being "prime" and "irreducible" are equivalent in the ring of Hurwitz integers (though not for non-integer primes). Proving this will require a Euclid's Lemma for the Hurwitz integers, which requires Bézout's Identity, which requires the Euclidean Algorithm, which requires our "perfectly lovely Division Algorithm," which requires a notion of divisibility, etc.

▶ **Prompt 9.2.2**   Show that for Theorem 9.2.1 it is sufficient to prove that:

- If $m, n \in \mathbb{N}$ can be written as a sum of four squares, then so can mn.
- Every odd (positive) prime can be written as a sum of four squares.

The multiplicativity of the problem, as shown in the above prompt, is our hint that factorization and norm calculations might be of use again here. So we turn to introducing these new numbers and their norms. It is convenient to embed some of our discussions in the language of matrices. As a warm-up exercise, it is instructive to see how we could view our standard arithmetic of complex numbers through the lens of matrix arithmetic.

▶ **Prompt 9.2.3**   Define two 2×2 matrices, I and J, as follows:

$$I = \begin{bmatrix} 1 & 0 \\ 0 & 1 \end{bmatrix} \quad \text{and} \quad J = \begin{bmatrix} 0 & 1 \\ -1 & 0 \end{bmatrix}.$$

Verify that $J^2 = -I$ and that for $a, b, c, d \in \mathbb{R}$ we have the matrix identity

$$(aI + bJ)(cI + dJ) = (ac - bd)I + (ad + bc)J,$$

analogous to the standard complex result $(a+bi)(c+di) = (ac-bd)+(ad+bc)i$.

This matrix interpretation of complex arithmetic permits the use of tools from linear algebra. For example:

▶ **Prompt 9.2.4** Take the determinant of both sides of the previous matrix identity to deduce the "Two Square Identity" for $a, b, c, d \in \mathbb{R}$:

$$(a^2 + b^2)(c^2 + d^2) = (ac - bd)^2 + (ad + bc)^2.$$

We now look toward sums of four squares instead of two, so we would like to have a four-dimensional analog of this identity.

Definition 9.2.5

Define 2×2 matrices $\mathbf{1}, \mathbf{i}, \mathbf{j}$, and $\mathbf{k}$ by

$$\mathbf{1} = \begin{bmatrix} 1 & 0 \\ 0 & 1 \end{bmatrix} \qquad \mathbf{i} = \begin{bmatrix} i & 0 \\ 0 & -i \end{bmatrix} \qquad \mathbf{j} = \begin{bmatrix} 0 & 1 \\ -1 & 0 \end{bmatrix} \qquad \mathbf{k} = \begin{bmatrix} 0 & i \\ i & 0 \end{bmatrix}.$$

We define the set of **quaternions** by

$$\begin{aligned}
Q &= \mathbb{R}[\mathbf{i}, \mathbf{j}, \mathbf{k}] \\
&= \{a\mathbf{1} + b\mathbf{i} + c\mathbf{j} + d\mathbf{k} : a, b, c, d \in \mathbb{R}\}. \\
&= \left\{ \begin{bmatrix} z & w \\ -\overline{w} & \overline{z} \end{bmatrix} : z, w \in \mathbb{C} \right\} \quad \text{(writing } z = a + bi \text{ and } w = c + di)
\end{aligned}$$ ◀

The definition provides two ways of interpreting quaternions. The first, like complex numbers, is an abstract expression of the form $a + b\mathbf{i} + c\mathbf{j} + d\mathbf{k}$ (where we abbreviate $a\mathbf{1} = a$), and the second as an explicit 2×2 matrix. We typically adopt the language of the first perspective, calling such a representation the "standard form" of the quaternion, and when we want to explicitly call attention to the matrix itself, we will reference its "matrix representation."

▶ **Prompt 9.2.6** Let $\alpha = (3 + 2\mathbf{i} - 4\mathbf{j} + \mathbf{k})$ and $\beta = -6(2 + 3\mathbf{i} - \mathbf{j} - 3\mathbf{k})$. Compute $\alpha - 6\beta$ and $\alpha\beta$. Report your result in both the standard form and the matrix representation.

Let's continue to get acquainted with this set.

▶ **Prompt 9.2.7** Prove that the quaternions satisfy a non-commutative integral domain condition: if $\alpha\beta = 0$ for some $\alpha, \beta \in Q$, then $\alpha = 0$ or $\beta = 0$.

▶ **Prompt 9.2.8** Prove that the quaternions form a vector space of dimension four over $\mathbb{R}$ with basis $\{\mathbf{1}, \mathbf{i}, \mathbf{j}, \mathbf{k}\}$.

▶ **Prompt 9.2.9** Show that $\mathbf{i}^2 = \mathbf{j}^2 = \mathbf{k}^2 = -1$ and verify the products $\mathbf{ij} = \mathbf{k}$, $\mathbf{jk} = \mathbf{i}$, and $\mathbf{ki} = \mathbf{j}$. Finally, show that reversing the order of the multiplication in these last three products changes the sign of the result.

These relationships are summarized neatly in the convenient diagram above: multiplying two distinct elements of $\{\mathbf{i}, \mathbf{j}, \mathbf{k}\}$ in the order depicted by the arrows (clockwise) gives the remaining element, and multiplying against the arrows (counterclockwise) gives the negative of that same result.

▶ **Prompt 9.2.10** Show (e.g., using matrix arithmetic) how to find a multiplicative inverse of a quaternion.

▶ **Prompt 9.2.11** Prove that Q forms a non-commutative ring with unity under matrix addition and multiplication, and all non-zero elements of Q have an inverse under multiplication. (We call such a ring, one satisfying all the conditions of a field except for commutativity, a **division ring**.)

The "squaring to -1" aspect of the quaternions is just the start of the analogies between Q and $\mathbb{C}$. For example, we can make sense of conjugates of quaternions, flipping the signs of the non-real parts:

Definition 9.2.12

If $\alpha = a + b\mathbf{i} + c\mathbf{j} + d\mathbf{k}$, define

$$\overline{a + b\mathbf{i} + c\mathbf{j} + d\mathbf{k}} = a - b\mathbf{i} - c\mathbf{j} - d\mathbf{k}.$$

◀

▶ **Prompt 9.2.13** What does the conjugate of a quaternion look like when written in its matrix representation?

▶ **Prompt 9.2.14** Prove that $\overline{\alpha\beta} = \overline{\beta}\,\overline{\alpha}$ for all $\alpha, \beta \in Q$.

Definition 9.2.15

The **norm** of a quaternion $\alpha = a + b\mathbf{i} + c\mathbf{j} + d\mathbf{j}$ is

$$N(\alpha) = \alpha\overline{\alpha} = a^2 + b^2 + c^2 + d^2.$$

◀

▶ **Prompt 9.2.16** Verify that the norm of a quaternion equals the determinant of its 2×2 matrix representation.

▶ **Prompt 9.2.17** Prove that for $\alpha, \beta \in Q$, we have $N(\alpha\beta) = N(\alpha)N(\beta)$.

This beautiful interpretation of the norm of a quaternion lays bare the application to our sums of four squares problem. To complete the connection, and prove the first of the two ingredients in Prompt 9.2.2, we need only restrict our attention to integer coefficients.

Definition 9.2.18

The **Lipschitz integers** L are the quaternions with integer coefficients:

$$L = \mathbb{Z}[i, j, k] = \{a + bi + cj + dk : a, b, c, d \in \mathbb{Z}\}. \qquad ◀$$

We can now complete the first phase of the outline in Prompt 9.2.2:

▶ **Prompt 9.2.19** Show that if integers m and n are each the sum of four (integer) squares, then so is mn.

For the second phase of our plan, we will need to perform our standard structural exploration of the new ring L: units, primes, factorizations, etc.

▶ **Prompt 9.2.20** Show that $L^{\times} = \{\pm 1, \pm \mathbf{i}, \pm \mathbf{j}, \pm \mathbf{k}\}$.

▶ **Prompt 9.2.21** Use norm arguments to find some irreducible elements of L.

The Lipschitz integers extend the Gaussian integers in the sense that $\mathbb{Z}[i]$ can be thought of as consisting of the Lipschitz integers with $c = d = 0$. As with Gaussian integers, the norm map serves a role greater than just forming expressions of sums of squares. Indeed, the norm of a Gaussian integer represents its *size*, and played a major role in generalizing the Division Algorithm from $\mathbb{Z}$ to $\mathbb{Z}[i]$ (and beyond). Let's see what we can do in L:

▶ **Prompt 9.2.22** Experiment with a quaternionic Division Algorithm:

1. Let $\alpha = 4 + 7\mathbf{i} + \mathbf{j} + 8\mathbf{k}$ and $\beta = 1 - \mathbf{i} + 2\mathbf{j} + 2\mathbf{k}$. Calculate $\alpha\beta^{-1}$.
2. Find a nearest Lipschitz integer to your quotient (i.e., round each coefficient to an integer), and use that to find γ and ρ such that $\alpha = \gamma\beta + \rho$.
3. Compute and then compare $N(\rho)$ and $N(\beta)$.
4. Use a different rounding to find a new γ and ρ and compare their norms.
5. What does this suggest about a Division Algorithm for the Lipschitz integers?

Oh, dear! And it seemed like such a nice norm, too. But, alas, we can't guarantee a small enough remainder, which is the driving force behind the Division Algorithms in $\mathbb{Z}$, $\mathbb{Z}[i]$, $\mathbb{Z}[\zeta_3]$, etc. However, inspired by our success with the Eisenstein integers, in which we exchanged the rectangular lattice geometry of $\mathbb{Z}[\sqrt{-3}]$ for the parallelogrammatic structure of $\mathbb{Z}[\zeta_3]$, we might be tempted to try to *adjust* the lattice a little so that the center of each cell is less than one unit away from any corner. The major obstacle we have here is that most of us can't visualize this lattice because it is four dimensional! Or, at least, not with *that* attitude.

Accordingly, begin by imagining a 4-dimensional integer lattice. A smallest "cube" would have a diagonal ranging from the origin $(0, 0, 0, 0)$, say, to the point $(1, 1, 1, 1)$. The length of that diagonal would be $\sqrt{N(1, 1, 1, 1)} = 2$, so the center is precisely 1 unit away from the vertices: it only *just* misses our goal! (Remember that we need the distance to be *strictly* less than 1). As a consequence, this center point is the *only* point that causes us a problem – so what if we include in our lattice this one extra point $(\frac{1}{2}, \frac{1}{2}, \frac{1}{2}, \frac{1}{2})$ (and its translated compatriots in the centers of other "cubes," of course)?

Definition 9.2.23

The **Hurwitz integers** are defined by

$$\mathbb{H} = \mathbb{Z}\left[\frac{1 + i + j + k}{2}, \, i, \, j, \, k\right]$$

$$= \{a + bi + cj + dk : a, b, c, d \in \mathbb{Z} \text{ or } a, b, c, d \in \mathbb{Z} + \tfrac{1}{2}\}. \qquad \blacktriangleleft$$

That is, we expand from the Lipschitz integers to also include quaternions whose coefficients are *all* $\frac{1}{2}$ more than an integer.

▶ **Example 9.2.24** Verify that $\frac{5}{2} + \frac{7}{2}i - \frac{9}{2}j - \frac{1}{2}k$ and $4 - i + 3j - 6k$ are both Hurwitz integers, but that $3 + \frac{1}{2}i + 3j - \frac{1}{2}k$ and $\frac{3}{2}i - \frac{1}{2}j + \frac{7}{2}k$ are not.

▶ **Prompt 9.2.25** Show the two definitions in Definition 9.2.23 are equivalent.

Structurally, $\mathbb{H}$ inherits most of its algebraic properties from Q — the principal consideration is verifying that its somewhat unorthodox definition makes it a ring at all.

▶ **Prompt 9.2.26** Prove that $\mathbb{H}$ is a ring with unity, but not a division ring.

▶ **Prompt 9.2.27** Prove that for all $\alpha \in \mathbb{H}$, we have $N(\alpha) \in \mathbb{Z}$. Deduce that the units of $\mathbb{H}$ are precisely the elements of norm 1.

Now for the point: $\mathbb{H}$ was constructed as an extension of L for the express purpose of making a Division Algorithm work well. Of course, since $\mathbb{H}$ is not commutative we cannot call it a Euclidean Domain, and in particular we don't get for free all the consequences of being a Euclidean Domain (e.g., being a UFD). Nevertheless, we will see that the Division Algorithm we do get is good enough to make the factorization arguments we need.

▶ **Prompt 9.2.28** Mimic earlier Division Algorithm proofs to prove that $\mathbb{H}$ satisfies a non-commutative version of the Division Algorithm: for $\alpha, \beta \in \mathbb{H}$ with $\beta \neq 0$, there exist $\chi, \rho \in \mathbb{H}$ such that $\alpha = \beta\chi + \rho$ and $0 \leq N(\rho) < N(\beta)$.

The proof is nearly identical to all of our prior encounters with Division Algorithms (consider $\alpha\beta^{-1} \in Q$ and how close this element must be to an element of $\mathbb{H}$), but the lack of commutativity requires extra care in maintaining the order of multiplications. Similarly, we will have to be careful in our non-commutative versions of divisibility (compare to Definition 3.1.14):

Definition 9.2.29

Given $\alpha, \beta \in H$, we say that α **left-divides** β if there exists $\gamma \in H$ such that $\alpha\gamma = \beta$, in which case we write $\alpha \mid_L \beta$. We say that α is a **left-divisor** of β. ◀

Though we will not use the analogous "right" notions[6], we would write $\alpha \mid_R \beta$ if we could write $\gamma\alpha = \beta$. In non-commutative rings, these are not necessarily equivalent!

▶ **Prompt 9.2.30** Show that $\alpha = 1+\mathbf{i}+2\mathbf{j}+3\mathbf{k}$ left-divides, but does not right-divide, $\beta = -16 + 5\mathbf{i} + 13\mathbf{j}$.

Definition 9.2.31

A **left gcd** $\delta = \mathrm{lgcd}(\alpha, \beta)$ of $\alpha, \beta \in \mathbb{H}$ is a Hurwitz integer such that δ left-divides both α and β and such that if γ is another left-divisor of both α and β, then γ left-divides δ. An element $\pi \in \mathbb{H}$ is a (left) **Hurwitz prime** if $\pi \mid_L \alpha\beta$ implies $\pi \mid_L \alpha$ or $\pi \mid_L \beta$. ◀

▶ **Remark 9.2.32** We adopt the same conventions as in Chapter 6 with regard to writing equalities like $\delta = \mathrm{lgcd}(\alpha, \beta)$ despite left gcds not being unique.

▶ **Prompt 9.2.33** Prove that the definition of left gcd is equivalent to the following: δ is a left gcd of α and β if and only if δ is a left divisor of both α and β and if γ is a left divisor of both, then $N(\gamma) \leq N(\delta)$.

[6] Arguably, the left notion *is* the right notion.

With divisibility and the Division Algorithm sorted out, we can progress down the path: the Linear Combination Lemma (Lemma 3.2.4) holds without issue in any ring as long as we are consistent about which side we are dividing on. As a consequence of this and the Division Algorithm, we get natural analogs of the Euclidean Algorithm and Bézout's Identity. Mimic prior instances of these proofs for the following prompt.

▶ **Prompt 9.2.34** Let $\alpha, \beta \in \mathbb{H}$. Prove that if $\alpha = \beta\chi + \rho$ for some $\chi, \rho \in \mathbb{H}$, then $\mathrm{lgcd}(\alpha, \beta) = \mathrm{lgcd}(\beta, \rho)$. Furthermore, if $\delta = \mathrm{lgcd}(\alpha, \beta)$, then there exist $\mu, \nu \in \mathbb{H}$ such that $\delta = \alpha\mu + \beta\nu$.

We keep insinuating that things go awry somewhere, and it's time for the hammer to drop—despite having a Division Algorithm, a Euclidean Algorithm, Bézout's Identity, and so on, it's just not true that $\mathbb{H}$ forms a unique factorization domain (even if the definition of UFD didn't require commutativity). There is a weaker notion of factorization that $\mathbb{H}$ does satisfy, but even better is that for our purposes, we only care about factorizations involving *integer* primes p. In this case, we have a very UFD-like result.

Theorem 9.2.35

Let $p \in \mathbb{Z}$ be an integer prime. Then the Hurwitz integer $p = p + 0\mathbf{i} + 0\mathbf{j} + 0\mathbf{k} \in \mathbb{H}$ is irreducible (in $\mathbb{H}$) if and only if it is prime (in $\mathbb{H}$).

We split the proof into two prompts.

▶ **Prompt 9.2.36** For the forward direction, assume $p \in \mathbb{Z}$ is irreducible in $\mathbb{H}$ and suppose $p \mid_L \alpha\beta$ for some $\alpha, \beta \in \mathbb{H}$. Show that if $p \nmid_L \alpha$, then $p \mid_L \beta$. (The proof is the same as that of the Prime Divisor Property, with extra care taken with regard to order of multiplication).

▶ **Prompt 9.2.37** Now assume that $p \in \mathbb{Z}$ is a Hurwitz prime and that $p = \alpha\beta$ for some $\alpha, \beta \in \mathbb{H}$. Show that if α is not a unit in $\mathbb{H}$, then β must be. Conclude that p is irreducible in $\mathbb{H}$.

This equivalence sure makes it seems like we're headed for the inevitable journey of deciding which primes are irreducible in $\mathbb{H}$ and which are not, perhaps governed by congruence conditions on p. But in the name of periodically subverting expectations, we present the following plot twist:

Theorem 9.2.38
For all integer primes p, the element p is reducible in $\mathbb{H}$.

By Theorem 9.2.35, it suffices to show that an integer prime p is never a Hurwitz prime. We will show that it is always possible to construct $\alpha, \beta \in \mathbb{H}$ such that $p \mid_L \alpha\beta$, but p left-divides neither α nor β. This is typically not too difficult on a case-by-case basis, as left-divisibility by $p \in \mathbb{Z}$ is straightforward.

▶ **Prompt 9.2.39** Let $p = 7$. Find $a, b, c, d \in \mathbb{Z}$ such that $p \mid (a^2 + b^2 + c^2 + d^2)$ but p divides neither $(a + b\mathbf{i} + c\mathbf{j} + d\mathbf{k})$ nor its conjugate. Conclude that $7 \in \mathbb{H}$ is not prime.

On the other hand, it is not clear how to systematize this process. We turn to Lagrange's simplification of this problem, showing that we can restrict our attention to factorizations of the form $1 + b^2 + c^2 = (1 + b\mathbf{i} + c\mathbf{j})(1 - b\mathbf{i} - c\mathbf{j})$. Since $p \nmid 1$ in $\mathbb{Z}$, p does not divide either of the factors on the right, so it suffices to show that for any prime p, there exist integers b and c such that $p \mid 1 + b^2 + c^2$.

▶ **Prompt 9.2.40** A warm-up for the general proof, focused on $p = 13$: For each of the 7 values of b and c in the range $0 \leq b, c \leq 6$, compute $-1 - b^2$ mod p and c^2 mod p. Which pairs (b, c) give $13 \mid 1 + b^2 + c^2$?

▶ **Prompt 9.2.41** Fix p an odd prime in $\mathbb{Z}$. Show that there must exist integers b and c such that $c^2 \equiv -1 - b^2 \pmod{p}$, and so $p \mid 1 + b^2 + c^2$. (It may help to prove that the values of c^2 must all be distinct for c in the range $0 \leq c \leq \frac{p-1}{2}$.)

▶ **Prompt 9.2.42** Connect the dots and give a careful proof of Theorem 9.2.38.

This is precisely the information we need to write primes as sums of four squares. In the Gaussian integers, we used that if we have a non-trivial factorization $p = \alpha\beta$, then $N(p) = p^2 = N(\alpha)N(\beta)$, and so $N(\alpha) = p$. Since writing p as a norm from $\mathbb{Z}[i]$ is equivalent to writing it as a sum of two squares, we're done. To replicate this argument for sums of four squares, we have one hurdle to jump: our recent structural results are about $\mathbb{H}$, but writing p as a sum of four integer squares is equivalent to writing p as a norm from L. Fortunately, this hurdle is only knee-high.

▶ **Prompt 9.2.43** Let p be an odd prime. Use the reducibility of p in $\mathbb{H}$ to show that p must be the norm of some Hurwitz integer.

If p is the norm of a Hurwitz integer $a + b\mathbf{i} + c\mathbf{j} + d\mathbf{k}$ for $a, b, c, d \in \mathbb{Z}$ we are done. To resolve the case for $a, b, c, d \in \mathbb{Z} + \frac{1}{2}$, we introduce an algebraic technique showing that any sum of four half-integer squares is also a sum of four *integer* squares. We begin with an illustrative example:

▶ **Prompt 9.2.44** Consider the representation

$$29 = \left(\frac{3}{2}\right)^2 + \left(\frac{5}{2}\right)^2 + \left(\frac{9}{2}\right)^2 + \left(\frac{1}{2}\right)^2.$$
$$= \left(1+\frac{1}{2}\right)^2 + \left(2+\frac{1}{2}\right)^2 + \left(4+\frac{1}{2}\right)^2 + \left(0+\frac{1}{2}\right)^2$$

Re-write each summand $a + \frac{1}{2}$ in the form $b \pm \frac{1}{2}$ with b even and argue this can be done in general.

The $\pm$'s occurring in the coefficients of the above re-writing turn out to be of relevance: Let ω be the Hurwitz integer of the form $\dfrac{\pm 1 \pm \mathbf{i} \pm \mathbf{j} \pm \mathbf{k}}{2}$ with signs chosen to match those found in Prompt 9.2.44, and notice that since $N(\omega) = 1$, we know that ω is a unit. In fact, the sixteen such elements, in conjunction with the eight units of L, comprise the 24 elements of $\mathbb{H}^{\times}$.

▶ **Prompt 9.2.45** With ω as above, verify the following computation:

$$\begin{aligned} 29 &= (\omega + (2 + 2\mathbf{i} + 4\mathbf{j} + 0\mathbf{k})) \cdot \overline{\omega} \cdot \omega \cdot (\overline{\omega} + (2 - 2\mathbf{i} - 4\mathbf{j} - 0\mathbf{k})) \\ &= (\omega\overline{\omega} + 2(1 + \mathbf{i} + 2\mathbf{j})\overline{\omega})(\omega\overline{\omega} + 2\omega(1 - \mathbf{i} - 2\mathbf{j})) \\ &= (1 + 2(1 + \mathbf{i} + 2\mathbf{j})\overline{\omega})(1 + 2\omega(1 - \mathbf{i} - 2\mathbf{j})) \\ &= \overline{(1 + 2\omega(1 - \mathbf{i} - 2\mathbf{j})}(1 + 2\omega(1 - \mathbf{i} - 2\mathbf{j})), \end{aligned}$$

showing that 29 is a norm of a Lipschitz integer. Which representation of 29 as a sum of four squares do you get from this computation?

Our final step is to show that this construction works in general. The key trick is the re-writing from Prompt 9.2.44.

▶ **Prompt 9.2.46** Show that all Hurwitz integer of the form $a + b\mathbf{i} + c\mathbf{j} + d\mathbf{k}$ for $a, b, c, d \in \mathbb{Z} + \frac{1}{2}$ can be written in the form

$$\frac{\pm 1 \pm \mathbf{i} \pm \mathbf{j} \pm \mathbf{k}}{2} + (a' + b'\mathbf{i} + c'\mathbf{j} + d'\mathbf{k})$$

with $a', b', c', d' \in 2\mathbb{Z}$ and some choice of $\pm$ signs.

▶ **Prompt 9.2.47** Mirror the computations in the example to show that every integer that is a norm from $\mathbb{H}$ is also a norm from L.

We have at last reached the punchline. Every positive prime p is a norm from L and so a sum of four squares. The following prompt should require nothing but organizing all of the work we've done thus far.

▶ **Prompt 9.2.48** Prove Theorem 9.2.1.

Finally, though we have this established as a theoretical result, there are still questions about explicit representations as a sum of four squares. We leave the following as an open-ended prompt to continue to investigate.

▶ **Prompt 9.2.49** Verify and experiment with the following computation:

$$
\begin{aligned}
30 &= 2 \cdot 3 \cdot 5 \\
&= (1^2 + 1^2)(1^2 + 1^2 + 1^2)(2^2 + 1^2) \\
&= N(1 + \mathbf{i})\, N(1 + \mathbf{i} + \mathbf{j})\, N(2 + \mathbf{i}) \\
&= \det\left(\begin{bmatrix} 1+i & 0 \\ 0 & 1-i \end{bmatrix} \begin{bmatrix} 1+i & 1 \\ -1 & 1-i \end{bmatrix} \begin{bmatrix} 2+i & 0 \\ 0 & 2-i \end{bmatrix} \right) \\
&= \det \begin{bmatrix} -2+4i & 3+i \\ -3+i & -2-4i \end{bmatrix} \\
&= 2^2 + 4^2 + 3^2 + 1^2.
\end{aligned}
$$

Does this calculation suggest any results on the *number* of ways of writing 30 (or more generally, n) as a sum of four squares?

Some parting thoughts:

▶ **Prompt 9.2.50** Quaternions rose to prominence in applied disciplines for their role in representing rotations in three dimensions. In this context, the *unit* quaternions have special significance as *versors*. Explore!

▶ **Prompt 9.2.51** We have explored quaternions over rings like $\mathbb{Z}$ and $\mathbb{R}$, but could just as well have considered rings like $\mathbb{Z}/(p)[i, j, k]$. Explore the algebra of these rings—are they division rings? Where are these rings used?

9.3 Exploration: Public Key Cryptography

It is certainly not a deep observation that much of today's commerce occurs over the internet. This is fairly remarkable in a number of ways, not the least of which is the implication that a buyer and seller may be literally thousands of miles apart but still can not only agree on the terms of the transaction, but also the *method* of transaction. In particular, it is rare that such a transaction involves the physical exchange of money, either via check or cash sent through the mail. Instead, modern financial transactions depend crucially on the principle of secrecy, that Person A can send Person B their credit card number, allowing Person B to tell Person A's credit card company to dedicate a certain amount of funds to Person B's store. A crucial aspect of this discussion is the inherent security risks in this transaction — the internet, being the series of tubes that it is, transmits that credit card number through any number of waypoints, each of which provides an opportunity for an unwanted eavesdropper[7] to intercept the number and use it for their own nefarious purposes.

So here's the big question: how do you secretly transmit numbers? It is tempting to assume that these concerns should be left to the computer scientists, and while it is true that they are a good and noble people who have much to say on this subject, even they turn to us mathematicians for solutions here. To underscore how little technology is needed, consider a more local example: suppose you would like to transmit a secret numerical message to a stranger on the other side of the room, using only your voice to communicate. You have no prior shared memories to rely upon (no "the first digit is the number of times we had spaghetti last week"), no shared phone numbers through which you can communicate more privately, and, analogous to the general insecurity of the internet, you have to be conscious of the fact that everyone in the room can hear everything you have to say. Yelling "I've posted the information on `darkwebmathsecrets.com` and the access password is `hunter2`, so go read it there" would convey the information to your intended recipient, but also to everyone else in the room.

▶ **Prompt 9.3.1** How can we securely transmit our secret information across the crowded room using only our voice?

With any luck, you have convinced yourself this is entirely impossible, in which case you are in for a breathtakingly pleasant surprise. It is, without exaggeration, completely and totally a miracle of mathematics that it is in fact possible to achieve this task. A solution was first discovered in the twentieth century, though it rests on mathematics developed centuries earlier. Until that point, all secure communications involved some *shared secret key*. Communicating parties had to privately meet and agree upon this shared information. There was no way to conduct secure communications without some *a priori* private communication.

The amazingly simple solution outlined below is the so-called RSA protocol, acronymed after the mathematicians (Ronald Rivest, Adi Shamir, and Leonard Adleman) who in the 1970s first made a protocol of this type public. It is an example of a *public key* cryptosystem, a cryptosystem based on a presumed *one-way function*—an easy-to-compute invertible function whose inverse is difficult to compute, by which we mean any algorithm that attempts to compute the inverse in a "reasonable" amount of time (i.e., less than the age of the universe), will fail for all but a set of values having measure zero in the space of all possible inputs. Public key cryptosystems are

---

[7] Is there any other kind?

based on one-way functions having some sort of trapdoor—additional information which makes calculation of the inverse feasible. This trapdoor information is often called the *private key*. Despite years of research, it is still not known whether one-way functions exist. A proof of the existence of one-way functions would simultaneously solve the famous $\mathcal{P} = \mathcal{NP}$ problem from complexity theory.

9.3.1 Public Key Encryption: RSA

The presumed one-way function upon which RSA is built is modular exponentiation. While there are algorithms that compute x^m mod n efficiently, current methods for calculating the m-th root of a number mod n are computationally infeasible for many large values of n, without knowing some additional trap door information about n. Take a moment to review the various results of Chapter 4 used in the implementation of RSA: the computation of the Euler φ function, our knowledge of units in modular worlds, and Euler's Theorem for computing modular exponentiation.

Definition 9.3.2

Suppose Alicia wants to send a secret numerical[8] message M to Robert. We assume that any information transmitted from Alicia to Robert is interceptable by the public, or even by nefarious interloper Evelyn. The **RSA protocol** for doing so is as follows:

1. **Set Up:** Alicia begins with her secret integer M (the Message). Robert chooses two gigantic primes p and q and forms the number $n = pq$. Robert chooses some $e \in \mathbb{Z}/(\varphi(n))^\times$ (e for "encryption") and publishes n and e, but *not* the primes p or q, in some public place that Alicia can access (as can Evelyn and anyone else motivated to do so).

2. **Encryption:** Alicia encrypts her secret message M using modular exponentiation. Assume $\gcd(M, n) = 1$. If not, Alicia can always pad her message M with irrelevant information (trailing digits) to make this the case. Let C (for "code") be defined by

$$C = M^e \bmod n.$$

Alicia broadcasts C over public channels to Robert.

3. **Decryption:** Robert uses his trap door knowledge of p and q to compute a secret decryption key d such that

$$C^d \equiv (M^e)^d \equiv M \pmod{n}.$$

---

[8] A text-based message could be converted to a numerical message in a variety of ways. Among more sophisticated ways, we could simply translate $A \to 1$, $B \to 2$, $C \to 3$, etc.

The RSA Cryptosystem

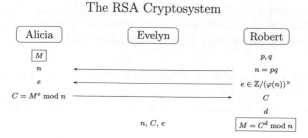

Fig. 9.2 The RSA Cryptosystem in picture form

This uniquely determines the integer M assuming we have chosen p and q (and hence n) to be sufficiently large. ◀

▶ **Prompt 9.3.3** What is the decryption key d and what algorithm will Robert use to compute it? (*Hint:* Consider corollaries to Euler's Theorem.) Why was it important that $\gcd(M, n) = 1$ and $e \in \mathbb{Z}/(\varphi(n))^\times$? How will Robert calculate $\varphi(n)$ and find an $e \in \mathbb{Z}/(\varphi(n))^\times$?

Figure 9.2 shows the steps, keeping track of who has access to which pieces of information. Most importantly, anything that the protocol has us transmit between Alicia and Robert becomes public information accessible to Evelyn.

The main discussion point is why this method is secure, and so in particular, why Evelyn can't deduce M from the information made public: n, C, and e. The argument is as follows. Surely Evelyn could deduce M if she had access to d, the same way that Robert did. Now in turn, Evelyn could compute d if she had access to $\varphi(n)$, which in turn she would have access to if she knew p and q. And herein lies the proverbial rub: it is very difficult to factor $n = pq$ into its two unknown prime factors. This is true even if Evelyn knows precisely the protocol that Alicia and Robert are employing – knowing *that* n is the product of two large primes does not significantly help in deducing what those two primes *are*. We will discuss some more implementation questions momentarily, but we first want to emphasize how straightforward this process is to accomplish in practice. In the following prompt, you will run through the steps with a couple of small, manageable primes.

▶ **Prompt 9.3.4** Suppose Ben wants to send the secret message $M = 15$ to Jerry[9]. How should they proceed? Most of the work can be done up front by Jerry (and indeed, can be reused for several secure communications). Suppose Jerry chooses secret primes $p = 31$ and $q = 43$ and computes their product, $n = 1333$. Take it from here.

---

[9] Clearly $M = 15$ is the 15th ice cream prototype flavor, *Euler's Modular Mocha Mania*.

- Find a value of $e \in \mathbb{Z}/(\varphi(n))^{\times}$ and the corresponding decryption key d.
- What information will Jerry broadcast? What code C will Ben calculate?
- Go through the calculations Jerry will do to decrypt C.

▶ **Prompt 9.3.5** Set up a more professional-grade RSA public encryption scheme (in that it could not be done by hand or with a cheap calculator). You can start from scratch or use the Chapter 9 Python worksheet "Prompt 9.3.5 RSA," which outlines the process.

- Pick two large-ish primes p and q, each on the order of 200-digits (the internet can help you out here).
- With a computer, calculate n and find an encryption key $e \in \mathbb{Z}/(\varphi(n))^{\times}$.
- Calculate the corresponding decryption key d. What algorithm have you (or your computer) used to calculate d?
- Send n and e to your friend and instruct them to encrypt their phone number M as $C = M^e \bmod n$. Your friend might need some additional instruction on how to carry out this calculation. What is an efficient algorithm for performing modular exponentiation?
- Upon receipt of C, decrypt the number and give your friend a call!

Again, the security of the system seems to rest on the claim that factoring a very large number n is very hard.[10] Let's consider some of the implementation issues. The name of the game here is speed. No one would claim that breaking such a system is *impossible* (outside of companies that sell this type of security, perhaps), but rather only that it would require a prohibitively large amount of time and computing power. More legitimately, the implicit claim is that the resources demanded of Alicia and Robert (or Ben and Jerry) to complete this transaction are far fewer than those required to break in from the outside. Thus a careful accounting of implementation issues needs to address both the difficulty of cracking but also the ease of implementing.

How big of a prime do we need? Are there enough?

Both technology and our theoretical understanding of prime factorization are constantly improving, and with it the suggested sizes of the primes p and q to ensure security. For the sake of this discussion, let us suppose that best practices dictate that p and q be 500 digits long. It's easy to be skeptical about the ubiquity of such primes given our understanding that primes get sparser and sparser as we move down the number line. But!

---

[10] and that factoring a very very large number is very very hard, etc.

▶ **Prompt 9.3.6** Using the Prime Number Theorem (Theorem 3.5.5), show that our fears are unfounded. Approximately how many 500-digit primes are there?

How do we find our primes? How fast is setting up RSA?

The remaining elephant in the room is the finding of the primes p and q. It is likely obvious that it is not best practice to set up your encryption system by first searching "500-digit primes" on the internet. So how do we generate our primes p and q?

Most algorithms have us chose a random large number and then check it for primeness. The odds of success can be improved by starting with a product of primes and adding 1. Section 3.5 had us think through some algorithms for testing primeness, and though we made great improvements over the "divide by everything" algorithm, we still had nothing practicable on the scale of 500-digit numbers. And after all, if we're going to claim that Evelyn's great hurdle is not being able to factor n into p and q, then surely we can't simultaneously claim that finding and testing p and q for primeness is trivial.

As it turns out, testing for primeness and actually factoring are of significantly different difficulty levels: there are much faster ways of testing *whether* a number is prime than of finding an explicit factorization. Among many other ways, here's one of relevance to material covered in Chapter 4: we've seen that if p is prime, then by Fermat's Little Theorem,
$$2^{p-1} \equiv 1 \pmod{p}.$$
Therefore, if we're given n and we find that $2^{n-1} \not\equiv 1 \pmod{n}$, then n is not prime. (Note that there are ways to compute $2^{n-1} \bmod n$ quickly, as in Example 4.4.13). If it does work out to be 1, then we test $3^{n-1} \equiv 1 \pmod{n}$, and so on. None of these individually guarantees primeness, but they detect most composites very quickly. This test, called the *Fermat primality test*, ends up being a *probabilistic* primality test in that after a given number of such modular computations evaluating to be 1, we conclude with a very high *probability* that the number is prime. Still, the process is non-trivial (we may have to run through many n's before finding one that passes all the tests), and not foolproof.

▶ **Prompt 9.3.7** If you didn't do it before, go back and complete Exercise 4.61.

▶ **Prompt 9.3.8** Research the Miller-Rabin primality test, and other primality tests commonly used in encryption protocols.

How hard is encryption for Alicia?
How hard is decryption for Robert?

Tremendously easy. As you have already surmised, finding d is just applying our old friend, the well-known and efficient algorithm discussed in Section 3.2. Modular exponentiation is similarly very fast using algorithms discussed in Chapter 4.

How hard is decryption for an evil interloper?

This is the crux of the matter—factorization of gigantic numbers is much harder than any other step addressed thus far. Some comments on this front: first, note that factorization depends on more parameters than just the length of the number (what's the prime factorization of the gigantic number 2^{1000})? Relatedly, since any naive factorization attempt begins with trial division of small primes, a number with small factors gets factored more quickly than a number of the same size with no small prime factors. In some sense, then, the difficulty of factoring a number depends on the size of its smallest prime factor. This explains the choice of the form of n in the RSA protocol—of all the possible factorizations of a 1000-digit number n, the type with the largest smallest prime factor is of the form $n = pq$ with p and q being 500-digit primes. There is no known *proof* that factorization of such numbers is hard, and indeed a revolutionarily fast algorithm to factor 1000-digit numbers would undermine a good deal of existing encryption techniques (and likely win several million dollars in prize money).

▶ **Prompt 9.3.9**   As practical testimony to the difficulty of factorization, look up the current best results for the RSA Challenge, and the running times for the best-known algorithms for factoring a k-digit number.

Are there other vulnerabilities?

This, of course, takes us to the broader realm of cryptography as a subject in its own right, but there are a few quick thoughts particularly meriting discussion. For example, we don't *know* that cracking RSA is as difficult as factoring, even though that does provide the one obvious way to break through the mathematical difficulties of taking roots mod n. Another attack on RSA is not mathematical but social— what if after Alicia and Robert completed their transaction, Evelyn were to pretend to be Robert and solicit secrets from Alicia? Another use of the ideas in the section is in the setting up of *digital signatures* to verify that the authors of documents (quotes, payments, etc.) are indeed who they claim to be. One protocol works as follows: we simply ask this "Robert" to send us, say, the value $s = 2^d$ mod n. It would be very hard for Alicia, or Evelyn, to deduce d from this (important, as Robert would not want to make d public), but it's easy for Alicia to verify that the d value is correct by computing s^e mod n, which should evaluate to 2 if the sender genuinely knew d.

Finally, we note that even given efficient implementation algorithms, it would still be rather time-consuming to hold a complete secret conversation using an RSA encryption protocol for each message sent. In practice, *private key cryptosystems*, in which two parties communicate by using some common knowledge (say, a shared secret password), are much more efficient. Typically one uses public key cryptography to establish a shared secret key which is subsequently used in a more convenient and efficient private key protocol. The following subsection covers such a public key exchange system based on elliptic curves.

9.3.2   Elliptic Curve Cryptography

We have encountered elliptic curves in a wide variety of contexts thus far – in Chapter 2 in regard to a stacking problem, in Chapter 4 in the study of modular arithmetic, in Chapters 5 and 6 as targets for applying our theory of unique factorization, and in Chapter 7 as a crucial step in our proof of quadratic reciprocity. But in addition to their role in developing this textbook's principle themes, elliptic curves have many applications outside the abstract study of number theory. The most significant of these is an infrastructure for cryptographic protocols based on the arithmetic of elliptic curves.

In 1985, Neal Koblitz and Victor Miller independently came up with the idea of using elliptic curves to create public-key cryptosystems. *Elliptic Curve Cryptography* (ECC) was born, based on the algebraic structure of elliptic curves over finite rings like $\mathbb{Z}/(p)$. What the set $\mathbb{Z}/(p)^{\times}$ did for RSA the set $E(\mathbb{Z}/(p))$ does for elliptic curve cryptography, in that the size and algebraic structure of the set determines the difficulty of breaking the overlying cryptosystem. It is believed that the same level of security afforded by an RSA-based system with large p and q can be achieved using a much smaller elliptic curve group. This reduces storage and transmission requirements, making implementation of the public-key cryptosystem more efficient and less costly.

Addition on Elliptic Curves

In Section 2.3 we explored the idea of using two known rational points on an elliptic curve to find a third such point. This process, generalizing the chord method of Diophantus, leads us to a new notion of the *sum* of two points on an elliptic curve and elevates the set of points on the curve from a simple set to a glorious group. Before giving the formal definition, let us illustrate the process with an example familiar from Section 2.3.

▶ **Prompt 9.3.10**   Let E be the elliptic curve in Figure 9.3, defined by

$$y^2 = x^3 - 36x.$$

Since it's not entirely obvious from the picture, find both algebra-based and calculus-based arguments that the dashed line through $(-3, 9)$ and $(-2, -8)$ must intersect the curve in a third point.

▶ **Prompt 9.3.11**   Find the third point of intersection described by the previous prompt. From this, find the sum $P \oplus Q$ of $P = (-3, 9)$ and $Q = (-2, -8)$, defined to be the point obtained by reflecting this third point of intersection over the x-axis. Why must that reflection also be on the curve?

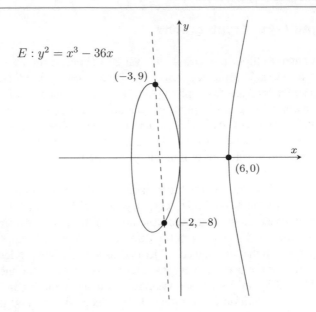

$E : y^2 = x^3 - 36x$

$(-3, 9)$

$(6, 0)$

$(-2, -8)$

Fig. 9.3 The elliptic curve E defined by $y^2 = x^3 - 36x$

There is much to say about this bizarre construction, and there are quite a few special cases that need addressing. For one, it's not clear from the above prompt how to add a point to itself! For this operation to define a group law, we need to make sure that the construction works for every pair of points, that there is some notion of an identity point, and that for each point in the set there is an additive inverse also in the set.

The good news is that all of these problems are handled by only two conventions—the use of *multiplicity* of intersections, and the adoption of a *point at infinity*. The latter of these is difficult to completely motivate without a substantive digression into the language of projective geometry. We offer instead just a glimpse of whence comes this idea. Recall our frequent interplay between rational points on the unit circle and Pythagorean triples (the equations $x^2 + y^2 = 1$ and $x^2 + y^2 = z^2$) obtained by clearing or introducing denominators, and that any scalar multiple of a Pythagorean triple corresponds to the same point on the unit circle.

▶ **Prompt 9.3.12** Instead of solutions (x, y) to the equation $y^2 = x^3 - 36x$, consider solutions $(x, y, z) \neq (0, 0, 0)$ to the equation $y^2 z = x^3 - 36x z^2$. Show that for every solution $P_0 = (x_0, y_0, z_0)$ to the latter with $z_0 \neq 0$, the point $P = (\frac{x_0}{z_0}, \frac{y_0}{z_0})$ is a solution to the former, and that all non-zero scalar multiples of P_0 correspond to the same point P. Finally, when $z_0 = 0$, show that up to scalar multiples, there is precisely one extra solution to the second equation.

The prompt suggests that to complete the picture of the points on E, we should include one more point, something akin to $(\frac{0}{0}, \frac{1}{0})$. Of course, this is nonsense as written, but it suggests a solution to our problems, and boy does it deliver. We consider every elliptic curve to possess one additional *point at infinity*, denoted $\mathcal{O}$ and drawn somewhere high above the graph for visualization purposes. We adopt the convention (also made sensible in terms of projective geometry) that $\mathcal{O}$ lies on *every vertical line* (i.e., on every line of slope infinity).

We can now resolve all of our point-addition dilemmas. When $P = Q$, we interpret "the line through P and Q" as the tangent line to the curve through P (the line which intersects the curve at P with multiplicity 2), and then as before find the third point of intersection and reflect over the x-axis. When P and Q are distinct but share the same x-coordinate, the line through P and Q is a vertical line, and so by our convention also contains the point $\mathcal{O}$ on the curve. In this case, we set $P \oplus Q = \mathcal{O}$ (with the convention that the reflection of $\mathcal{O}$ over the x-axis is $\mathcal{O}$ itself[11].

▶ **Prompt 9.3.13** Explain how, with these conventions, $\mathcal{O}$ serves as an identity for this operation; i.e., $\mathcal{O} \oplus P = P \oplus \mathcal{O} = P$ for every point P. Explain the claim that in Figure 9.4 we have $Q = -P$.

▶ **Prompt 9.3.14** Find the coordinates of the point $2P$ in Figure 9.4. It is probably wise to recall implicit differentiation to find the tangent line.

Before turning to an algebraic perspective on the same topic, convince yourself that you know the conventions well enough to add any two points on an elliptic curve. While one can be forgiven for thinking this construction seems awfully *ad hoc*, rest assured that it becomes quite natural the more one learns about algebraic and projective geometry. Here is one step toward a more natural geometric interpretation:

▶ **Prompt 9.3.15** Show that for all cases, our definition of point-addition is equivalent to the following: We have $P \oplus Q \oplus R = \mathcal{O}$ if and only if P, Q, and R are the three points of intersection (counting multiplicity) of some line with the curve.

While the geometric definition of addition on elliptic curves is visually pleasing, it is difficult to implement on a computer. Fortunately, a combination of elementary geometry and differential calculus provides explicit formulas for the addition of elliptic curve points.

---

[11] After all, who among us could dispute that $(\frac{0}{0}, \frac{1}{0}) = (\frac{0}{0}, -\frac{1}{0})$?

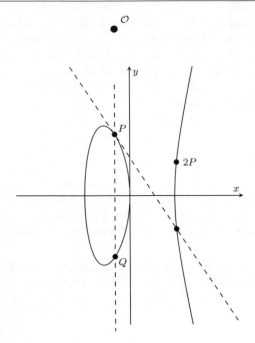

Fig. 9.4 The elliptic curve E defined by $y^2 = x^3 - 36x$

Theorem 9.3.16 (Point Addition Formulas)

Let E be the elliptic curve defined by $y^2 = x^3 + sx + t$, and let $P_1 = (x_1, y_1)$ and $P_2 = (x_2, y_2)$ be two non-$\mathcal{O}$ points on E. We construct $P_1 \oplus P_2$ as follows:

- If $P_1 = -P_2$ then $P_1 \oplus P_2 = \mathcal{O}$.
- Otherwise, define λ by

$$\lambda = \begin{cases} \dfrac{y_2 - y_1}{x_2 - x_1} & \text{if } P_1 \neq P_2 \\[2ex] \dfrac{3x_1^2 + s}{2y_1} & \text{if } P_1 = P_2 \end{cases}$$

Then $P_1 \oplus P_2 = (x_3, y_3)$ where

$$x_3 = \lambda^2 - x_1 - x_2 \quad \text{and} \quad y_3 = \lambda(x_1 - x_3) - y_1.$$

▶ **Prompt 9.3.17** Show that the formulas in the theorem concur with the geometric definition, i.e., that the point (x_3, y_3) is in each case the desired reflection of the third point of intersection.

Perhaps the most remarkable aspect of the theorem is its ambiguity in what sorts of numbers the coordinates x_1, x_2, y_1, and y_2 are (and for that matter, the values of s and t). Since the geometric definition of point-addition agrees with the algebraic one, and since the algebraic comes from polynomial arithmetic, we can conclude that the formulas continue to hold in any ring for which they make sense. If P_1 and P_2 have rational coordinates, then so does $P_1 \oplus P_2$, and likewise for real, complex, and $\mathbb{Z}/(p)$ solutions (for odd prime p). Looking at the formulas for λ in Theorem 9.3.16, we must only take care to exclude rings in which $(x_2 - x_1)$ might not be invertible even if $x_1 \neq x_2$ (for example, the sum of points in $E(\mathbb{Z})$ or $E(\mathbb{Z}/(6))$ would crash the algorithm if $x_2 - x_1 = 3$), and likewise throwing out rings like $\mathbb{Z}/(2)$ where 2 is not invertible. We synthesize the section so far with the following result:

▶ **Prompt 9.3.18** Let F be a field in which 2 is invertible. Prove that the set $E(F)$ of points on an elliptic curve, together with the point at infinity, forms an abelian group under $\oplus$. (Associativity is somewhat terrible, so we will settle for a carefully-sketched diagram over $\mathbb{R}$).

While our previous encounters with elliptic curves had us interested in integer or rational solutions, we will focus now on solutions involving modular arithmetic. For the rest of the section, p will be an odd prime and E an elliptic curve over the field $\mathbb{Z}/(p)$. That is, we consider the group $E(\mathbb{Z}/(p))$ where E is defined by an equation of the form $y^2 = x^3 + sx + t$ with $s, t \in \mathbb{Z}/(p)$ satisfying $4s^3 + 27t^2 \neq 0$ (see Definition 2.3.3). One pleasant benefit of working over finite rings is that for any given p and E, finding $E(\mathbb{Z}/(p))$ is (among other solution trajectories) a simple brute force search.

▶ **Prompt 9.3.19** Consider the elliptic curve E over $\mathbb{Z}/(13)$ defined by $y^2 = x^3 + 3x + 8$. Find the 9 points of $E(\mathbb{Z}/(13))$. (Hint: It may help to pre-compute the Legendre symbols $\left(\frac{a}{13}\right)$ for $a \in \mathbb{Z}/(13)$ first. Don't forget to count $\mathcal{O}$!)

Again we marvel quickly at Theorem 9.3.16, permitting us to add points on this elliptic curve without having the notions of implicit differentiation, tangent lines, or even slopes in the world of modular arithmetic. Note that we can now make sense of the notation nP for a point P on an elliptic curve: $2P$ is short for $P \oplus P$, $-3P$ is the reflection of $3P = 2P \oplus P$, etc. We stress that the notation nP does *not* mean to scalar multiply P's coordinates by n.

▶ **Prompt 9.3.20** With E as above, find $(9, 7) \oplus (1, 8)$, $2(9, 7)$, and $-3(9, 7)$.

▶ **Prompt 9.3.21** Let P be a point on an elliptic curve defined over $\mathbb{Z}/(p)$. Show that there exists a smallest positive integer n such that $nP = \mathcal{O}$. Such an n is called the *order* of P on E and denoted $|P|$.

Elliptic Curve Cryptosystems

Recall that the fundamental basis for the success of a public key cryptosystem rests on some underlying presumed one-way function. In the context of elliptic curves, the difficult inverse problem is the following:

Definition 9.3.22

Suppose E is an elliptic curve, $P \in E(\mathbb{Z}/(p))$, and suppose $Q = nP$ is an integer multiple of P. The **Elliptic Curve Discrete Logarithm Problem** (ECDLP) is the problem of finding n given P and Q. ◄

It is important to note that there will typically be infinitely many such n, since by Prompt 9.3.21, if $nP = Q$, then for all $k \in \mathbb{Z}$ we have

$$(n + k|P|)P = nP \oplus k|P|P = Q \oplus \mathcal{O} = Q.$$

We typically therefore ask for the smallest such n, or better, simply recognize that we should be asking for this value mod $|P|$.

Definition 9.3.23

Let E be an elliptic curve E and let $\langle P \rangle$ be the cyclic subgroup of $E(\mathbb{Z}/(p))$ generated by a point $P \in E(\mathbb{Z}/(p))$. The **Elliptic Curve Discrete Logarithm Function** (for E and P) is the function

$$\log_P : \langle P \rangle \to \mathbb{Z}/(|P|)$$

defined by $\log_P(nP) = [n]$ for each $0 \le n < |P|$. ◄

▶ **Prompt 9.3.24** Figure out what Definition 9.3.23 says, and prove the familiar log-like rule that for all $Q_1, Q_2 \in \langle P \rangle$,

$$\log_P(Q_1 \oplus Q_2) = \log_P(Q_1) + \log_P(Q_2).$$

Explain why this would be false if we took $\log_P$ to have codomain $\mathbb{Z}$.

▶ **Prompt 9.3.25** Compute the cyclic subgroups generated by $(1, 5)$ and $(9, 6)$ inside $E(\mathbb{Z}/(13))$ for the curve of Prompt 9.3.19. Explain why $\log_{(1,5)}(9, 6)$ makes sense but $\log_{(9,6)}(1, 5)$ doesn't.

▶ **Prompt 9.3.26** Propose a definition for the discrete logarithm problem in an arbitrary finite abelian group G. Show that this problem is comparatively trivial for the groups $G = \mathbb{Z}/(n)$.

Any claims of difficulty for the ECDLP are rendered moot if we cannot convince ourselves of the ease of the "easy" direction, i.e., that the calculation of nP from n and P is comparatively easy. Analogous to the modular exponentiation methods we encountered in Example 4.4.13, a prominent algorithm here is to write n as a sum *or difference* of powers of 2:

$$n = n_0 + n_1 \cdot 2 + n_2 \cdot 2^2 + \cdots + n_r \cdot 2^r,$$

with each $n_i \in \{-1, 0, 1\}$. Then to compute nP one computes the quantities

$$Q_0 = P, \quad Q_1 = 2Q_0, \quad Q_2 = 2Q_1, \quad \ldots \quad Q_r = 2Q_{r-1},$$

after which the desired value nP is

$$nP = n_0 Q_0 \oplus n_1 Q_1 \oplus \cdots \oplus n_r Q_r.$$

▶ **Prompt 9.3.27** Suppose $n = 349$ and we want to compute nP for some point P on an elliptic curve. How many point additions would you need to enact using the binary expansion of n, i.e., writing n as

$$n = n_0 + n_1 \cdot 2 + n_2 \cdot 2^2 + \cdots + n_r \cdot 2^r,$$

with each $n_i \in \{0, 1\}$? Can you express n as the sum or differences of powers of 2 (i.e., taking each n_i from the set $\{-1, 0, 1\}$) in a way that leads to fewer point additions in the calculation of nP?

▶ **Prompt 9.3.28** Provide an estimate or upper bound for how many elliptic curve point additions/doublings are generally necessary to compute nP using the algorithm described above with $n_i \in \{-1, 0, 1\}$.

▶ **Prompt 9.3.29** Research: How long *does* it take for a computer to sum two points on an elliptic curve? Does it depend on the curve? If so, how? What are the "best" curves?

▶ **Prompt 9.3.30** Consider the following thought experiment to convince yourself of the relative difficulty of our two main calculations. Suppose $n \approx 10^6$, and compare the number of point additions needed to compute:

(a) $Q = nP$ given n and P.
(b) kP for all $1 \leq k \leq n$, in order to find n from P and Q.

Elliptic Curve Diffie-Hellman Key Exchange

The following protocol allows two parties, named Alicia and Robert, to establish a *shared*, secure, secret key while conducting all communication over insecure lines.

1. Alicia and Robert publicly agree upon a very large odd prime p, an elliptic curve E over $\mathbb{Z}/(p)$, and a point $P \in E(\mathbb{Z}/(p))$ of large order.
2. Alicia chooses a secret integer a and computes $Q_a = aP$. Likewise, Robert chooses a secret integer b and computes $Q_b = bP$. The values a and b are known as the private keys and are not public knowledge. Alicia and Robert then transmit Q_a and Q_b to one another.
3. Upon receiving Q_b, Alicia uses her private key to calculate $K = aQ_b$, and Robert similarly computes $K' = bQ_a$. Now, these two points are the same, as

$$K = aQ_b = a(bP) = b(aP) = bQ_a = K'.$$

Thus, Alicia and Robert share the secret key K. If they want a single number for a key, they could agree to use its x-coordinate.

Figure 9.5 shows the steps, keeping track of who has access to which pieces of information. Most importantly, anything that the protocol has us transmit between Alicia and Robert becomes public information accessible to Evelyn.

As with our discussion of RSA, the principal concern is whether or not an eavesdropping Evelyn can also determine K. Evelyn has access to a lot of information! She has p, P, E, Q_a, and Q_b all at her disposal. Yet the only obvious way for her to proceed would be to solve the ECDLP. If Evelyn can solve the ECDLP, she can deduce a from P and Q_a and use it to calculate the shared secret key $K = aQ_b$. While this is the most obvious approach, in principal there might be some *other* way for Evelyn to discover K from the available information, and the *Diffie-Hellman problem* is exactly that question.

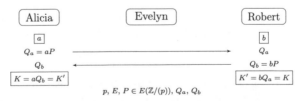

Fig. 9.5 Elliptic Curve Diffie-Hellman Key Exchange in picture form

---

Definition 9.3.31

Let $E(\mathbb{Z}/(p))$ be an elliptic curve over $\mathbb{Z}/(p)$ and let $P \in E(\mathbb{Z}/(p))$. The **Elliptic Curve Diffie-Hellman Problem** (ECDHP) is the problem of computing the value of abP from the known values of aP and bP. ◄

▶ **Prompt 9.3.32** It is not unreasonable to wonder whether the problems ECDLP and ECDHP are in fact equivalent. Research what is known about the relationship between these two problems. What are the best-known algorithms for attempting each?

The Elliptic Curve Diffie-Hellman Key Exchange does *not* replicate the encryption functionality of RSA. Whereas in Diffie-Hellman, Alicia and Robert work together to generate a new secret key from scratch, in RSA we took an existing secret key or message and found a way to securely transmit it over open channels. In Section 9.3.3 we take a look at an elliptic curve public key *encryption* scheme.

9.3.3 Elliptic ElGamal Public Key Cryptosystem

In the *Elliptic ElGamal Public Key Cryptosystem*, Alicia wants to convey a given secret message to Robert who has set up and announced a public key. Alicia encrypts her message using this public key and then transmits her encrypted message over an open channel to Robert. Robert is able to decrypt the message using his secret key, while an eavesdropper with access to Robert's public key and Alicia's encrypted message cannot easily do so.

▶ **Remark 9.3.33** In the following protocol, we will have Alicia convert her message to a point on Robert's elliptic curve. As with RSA, there is a logistical detail as to how actually one does the conversion. If nothing else, Alicia could convert her message to an integer, and use that as the x-coordinate of a point on the elliptic curve – if no such point exists, she could tweak her message or ask Robert to choose a new curve or prime.

Here is the protocol:

1. Robert selects and makes public a large prime p, an elliptic curve E over $\mathbb{Z}/(p)$, and a point $P \in E(\mathbb{Z}/(p))$. He then selects a private key $a \in \mathbb{N}$ and sends $Q_a = aP$ (but *not* a) to Alicia.
2. Alicia begins with her secret message $M \in E(\mathbb{Z}/(p))$ (see Remark 9.3.33). She chooses an integer k and computes both $C_1 = kP$ and $C_2 = M \oplus kQ_a \in E(\mathbb{Z}/(p))$ efficiently. She sends the pair (C_1, C_2) of points to Robert.

Fig. 9.6 Elliptic Curve ElGamal Public Key Cryptosystem in picture form

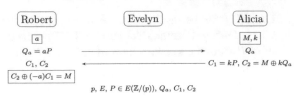

The ElGamal Public Key Cryptosystem

3. To decrypt (C_1, C_2), since Robert possesses the integer a, he can easily compute $C_2 \oplus (-a)C_1$:

$$C_2 \oplus (-a)C_1 = (M \oplus kQ_a) \oplus (-a)kP$$
$$= M \oplus (kaP \oplus -kaP)$$
$$= M \oplus (ka - ka)P$$
$$= M,$$

which recovers the secret message M.

The process is depicted in Figure 9.6.

Again the secrecy hinges primarily on the inability of an eavesdropper to deduce a from aP. Like with Diffie-Hellman, and indeed like RSA, there are no *proofs* of the security of the elliptic curve discrete logarithm problem, and we instead rely on the *presumed* difficulty of the problem. That said, ECC has recently gained in prominence and was endorsed by the National Security Agency, making it a recommended algorithm for protecting government information classified up to top secret.

One cannot help but feel impressed with the almost-mystical pervasiveness that elliptic curves seem to have in modern mathematics, and we are left to wonder the extent elliptic curves were *necessary* to make these protocols work. There are protocols based on other families of curves (e.g., *hyper*elliptic curve cryptography), but the setting is less natural and not particularly used in practice.

▶ **Prompt 9.3.34** Hyperelliptic curves are generalizations of elliptic curves, with the corresponding group structure given by the *Jacobian*. Research the definition of the Jacobian for a hyperelliptic curve defined over a finite field. What is the discrete logarithm problem in this setting?

▶ **Prompt 9.3.35** Cryptosystems based on the discrete logarithm problem may be vulnerable to certain attacks based on the order of the corresponding finite group. Research the Pollard-ρ and Pohling-Hellman attacks. What type of group orders are susceptible to these attacks?

▶ **Prompt 9.3.36** An elliptic curve E over $\mathbb{Z}/(p)$ is called *anomalous* if the order of the group $|E(\mathbb{Z}/(p))|$ is p. In such settings, the discrete logarithm problem can be solved by first lifting the curve to the p-adics, i.e., considering the group $E(\mathbb{Q}_p)$. Research this lifting. Research the Smart Anomalous attack on the discrete logarithm problem.

9.4 Exploration: Units of Real Quadratic Fields

When d is a negative square-free integer the units of the ring of integers of $\mathbb{Q}(\sqrt{d})$ are easily dispatched, in that with the notable exceptions of $d = -1$ and $d = -3$ (see Lemma 6.3.7), the units are simply ± 1. When d is positive, however, the situation can be pleasantly more intricate. Most notably, we have (by Corollary 6.3.9) infinitely many units in $\mathbb{Z}[\sqrt{2}]$. Developing the full theory of real quadratic units requires addressing several variants at once—we must decide if we want to restrict our attention to positive units, or to units of norm 1, to consider units of $\mathbb{Z}[\sqrt{d}]$ vs. the full ring of integers, etc. To make a semi-arbitrary set of such choices, we will focus on trying to find all units of $\mathbb{Z}[\sqrt{d}]$ with norm 1, in order to cleanly make a connection to a class of Diophantine equations, and periodically check on how we would accommodate variant questions as we go. Assume that d is positive and square-free for the entirety of this section, and as a starting point:

▶ **Prompt 9.4.1** Show that $x + y\sqrt{d}$ is a unit of $\mathbb{Z}[\sqrt{d}]$ if and only if

$$x^2 - dy^2 = \pm 1.$$

▶ **Prompt 9.4.2** Let $R = \mathbb{Z}[\sqrt{d}]$ for $d > 0$. Show that the units $R^\times$ of R form a group under multiplication and that the following sets are subgroups of $R^\times$:

- The positive[12] elements of $R^\times$.
- Elements of $R^\times$ of norm 1.
- For each $\alpha \in R^\times$, the elements $\{\alpha^n : n \in \mathbb{Z}\}$.
- For each $\alpha \in R^\times$, the elements $\{\pm\alpha^n : n \in \mathbb{Z}\}$.

These two results combined provide the intriguing prospect that group theory could have sway over the solutions to certain Diophantine equations. Once we have found one solution (a, b), any element in the cyclic subgroup of units generated by $\alpha = a + b\sqrt{d} \in \mathbb{Z}[\sqrt{d}]$ will provide another solution.

▶ **Prompt 9.4.3** Let $\alpha = 2 + \sqrt{3} \in \mathbb{Z}[\sqrt{3}]$. Show that $\alpha \in \mathbb{Z}[\sqrt{3}]^\times$ and compute α^n for $-2 \leq n \leq 4$. Find 14 integer solutions to $x^2 - 3y^2 = 1$.

---

[12] This may be the clearest benefit to working in real fields. Complex lattices are great and all, but it sure feels nice to regain the use of the words positive and negative.

Though prime factorization in real quadratic fields is not the focus of the section, recall that units played a central role in any careful statement of a Fundamental Theorem of Arithmetic.

▶ **Prompt 9.4.4** The ring $\mathbb{Z}[\sqrt{3}]$ turns out to be a unique factorization domain. How is it, then, that we can have such seemingly distinct factorizations as

$$(2\sqrt{3} - 1)(2\sqrt{3} + 1) = 11 = (250\sqrt{3} + 433)(250\sqrt{3} - 433)?$$

▶ **Prompt 9.4.5** It seems plausible that as a novice number-theorist starting way back in Chapter 1, you would have found it difficult to find many examples of a square number that is 1 more than 15 times another square number. Find some such examples now, enough that you feel convinced that your past self would be suitably impressed.

It is striking that questions about real quadratic units can be rephrased as questions about solutions to Diophantine equations, and this perspective allows us to view our investigation in the broader historical context in which these equations were tackled. Equations of the form $x^2 - dy^2 = 1$ have been studied for literally thousands of years. Specific examples can be found in the works of Pythagoras and Archimedes, and such equations were studied more systematically by the Indian mathematician Brahmagupta in the seventh century and his intellectual descendant Bhaskara II in the twelfth century. In the mid-seventeenth century William Brouncker became the first European mathematician to find non-trivial solutions, having largely rediscovered the ideas of Brahmagupta and Bhaskara. Euler's contributions in the eighteenth-century included a complete solution technique as well as giving these equations the moniker "Pell equations"—likely misattributing the work of Brouncker[13]. In the spirit of correcting past mistakes, we refer to these equations as *Brahmagupta-Bhaskara-(Pell) equations*[14].

▶ **Prompt 9.4.6** Research: The most infamous of all Brahmagupta-Bhaskara equations is likely the *Archimedes Cattle Problem*. The problem illustrates both how such problems can "naturally arise" and how surprising their solutions can be. Describe the problem, its translation to a Brahmagupta-Bhaskara equation, and the eventual resolution.

The early solutions to Brahmagupta-Bhaskara equations were phrased in Brahmagupta's language of *composition*:

---

[13] and continuing the notorious tradition of mathematical naming practices

[14] The "Pell" is silent.

> **Theorem 9.4.7 (Brahmagupta's Composition Formula)**
> If (x_1, y_1) is a solution to $x^2 - dy^2 = a$ and (x_2, y_2) is a solution to $x^2 - dy^2 = b$,
> then (x_3, y_3) is a solution to $x^2 - dy^2 = ab$, where
>
> $$(x_3, y_3) = (x_1 x_2 + dy_1 y_2, x_1 y_2 + x_2 y_1)$$

▶ **Prompt 9.4.8** Find an integer solution to $x^2 - 6y^2 = 57$ by composing easily-found solutions to $x^2 - 6y^2 = 3$ and $x^2 - 6y^2 = 19$.

In modern language, this is just expressing that the norm is multiplicative: The product of an element of $x_1 + y_1\sqrt{d}$ of norm a and an element $x_2 + y_2\sqrt{d}$ of norm b gives the element $(x_1 x_2 + dy_1 y_2) + (x_1 y_2 + x_2 y_1)\sqrt{d}$ of norm ab. Of course, when $a = b = 1$ we have the special case that a product of units of norm 1 is again a unit of norm 1. Our first dip into units of norm equal to negative one provides another instance:

▶ **Prompt 9.4.9** Find an easy solution to $x^2 - 17y^2 = -1$ and use this to find a unit of $\mathbb{Z}[\sqrt{17}]$ of norm 1.

The realization that studying $x^2 - dy^2 = 1$ can be studied by working with $x^2 - dy^2 = n$ for various n is a liberating one, and Brahmagupta himself put it to good use, being the first to find a solution to $x^2 - 92y^2 = 1$. We rephrase Brahmagupta's ingenious composition solution in the language of modern algebraic number theory.

▶ **Prompt 9.4.10** Verify that $\alpha = 10 + \sqrt{92} \in \mathbb{Z}[\sqrt{92}]$ has norm 8, so that $\frac{\alpha^2}{8}$ has norm 1. This number is not an element of $\mathbb{Z}[\sqrt{92}]$, but its square is. Compute it and thus provide a solution to $x^2 - 92y^2 = 1$.

Ancient mathematicians were also interested in Brahmagupta-Bhaskara equations for their role in understanding irrational numbers. The first proof of the existence of irrational numbers is commonly attributed to the ancient Greek mathematician Hippasus, where legend has it that the Pythagoreans were so incensed by the discovery that they threw Hippasus from a boat and he drowned at sea. Whether or not this is apocryphal, it's certainly true that the ancient Greeks wanted to understand numbers like $\sqrt{2}$ better[15]. We begin by noting that the irrationality of $\sqrt{d}$ is equivalent to the non-existence of integral solutions to the equation $x^2 - dy^2 = 0$, as a solution

---

[15] If for no other reason than to avoid being thrown off a boat.

would give $\frac{x}{y} = \pm\sqrt{d}$. But the closely related Brahmagupta-Bhaskara equation, $x^2 - dy^2 = 1$, seems to frequently have an infinitude of solutions.

▶ **Prompt 9.4.11** Prove that if (x_n, y_n) is an infinite family of solutions to the Brahmagupta-Bhaskara equation $x^2 - dy^2 = 1$ with $y_n \to \infty$ as $n \to \infty$, then $\lim\limits_{n \to \infty} \frac{x_n}{y_n} = \sqrt{d}$.

So for example, finding more and more units of $\mathbb{Z}[\sqrt{2}]$ gives us better and better rational approximations to $\sqrt{2}$, a pretty reasonable interpretation of the ancient Greek goal of "understanding $\sqrt{2}$."

▶ **Prompt 9.4.12** For each of the solutions (x, y) you found in Prompt 9.4.3 (with $y \neq 0$), compute the ratio $\frac{x}{y}$. Do you find any good approximations to $\sqrt{3}$? How good? Can you do better?

▶ **Prompt 9.4.13** Find a rational approximation to $\sqrt{92}$ accurate to 8 decimal places. Note that the approach of Prompt 9.4.11 permits verifying the number of digits of decimal accuracy even without the exact decimal expansion.

The key element in all of our calculations thus far has been to eyeball, or just be given, one starting solution and then generate new ones by composition/multiplication. To systematize the approach we return to structural results about the group $\mathbb{Z}[\sqrt{d}]^\times$. Continuing to focus on units of norm 1, note that we can restrict our attention to positive such units, since u has norm 1 if and only if $-u$ does. We have already observed that such units form a group, and the major strengthening of this result is that this group is always *infinite* and *cyclic*. First, the cyclicity result:

Theorem 9.4.14
The group of positive units in $\mathbb{Z}[\sqrt{d}]$ of norm 1 is cyclic.

▶ **Prompt 9.4.15** Prove the theorem by establishing the following steps:

(a) Argue that if $\mathbb{Z}[d]^\times = \{\pm 1\}$, we are done. Else, show that...
(b) every positive unit $x + y\sqrt{d} > 1$ of norm 1 must have $x, y > 0$, and hence there must be a smallest unit $u > 1$ of norm 1.
(c) Get a contradiction to (b) if there were a positive unit v not in $\langle u \rangle$.

The obvious solutions $(x, y) = (\pm 1, 0)$ to the Brahmagupta-Bhaskara equation $x^2 - dy^2 = 1$ are called the **trivial solutions**. The existence of *any* non-trivial solution (x, y) implies the existence of the unit $u_d = x + y\sqrt{d} \in \mathbb{Z}[\sqrt{d}]^\times$ described in the prompt above (the smallest unit of norm 1 greater than 1 with $x, y > 0$), which corresponds to the **fundamental solution** of the equation. So if any non-trivial solution exists, the units of norm 1 are precisely $\pm \langle u_d \rangle$, and since $u_d > 1$, this group is infinite. In other words, if there is one non-trivial solution to the Brahmagupta-Bhaskara equation, there are infinitely many. The remaining questions are whether or not there always exists a non-trivial solution and how to find and recognize the fundamental one. For example, in Example 9.4.3 we encountered the unit $2 + \sqrt{3}$, and we can verify that neither of the smaller candidates, $\sqrt{3}$ nor $1 + \sqrt{3}$, are units. So in the prior notation, $u_3 = 2 + \sqrt{3}$. More generally, we see that given an explicit unit of $\mathbb{Z}[\sqrt{d}]$, there is a finite algorithm for testing whether it is the unit u_d promised by the theorem (namely, brute force search over smaller positive coefficients). Now, what if we don't have an initial solution? This turns out to be harder than one might have expected, as illustrated by the example $d = 61$, tackled first by Bhaskara II. Here the smallest non-trivial solution to $x^2 - 61y^2 = 1$ was found to be

$$(x, y) = (1766319049, 226153980).$$

This example is indeed the fundamental solution, and so definitively refutes any thought that there's *clearly* always a non-trivial solution, or that they are easy to find. As a starting point for our proof that non-trivial solutions always exist, the following question shows that there's mathematics of interest to the question even if we temporarily move away from units.

▶ **Prompt 9.4.16**   What is the *smallest* positive element of $\mathbb{Z}[\sqrt{d}]$?

For $\alpha \in \mathbb{R}$, let's adopt the notation $\alpha \bmod 1 = \alpha - \lfloor \alpha \rfloor$, the decimal part of α. For example, $5\sqrt{2} = 7.07106...$, so $5\sqrt{2} \bmod 1 = 5\sqrt{2} - 7 \approx 0.07106....$, a smallish positive element of $\mathbb{Z}[\sqrt{2}]$.

▶ **Prompt 9.4.17**   Use a computer to tabulate/experiment with the numbers $n\sqrt{2} \bmod 1$. How do these numbers help address Prompt 9.4.16? How small of a positive element of $\mathbb{Z}[\sqrt{2}]$ can you find?

The inevitable conclusion of any serious experimentation is a prediction that we seem to be able to find arbitrary small elements of $\mathbb{Z}[\sqrt{2}]$ (and more generally, $\mathbb{Z}[\sqrt{d}]$). In fact, for *any* irrational number α, the values $n\alpha \bmod 1$ densely permeate the unit interval $[0, 1]$. As a specific consequence we will learn that between any two distinct real numbers there exists an element of $\mathbb{Z}[\sqrt{2}]$! We will establish the claims of this paragraph and then show the link to the existence of a fundamental solution when the irrational number in question is $\alpha = \sqrt{d}$. We need one more tool, a fundamental ingredient to many combinatorial arguments:

Theorem 9.4.18 (Dirichlet's Pigeonhole Principle)

- If you have $n + 1$ objects of n possible types, there must be at least two objects of the same type.
- Further, if you instead have infinitely many objects of finitely many types, there must be an infinite subcollection of your objects that all have the same type.
- Further still, if you have uncountably many objects of countably many types, there must be an uncountable subcollection of your objects that all have the same type.

Lejeune Dirichlet was certainly not the first human in history to have made deductions of this kind, though he was one of the first to explicitly identify the argument (and its myriad generalizations) as an essential tool of mathematical proof, and certainly the first to invoke pigeons to do so. The application of the theorem typically takes the form of identifying what one's "objects" are, what one's "types" are, and then concluding that two (or infinitely many) objects must be of the same type. Try out this approach on some warm-up problems:

▶ **Prompt 9.4.19** For each $n \in \mathbb{N}$, show that among any $n + 1$ integers there must exist two numbers whose difference is divisible by n.

▶ **Prompt 9.4.20** Prove that for each $n \in \mathbb{N}$, there is a multiple of n whose digits are only 0s and 7s. (Hint: Apply the previous problem to the "objects" that are strings of 7s).

▶ **Prompt 9.4.21** Prove that if you choose any $n + 1$ distinct numbers from $\{1, 2, 3, \ldots, 2n\}$ then at least one must divide another. (Hint: Let the "type" of k be $k/2^{v_2(k)}$).

And now, the big one:

▶ **Prompt 9.4.22** Prove Dirichlet's Approximation Theorem: given a positive integer n and any irrational number α, there exist integers a, b with $0 < b < n$ such that

$$|a - b\alpha| < \frac{1}{n}.$$

Hint: Consider the $n + 1$ objects $k\alpha \bmod 1$ for $0 \leq k \leq n$ and "type" them according to the location in the unit interval.

Dirichlet's Approximation Theorem has a number of interesting consequences and applications beyond the proof of the existence of non-trivial solutions to the Brahmagupta-Bhaskara equation. The following corollary and problem are just a sample.

▶ **Prompt 9.4.23** Let α be an irrational number. Prove that between every two distinct real numbers, there is an element of $\mathbb{Z}[\alpha]$.

▶ **Prompt 9.4.24** A party trick sure to impress. Step 1: Have a party-goer tell you their phone number. Step 2: Prove to them that there exists a power of 2 whose leading digits are precisely their number. Step 3: Profit. (Hint: $\log_{10}(2)$ is irrational).

More pertinently, Dirichlet's Approximation Theorem, combined with a substantial use of the Pigeonhole Principle, resolves our fundamental question about units.

Theorem 9.4.25

When d is a square-free positive integer, the Brahmagupta-Bhaskara equation $x^2 - dy^2 = 1$ has a non-trivial solution (a, b).

▶ **Prompt 9.4.26** Prove the theorem as follows:

(a) Use Dirichlet's Approximation Theorem to show that there exist infinitely many numbers $a + b\sqrt{d}$ such that $|a - b\sqrt{d}| < \frac{1}{b}$. These numbers will be our "objects".

(b) Let the "type" of an object $\alpha = a + b\sqrt{d}$ be the triple $(N(\alpha), a \bmod N(\alpha), b \bmod N(\alpha))$. Show there are only finitely many possible types for our objects, and hence at least two of the same type.

(c) Show that the quotient of the two objects of the same type is a non-trivial unit of $\mathbb{Z}[\sqrt{d}]$.

Consequently, every $\mathbb{Z}[\sqrt{d}]$ with $d > 0$ has infinitely many units. It is not quite true, however, that *every* unit is a power of the one corresponding to the fundamental solution of the Brahmagupta-Bhaskara equation, as we have lazily avoided dealing with units of norm -1. For example, $3 + 2\sqrt{2}$ is the smallest positive unit of norm 1 in $\mathbb{Z}[\sqrt{2}]$, but is itself the square of the smaller unit $1 + \sqrt{2}$ of norm -1. We can continue our blissful ignorance of this hiccup for a little bit longer by imposing one condition:

▶ **Prompt 9.4.27** Suppose d is divisible by a prime p with $p \equiv 3 \bmod 4$. Then prove that $\mathbb{Z}[\sqrt{d}]$ has no units of norm -1. (Hint: Work mod p).

As a consequence, if for such a d we let u_d be the unit provided by Prompt 9.4.15, then it is reasonable to call this unit the *fundamental unit* of $\mathbb{Z}[\sqrt{d}]$, and we then have the following.

Theorem 9.4.28

For d as in the previous prompt, let u_d be the fundamental unit of $\mathbb{Z}[\sqrt{d}]$. Then

$$\mathbb{Z}[\sqrt{d}]^{\times} = \{\pm u_d^n : n \in \mathbb{Z}\}.$$

Finally, we close up shop by letting you close up shop. We leave two glaring questions on the table for you to resolve by experimentation or research:

▶ **Prompt 9.4.29** The *negative Brahmagupta-Bhaskara equation* $x^2 - dy^2 = -1$ has no solutions if d is divisible by any prime congruent to 3 mod 4. Is the converse true? If not, what *is* known about the set of d's for which there is a solution?

▶ **Prompt 9.4.30** Even in the best cases, we have typically only proven the *existence* of a fundamental unit. Since we can't all have Bhaskara's ingenuity, how do we go about finding them with modern insights? What do continued fractions have to do with anything?

9.5 Exploration: Ideals and Ideal Numbers

Our approach to solving Diophantine equations using algebraic number theory has focused predominantly on the power of leveraging unique factorization. The theory of Euclidean domains, of norms and primality, of units and associates, all went to stating and proving a version of the Fundamental Theorem of Arithmetic that would apply to increasingly exotic rings. These developments allowed us to classify Pythagorean triples, find integer points on elliptic curves, and explore the behavior of primes in larger rings of integers. Missing so far, however, has been the inevitable follow-up question of what we're supposed to do when we need to work in a ring that does *not* provide the many fruits of the unique factorization tree. This Exploration will crack open the door to the vast expanse of abstract mathematics that has been developed in an attempt to tackle such problems.

Recall from Section 6.5 that the ring $\mathbb{Z}[\sqrt{-26}]$ is not a unique factorization domain, and our attempts in Example 6.1.2 to solve the Diophantine equation $x^3 = y^2 + 26$ were consequently stymied. Recall too that we have fixed unique factorization once before. The apparent failure of unique factorization

$$(1 + \sqrt{-3})(1 - \sqrt{-3}) = 4 = 2 \cdot 2$$

in $\mathbb{Z}[\sqrt{-3}]$ was rectified by the observation that these factors are all associates of one another once we move from $\mathbb{Z}[\sqrt{-3}]$ to the *correct* ring, the full ring of Eisenstein integers. This approach fails for our current example, however, as $\mathbb{Z}[\sqrt{-26}]$ is already the full ring of integers of $\mathbb{Q}[\sqrt{-26}]$ (Theorem 6.3.5). That is, we cannot rectify the problematic factorizations

$$(1 + \sqrt{-26})(1 - \sqrt{-26}) = 27 = 3 \cdot 3 \cdot 3$$

in the same fashion, as $\frac{1+\sqrt{-26}}{3}$ is not an algebraic integer (and even if it were, it's not clear that this would help, as even the *number* of irreducible factors differs on the two sides of this factorization).

One of the first attempts to address this situation comes from the mid-nineteenth century German mathematician Ernst Kummer, who, much like how we formally introduced a new number i when we need a square root of -1, introduced the notion of new *ideal numbers* to better understand these multiple factorizations. *If*, he postulated, we could construct from thin air numbers like $\gcd(3, 1 \pm \sqrt{-26})$ in the ring $\mathbb{Z}[\sqrt{-26}]$, then we could use these to resolve multiple factorizations. This is precisely how we proceed in $\mathbb{Z}$, as whenever we have multiple factorizations of a given number, e.g.,

$$4 \cdot 6 = 24 = 3 \cdot 8$$

there is an immediate explanation that the factors on each side have non-trivial gcd with one another.

Kummer's attempt did not prove sufficiently workable to become the mainstream approach (though Kummer put it to great use resolving some cases of Fermat's Last Theorem), but has instead the more substantive legacy of serving as a springboard for Dedekind to redevelop the theory into the language of *ideals*. These objects, the focus of this section, have become a staple of modern algebra, and as we will see through a series of heartfelt literary quotations have greatly inspired enthusiasts of the language arts as well.

We must have ideals...life would be a sorry business without them.
  – Lucy Maud Montgomery, *Anne of Green Gables*

Definition 9.5.1

Let R be a commutative ring with unity (*hypotheses we shall assume about R for the rest of the section*). An **ideal** of R is a non-empty subset I of R that satisfies the following two properties:

- For all $a, b \in I$, we have $a - b \in I$.

- For all $a \in I$ and $r \in R$, we have $ra \in I$.

◀

These are rather minimalist axioms, though many other basic properties follow as a consequence.

▶ **Prompt 9.5.2** Argue that the set of multiples of 6 $S = \{6n : n \in \mathbb{Z}\}$ is an ideal in $\mathbb{Z}$ but that the set $T = \{6n + 1 : n \in \mathbb{Z}\}$ is not.

▶ **Prompt 9.5.3** Prove that an ideal must be closed under addition and multiplication, and must contain the additive identity 0 of the ring.

▶ **Prompt 9.5.4** Prove that in any ring R, both R itself and the set $\{0\}$ are ideals of R. An ideal is **proper** if it not equal to R itself.

▶ **Prompt 9.5.5** Prove that if I contains at least one unit of R, then $I = R$.

One collection of ideals is easy to come by in any ring.

▶ **Prompt 9.5.6** Prove that for each element $r \in R$, the set

$$(r) = \{ar : a \in R\}$$

is an ideal of R, called the **principal ideal generated by** r. Any ideal of this form is called **principal**. Note that R itself is the principal ideal (1).

That is, the set of all ring multiples of any fixed element of the ring forms an ideal. Such sets are familiar to us from earlier chapters, e.g., as the evenly spaced multiples of 6 in the number line representation of $\mathbb{Z}$ (and Prompt 9.5.2) and the lattice in $\mathbb{Z}[i]$ consisting of multiples of $2 + i$, and are of interest in every ring:

▶ **Prompt 9.5.7** Prove that the set $\{f \in \mathbb{R}[x] : f(3) = 0\}$ is an ideal of $\mathbb{R}[x]$, and decide if it is principal.

▶ **Prompt 9.5.8** Prove that for $r, r' \in R$, we have $(r) = (r')$ if and only if r and r' are associates in R. That is, ideal generators are only determined up to units.

▶ **Prompt 9.5.9** Show that for all $a, b \in R$, we have

$$b \mid a \iff a \in (b) \iff (a) \subseteq (b).$$

To repeat a common mantra for this relationship, "to divide is to contain."

▶ **Prompt 9.5.10** Prove that any ideal of $\mathbb{Z}$ that contains both 42 and 87 must also contain 3, and hence must contain the principal ideal (3).

More generally:

▶ **Prompt 9.5.11** Prove that every ideal in $\mathbb{Z}$ is principal; i.e., if I is an ideal of $\mathbb{Z}$ then $I = (a)$ for some $a \in \mathbb{Z}$. (Hint: If $I \neq (0)$, let a be the smallest positive integer in I, and prove that $I = (a)$.)

There is a clean generalization of this result: if we call R a **Principal Ideal Domain** (or PID) whenever *every* ideal of R is principal, then a similar argument to Prompt 9.5.11 shows that every Euclidean Domain is a Principal Ideal Domain. One can also prove that every Principal Ideal Domain is a Unique Factorization Domain, which provides an alternative approach to the Fundamental Meta-Theorem of Arithmetic (Theorem 6.5.3), contrasting to our previous direct proof that every Euclidean Domain is a UFD. Already having a perfectly good (better?) proof of this result, however, let us move on.

The Arithmetic of Ideals

The pinnacle of happiness lies in finding the sum total of one's ideals.
  —Anonymous[16]

We define the sum of two ideals to be the set of sums of their elements:

Definition 9.5.12

If I and J are ideals, define their **sum** by

$$I + J = \{i + j : i \in I, j \in J\}. \qquad \blacktriangleleft$$

▶ **Prompt 9.5.13** Prove that the sum of (finitely many) ideals of R is again an ideal of R.

Adding together principal ideals provides a construction for potentially non-principal ideals. Given a finite list of elements $a_1, \ldots, a_n \in R$, define the **ideal generated by** $a_1, \ldots, a_n$ as

$$(a_1, a_2, \ldots, a_n) = (a_1) + (a_2) + \cdots + (a_n).$$

---

[16] Or at least, anonymous in the sense that we're not telling you which one of us said this.

► **Remark 9.5.14** If one were to work in the ring $\overline{\mathbb{Z}}$ of *all* algebraic integers, or other sufficiently exotic rings, one would eventually run across *infinitely*-generated ideals. But, following the advice of legendary cowboy Will Rogers, we can for the time being safely exclude such fantastical scenarios.[17]

► **Prompt 9.5.15** Consider the principal ideals $I = (2 + i)$ and $J = (6 + 4i)$ of $R = \mathbb{Z}[i]$. Show that $I + J = R$.

Call two ideals I and J **relatively prime** if $I + J = R$. A pleasant bonus of working with ideals is the ease with which they accommodate many of our existing ideas:

► **Prompt 9.5.16** For $a, b \in \mathbb{Z}$, show that $(a, b) = (\gcd(a, b))$. In particular, a and b are relatively prime if and only if (a) and (b) are.

For those who write their integer gcds more succinctly as simply (a, b), their analogous result would be the notationally amusing $(a, b) = ((a, b))$. In any case, ideal generation and gcds are intimately linked.

► **Prompt 9.5.17** Find and prove a result analogous to that of Prompt 9.5.16 for the intersection $(a) \cap (b)$ of two ideals of $\mathbb{Z}$.

► **Prompt 9.5.18** Prove that the ideal $(2, x)$ of $\mathbb{Z}[x]$ is not principal.

► **Remark 9.5.19** Continuing our discussion of Euclidean Domains, PIDs, and UFDs, one can show that $\mathbb{Z}[x]$ is a UFD, but since it is not a PID, it is also not a Euclidean Domain. Thus $\mathbb{Z}[x]$ furnishes a counter-example to the converse of the Fundamental Meta-Theorem of Arithmetic.

Recall that the premise of Kummer's notion of *ideal numbers* was to construct a gcd of 3 and $1 + \sqrt{-26}$ in a ring where no such element exists. Motivated by the fact that $(a, b) = (\gcd(a, b))$ for $a, b \in \mathbb{Z}$, the ideal in the following prompt seems of promising significance:

► **Prompt 9.5.20** Show that $(3, 1 + \sqrt{-26}) \subset \mathbb{Z}[\sqrt{-26}]$ is a non-principal ideal.

And on this topic of factorization, next up is ideal multiplication:

---

[17] "People love high ideals, but they got to be about 33-percent plausible."—Will Rogers

> **Definition 9.5.21**

Given ideals I and J of R, we define their product by

$$IJ = \{a_1 b_1 + a_2 b_2 + \cdots + a_n b_n : a_i \in I, b_i \in J\}.$$

In words, the product IJ is the set of all (finite) sums of products of an element in I with an element in J. ◄

▶ **Prompt 9.5.22** Prove that the product as defined above is itself an ideal.

▶ **Prompt 9.5.23** Prove that for $a, b \in R$, we have $(a)(b) = (ab)$.

▶ **Prompt 9.5.24** Let I and J be ideals of R. Prove that $IJ \subseteq I \cap J$ and that if I and J are relatively prime, then in fact $IJ = I \cap J$.

Once we have multiplication of ideals, we can talk about factoring, and once we have factoring, we can talk about primeness, which we can adapt from the notion of primeness of elements in rings. Recall that p is prime if whenever $p \mid ab$, we must have $p \mid a$ or $p \mid b$. In the language of ideals, this means p is prime if whenever $ab \in (p)$, we must have $a \in (p)$ or $b \in (p)$.

> **Definition 9.5.25**

A proper ideal P of a ring R is **prime** if for all $a, b \in R$, whenever $ab \in P$ it must be that either $a \in P$ or $b \in P$. ◄

▶ **Prompt 9.5.26** Prove that a principal ideal (a) of $\mathbb{Z}$ is prime if and only if a is prime in $\mathbb{Z}$.

▶ **Prompt 9.5.27** Prove that R is an integral domain if and only if the ideal (0) is prime.

▶ **Prompt 9.5.28** Prove that for a prime $p \in \mathbb{Z}$, the ideal (p) cannot be written in the form IJ for proper ideals I and J of $\mathbb{Z}$. Thus, the ideal (p) is "irreducible" as well as prime.

The arguments above hold for prime ideals of $\mathbb{Z}[i]$, as well. This development allows us to start thinking of prime factorization of *ideals* of a ring R, rather than prime factorization of its *elements*.

▶ **Prompt 9.5.29** Verify the identity of ideals in $\mathbb{Z}$

$$(30) = (2)(3)(5),$$

and likewise the following identity of ideals in $\mathbb{Z}[i]$:

$$(74) = (1+i)^2(6+i)(6-i).$$

It is interesting to note that the first of these identities remains valid even if we drop all the parentheses and think of it as a factorization of integers, whereas the second is true only with the unit $-i$ included in the factorization. This highlights a pleasant feature of the ideal-theoretic approach—we can talk about factorization of ideals into prime ideals without having to obsess over units (recall Prompt 9.5.8), one of the major caveats we had to consider when articulating a precise Fundamental Theorem of Arithmetic. Far better still, we will soon state a rather advanced result (Theorem 9.5.34) that ideal factorization continues to work well even in cases we've run in to where unique factorization fails. Let us recall what this looks like.

The Failure (?) of Unique Factorization in $\mathbb{Z}[\sqrt{-26}]$

Failure comes only when we forget our ideals.
   – Jawaharlal Nehru

Consider the ring $\mathbb{Z}[\sqrt{-26}]$, where we had the failure of unique factorization

$$(1 + \sqrt{-26})(1 - \sqrt{-26}) = 27 = 3 \cdot 3 \cdot 3.$$

To briefly repeat ourselves, we consider this a failure of unique factorization because despite working in the full ring of integers of $\mathbb{Q}(\sqrt{-26})$, *and* having checked carefully that both 3 and $1 \pm \sqrt{-26}$ are irreducible but not prime, *and* knowing that their are no unit shenanigans at work (the units of $\mathbb{Z}[\sqrt{-26}]$ are just ± 1), we have two genuinely different factorizations of 27. Kummer had envisioned the existence of a number $\gcd(3, 1 + \sqrt{-26})$ to rescue us, but because these factors are each irreducible, their only common divisors are units. Alas. But! As developed above, it *is* reasonable to consider the ideal generated by any of these elements, and these ideals often tend to behave like gcd.

▶ **Prompt 9.5.30** Verify that in $\mathbb{Z}[\sqrt{-26}]$ we have

$$(3, 1 + \sqrt{-26}) = \{3a + b(1 + \sqrt{-26}) : a, b \in \mathbb{Z}\}$$

(Since we are asking for $a, b \in \mathbb{Z}$, this is not quite the *definition* of this ideal).

▶ **Prompt 9.5.31** Prove that in the ring $\mathbb{Z}[\sqrt{-26}]$, the ideals $I = (3, 1 + \sqrt{-26})$ and $J = (3, 1 - \sqrt{-26})$ are both prime.

▶ **Prompt 9.5.32** With I and J as in the previous prompt, verify all of:

$$IJ = (3) \qquad I^3 = (1 + \sqrt{-26}) \qquad J^3 = (1 - \sqrt{-26})$$

▶ **Prompt 9.5.33** With I and J as in the previous prompt, conclude that

$$(IJ)^3 = (27) = I^3 J^3,$$

and discuss the extent to which these calculations explain and resolve the failure of the unique factorization of 27 in $\mathbb{Z}[\sqrt{-26}]$.

That we have the ability to diagnose and partially rectify a failure of unique factorization using the language of ideals is a clarion call to continue to apply the tools of abstract algebra to study these number systems. In the final section, we peek ahead at the modern wonders of algebraic number theory and how they can be brought to bear upon the types of problems we have been considering.

What Lies Ahead

Ideals are like stars; you will not succeed in touching them with your hands. But like the seafaring man on the desert of waters, you choose them as your guides, and following them you will reach your destiny.
  – Carl Schurz

We close this Exploration with a survey of the landscape ahead of us, summarizing in broad strokes what a second course on algebra and number theory might provide for us. We begin by stating a result promised in the previous section, and then apply it to one final elliptic curve calculation that eludes our previous technique of exploiting unique factorization.

Theorem 9.5.34
Let K be a quadratic field and R its ring of integers. Every non-zero proper ideal of R can be written uniquely (up to order) as a product of non-zero prime ideals.

Particularly remarkable is there is no Euclidean-ness requirement for our ring, so the theorem applies even to rings for which unique factorization in the sense that we've been studying fails. For example, in the ring $\mathbb{Z}[\sqrt{-5}]$, where the factorization

$$2 \cdot 3 = 6 = (1 + \sqrt{-5})(1 - \sqrt{-5})$$

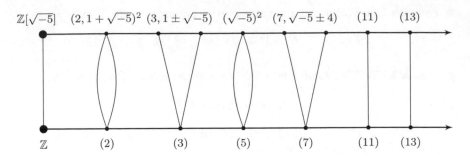

Fig. 9.7 The primes of $\mathbb{Z}$ and $\mathbb{Z}[\sqrt{-5}]$

demonstrates a failure of unique factorization, we can turn to the gcd-like ideals $(2, 1 + \sqrt{-5})$ and $(3, 1 + \sqrt{-5})$ for inspiration.

▶ **Prompt 9.5.35** Verify the following equalities of ideals in $\mathbb{Z}[\sqrt{-5}]$:

$$(2) = (2, 1 + \sqrt{-5})^2$$
$$(3) = (3, 1 + \sqrt{-5})(3, 1 - \sqrt{-5})$$

and hence *the* factorization of (6) in $\mathbb{Z}[\sqrt{-5}]$:

$$(6) = (2, 1 + \sqrt{-5})^2(3, 1 + \sqrt{-5})(3, 1 - \sqrt{-5})$$

The prompt is suggestive of a general phenomenon, that if we can understand how the prime ideals of $\mathbb{Z}$ factor in $\mathbb{Z}[\sqrt{-5}]$ then we will be able to compute prime factorizations of ideals of $\mathbb{Z}[\sqrt{-5}]$, restoring both a sense of unique factorization and how to accomplish it. Figure 9.7 shows the beginnings of a systematic inquiry into how this picture would look if we continued the work of the previous prompt. As it turns out, the prime ideals (2) and (5) in $\mathbb{Z}$ are the only ones that factor as perfect squares in $\mathbb{Z}[\sqrt{-5}]$, with the rest of the primes either splitting into two factors (e.g., (3) and (7)) or remaining prime in $\mathbb{Z}[\sqrt{-5}]$ (e.g., (11) and (13)). Also, as a final curiosity, the big fuzzy dot on the left of each axis could be thought of the last remaining prime ideal of each ring—the prime ideal (0)—adding substance to this otherwise ornamental component of Figures 5.9 and 7.5.)

Today's knowledge of number theory includes a relatively complete understanding of the patterns in Figure 9.7. We can describe precisely the pattern of which primes (p) below split into two prime ideals above, and which remain prime in $\mathbb{Z}[\sqrt{-5}]$, and the densities of those sets, all through the language of modular arithmetic and quadratic reciprocity. Moreover, there is a further hidden layer of algebraic structure to unmask:

▶ **Prompt 9.5.36** There are 5 non-principal ideals depicted in Figure 9.7. Verify for a sample of possible products the product of any two of these 5 ideals is always principal.

▶ **Prompt 9.5.37** Challenge: Prove that for any non-principal ideal I of the ring $\mathbb{Z}[\sqrt{-5}]$, the product $I \cdot (2, 1 + \sqrt{-5})$ is principal. Deduce that if I and J are both non-principal, then IJ is principal.

   This curious property about $\mathbb{Z}[\sqrt{-5}]$ does not hold for most other quadratic rings (it fails in $\mathbb{Z}[\sqrt{-26}]$, for example), and illustrates that this particular ring is not *so* far from being a UFD. The set of ideals of $\mathbb{Z}[\sqrt{-5}]$ manifests as a sort of $\mathbb{Z}/(2)$, where half of the ideals are principal ideals (P), half are non-principal (N), and we have a multiplicative structure as follows:

$$
\begin{array}{c|cc}
\times & P & N \\
\hline
P & P & N \\
N & N & P
\end{array}
$$

That is, much like the addition of even and odd integers, the product of two principal or two non-principal ideals is principal, while the product of a principal and a non-principal is again non-principal. In the ring $\mathbb{Z}[\sqrt{-26}]$, for contrast, the analogous description would be that of a $\mathbb{Z}/(6)$ structure, where every ideal is one of 6 different types. The interested reader should do a literature search for the *class group* and *class number* of a quadratic field.

Application: One Final Elliptic Curve

We make one concrete use of the ideal-theoretic interpretation of affairs, and in particular the recently-observed properties of $\mathbb{Z}[\sqrt{-5}]$, to improve our methods for solving Diophantine equations. Recall, for example, that when the ring of integers of $\mathbb{Q}[\sqrt{d}]$ is a unique factorization domain, we can successfully find all of the integer points on the elliptic curve $x^3 = y^2 + d$. In the absence of unique factorization, such efforts have proven ineffectual.

▶ **Prompt 9.5.38** Find all integers point on the elliptic curve $y^2 = x^3 - 5$.

   We've tackled several problems of this type before. But for this curve, wery early on in our process we run into a problem: we can factor

$$
x^3 = y^2 + 5 = (y + \sqrt{-5})(y - \sqrt{-5})
$$

as before, but we cannot then apply the Power Lemma to conclude that $y \pm \sqrt{-5}$ are themselves cubes, since we lack the luxury of unique factorization in $\mathbb{Z}[\sqrt{-5}]$.

But! We can consider instead the corresponding identity of ideals,

$$(x)^3 = (y + \sqrt{-5})(y - \sqrt{-5}),$$

and use the results of our Exploration to salvage the argument:

▶ **Prompt 9.5.39** Prove that the ideals $I = (y + \sqrt{-5})$ and $I' = (y - \sqrt{-5})$ are relatively prime. Then use Theorem 9.5.34 to argue as in the Power Lemma that $I = J^3$ for some ideal J. Finally, use the results of Prompt 9.5.37 to prove that such a J must be principal.

▶ **Prompt 9.5.40** Deduce from the previous problem that $y + \sqrt{-5}$ is the perfect cube an element of $\mathbb{Z}[\sqrt{-5}]$ and then finish off the solution to the Diophantine equation as in prior versions of this argument.

Full Circle: A Return to Fermat's Last Theorem

Chapter 2 kicked off our study of Diophantine equations with a discussion of Fermat's Last Theorem, studying the infamous equation

$$x^p + y^p = z^p.$$

Arguments like those employed in the previous section were a major advance in our understanding of these equations. Working in the cyclotomic rings $\mathbb{Z}[\zeta_p]$ rather than quadratic ones, Kummer observed that a solution to the equation would allow an ideal factorization

$$(z)^p = (x^p + y^p) = (x + y)(x + \zeta_p y)(x + \zeta_p^2 y) \cdots (x + \zeta_p^{p-1} y).$$

From this a suitable Power Lemma would guarantee that each $(x + \zeta_p^i y)$ was a p-th power of an ideal, and that each number $x + \zeta_p^j y$ was a p-th power in $\mathbb{Z}[\zeta_p]$...as long as we could be sure that it was impossible for the p-th power of a non-principal ideal in $\mathbb{Z}[\zeta_p]$ to be principal. Surprisingly, this property holds for some primes but not others, and in 1850 Kummer proved Fermat's Last Theorem for such primes (now called *regular* primes).

▶ **Prompt 9.5.41** What is currently known about regular primes? How many are there? What are the first few? Are there infinitely many? Are there more regular or irregular primes?

The story for quadratic fields is just as exciting as for cyclotomic ones. Despite our hard-fought victories in earlier chapters showing that the ring of integers of $\mathbb{Q}[\sqrt{d}]$ is a unique factorization domain for small negative values of d, it was conjectured by Gauss in 1798 that the list of such values was finite. That is, for nearly all negative

d, we lose unique factorization. The question remained unresolved until the 1950s–1960s when the remarkable punchline was concluded:

Theorem 9.5.42 (Baker-Heegner-Stark Theorem)
For negative square-free integers d, the ring of integers in $\mathbb{Q}(\sqrt{d})$ has unique factorization if and only if

$$d \in \{-1, -2, -3, -7, -11, -19, -43, -67, -163\}.$$

▶ **Prompt 9.5.43** Explore the fascinating history of this conjecture and proof. Did Gauss know the complete list? Who gave the first proof? The last proof? Who do you think deserves the most credit?

▶ **Prompt 9.5.44** The value 163 appearing at the end of the list ties to a remarkably vast swath of modern mathematics, leading to some commentators dubbing it the coolest number of all time. Explore.

Flipping signs, for positive values of d we know even less—dare we say embarrassingly less—than we do for negative d. The analogous, and remarkably different, conjecture of Gauss for this case reads as follows:

▶ **Conjecture 9.5.45** There are infinitely many integers $d > 0$ such that $\mathbb{Z}[\sqrt{d}]$ is a unique factorization domain.

Amazingly, this conjecture remains wide open.

▶ **Prompt 9.5.46** Explore modern perspectives on this conjecture. What do computer calculations suggest? Why should there be a difference between positive and negative values of d?

9.6 Conclusion: The Numberverse, Redux

We asked in Chapter 1 for a definition of *number*, and we encourage the reader to see how, if indeed at all, their perspective on this question has evolved over the course of the book. It seems safe to say that outside the most abstract of mathematical circles, the notion of numberhood ends at the introduction of complex numbers (and often before that). But in this text we have journeyed through quite a few potential expansions of the notion, including Hamilton's quaternions, Hensel's p-adic numbers, Kummer's ideal numbers, and perhaps even ideals, matrices, or polynomials.

Further, there are ample notions of numberhood that extend beyond the reach of this book, including such intriguing-sounding constructions as *surreal* numbers, or *infinitesimals*, or *octonions*, or *transfinite* numbers, etc. Or perhaps the truest numbers are simply the friends we made along the way.

▶ **Prompt 9.6.1** Reflect.

Number Systems

I.1 Introduction

In Chapter 1 we ask the question "What is a number?" and proceed to discuss increasingly exotic classes of objects that might fall under that heading. This approach leaves unaddressed, however, some rather deep questions about the nature of more fundamental classes of numbers, and in particular the natural numbers $\mathbb{N} = \{1, 2, 3, \ldots\}$. Of course there is a sense in which we have understood these objects ever since early childhood, but it was not until the 19th century that mathematicians attempted their formal construction from axiomatic principles. Since number theory as a whole rests on the foundation provided by $\mathbb{N}$, we would be remiss if we did not include at least some discussion of how one arrives at these fundamental objects.

The construction of the natural numbers occupies the first section of this appendix. One of the most significant consequences of this formal construction is a clear justification for perhaps the most "natural" proof technique in number theory, that of mathematical induction. Section I.3 explores the connections among the construction of the natural numbers, induction, and the Well-Ordering Principle.

▶ **Remark I.1.1** A note worth addressing before we begin: there is a rift in the mathematical community, rending otherwise peaceable math departments in twain. The cause of this tumultuous divide? Whether or not to consider 0 a natural number. To the logicians attempting to axiomatize them, it is convenient to start with 0 so that one doesn't have to return and construct it later. To the number theorists, 0 is a persistently special case that would constantly need to be excluded from statements of the form "for all natural numbers n..." Now it is convenient to start with 1 to avoid this and use the phrase *non-negative integers* if one wants to include 0. Moderate appeasers will attempt to assuage both sides by distinguishing *natural* from *whole* numbers, but we have little time for such niceties here. For us, 0 will be decidedly unnatural. But in a good way.

© Springer Nature Switzerland AG 2022
C. McLeman et al., *Explorations in Number Theory*, Undergraduate Texts in Mathematics,
https://doi.org/10.1007/978-3-030-98931-6

I.2   Construction of the Natural Numbers

In 1889, Giuseppe Peano published *Arithmetices principia, nova methodo exposita*
(The Principles of Arithmetic, Presented by a New Method), simplifying an axiomatic
foundation for the natural numbers given by Richard Dedekind in his 1888 work *Was
sind und was sollen die Zahlen?* (What are numbers and what should they be?) The
goal behind such an axiomatization was to produce a list of postulates such that
any mathematical structure satisfying the postulates would in essence have to *be* the
natural numbers, and then to deduce all fundamental arithmetic laws solely from the
postulates. Peano's solution to this problem consists of only four such postulates, now
called Peano's Postulates, and one could say that they are the keys to constructing
the natural numbers (if one wished to speak in such black-and-white terms).

Definition I.2.1

Peano's Postulates for a set N are the following:

1. There is an element $1 \in N$.
2. There exists a one-to-one function $\sigma : N \to N$.
3. The element 1 is not in the image of σ.
4. Suppose S is a subset of N with the following two properties:

   - $1 \in S$
   - If $n \in S$, then $\sigma(n) \in S$.

   Then $S = N$.

For reasons that will soon be clear, we will call σ the **successor function**, and the
last of these is the **Principle of Mathematical Induction.** ◄

Our intuitive understanding of the set $\mathbb{N} = \{1, 2, 3, \ldots\}$ shows Peano's Postulates
to be satisfied for this set when σ is taken to be the "next number" function. That is,
we *define* the symbol 2 to be the successor $\sigma(1)$ of 1, and then 3 to be the next number
after 2, and so on. Here the postulates merely insist that $1 \in \mathbb{N}$, that no two distinct
natural numbers have the same next number, that 1 is not the next number after any
element of $\mathbb{N}$, and finally, that induction works in $\mathbb{N}$. This says that any subset of $\mathbb{N}$
that contains 1 and is closed under taking successors must be all of $\mathbb{N}$, and indeed if
$1 \in S$ and S is closed under successors, then we must have $\sigma(1) = 2 \in S$, and then
$\sigma(2) = 3 \in S$, etc., filling up all of $\mathbb{N}$.

It is a theorem of mathematical logic (difficult primarily only in figuring out exactly what rules there are to prove things at this level of abstraction) due to Dedekind
that there is essentially a *unique* set that satisfies these postulates. Namely, if N is any
set satisfying Peano's postulates, then we have $1 \in N$ by the first postulate, and then
we could define the symbol 2 to mean the element $\sigma(1)$ of N, then define $3 = \sigma(2)$
and $4 = \sigma(3)$, etc. Applying induction shows that we can name all elements of

N in this way, and so N differs from $\mathbb{N}$ in names only. Thus for the remainder of the section we can continue using $\mathbb{N} = \{1, 2, 3 \ldots\}$ as *the* set that satisfies Peano's Postulates.

▶ **Remark I.2.2** A consequence of Dedekind's uniqueness result is that no other set of numbers can satisfy all four postulates. It is worth pausing to think of what goes wrong for $\mathbb{N} \cup \{0\}$, $\mathbb{Z}$, $\mathbb{Q}$, the set $\{1, 2, 3, 4, 5\}$, etc.

We next show how the familiar rules of arithmetic in $\mathbb{N}$ follow directly from the four basic postulates. We begin by observing that $\mathbb{N}$ must consist precisely of 1 and its successors.

Theorem I.2.3

If $n \in \mathbb{N}$ and $n \neq 1$, then $n = \sigma(m)$ for some $m \in \mathbb{N}$.

The proof, like most in this section, will show something to be true for all elements of n by invoking induction, showing that the set of elements with some property both includes 1 and is closed under taking successors.

Proof Let $S = \{1\} \cup \{n \in \mathbb{N} : n = \sigma(m) \text{ for some } m \in \mathbb{N}\}$. By definition, we have $S \subseteq \mathbb{N}$, $1 \in S$, and for any $k \in S$, it is also the case that $\sigma(k) \in S$. Thus, by Postulate 4, $S = \mathbb{N}$. That is, *every* element of $\mathbb{N}$ other than 1 is a successor of some other element. ☐

This technique applies to definitions as well. For example, to define addition on $\mathbb{N}$, we will define "adding 1" as applying the successor function and then, by way of induction, show how from a definition of "adding m" we can also define "adding $\sigma(m)$." By the previous theorem, we will thus have defined adding n for all $n \in \mathbb{N}$.

Definition I.2.4

Let $a \in \mathbb{N}$. Then for $n \in \mathbb{N}$ we define the sum $a + n$ as follows: if $n = 1$, we set $a + n = \sigma(a)$. Otherwise, $n = \sigma(m)$ for some $m \in \mathbb{N}$ and then we set $a + n = \sigma(a + m)$. ◀

So adding 1 is the same as looking at the successor, and since $2 = \sigma(1)$ by definition, we find that $a + 2 = \sigma(a + 1) = (a + 1) + 1$. Thus adding two is just adding one twice, and generally, the definition just says that to add $m + 1$, you simply add m and then look at the next number. While it would be obscene to belabor the point further, we note that the evaluation of the classical sum $2 + 2$ is now a mere

5-step derivation:

$$2 + 2 = 2 + \sigma(1) = \sigma(2 + 1) = \sigma(\sigma(2)) = \sigma(3) = 4.$$

Ta-da! We will not fully develop all of the standard laws of arithmetic here, but as a sample derivation, let's take a look at the associative law:

Lemma I.2.5

The associative law of addition: for any $a, b, c \in \mathbb{N}$, we have

$$(a + b) + c = a + (b + c).$$ ◄

Proof Again we invoke induction: let S be the set of all $c \in \mathbb{N}$ such that $(a+b)+c = a+(b+c)$ for all $a, b \in \mathbb{N}$. First we show that $1 \in S$: observe that $(a+b)+1 = \sigma(a+b)$ from the definition of addition, and $a + (b + 1) = a + \sigma(b) = \sigma(a + b)$ for the same reason. Thus $(a + b) + 1 = a + (b + 1)$, so $1 \in S$. Next we show that if $c \in S$, then $\sigma(c) \in S$ as well. We compute

$$
\begin{aligned}
(a + b) + \sigma(c) &= \sigma((a + b) + c) && \text{definition of } + \\
&= \sigma(a + (b + c)) && \text{since } c \in S \\
&= a + \sigma(b + c) && \text{definition of } + \\
&= a + (b + \sigma(c)) && \text{definition of } +,
\end{aligned}
$$

showing that $\sigma(c) \in S$ as well. This shows that $S = \mathbb{N}$ by induction, proving the result. □

We will only prove one more fundamental arithmetic property in $\mathbb{N}$, the commutative law of addition. This will again be a fundamental application of induction, for which the following lemma will serve as a base case.

Lemma I.2.6

For any $a \in \mathbb{N}$, we have $a + 1 = 1 + a$. ◄

Proof Proof by induction left to the reader. □

Lemma I.2.7

For any $a, b \in \mathbb{N}$, we have $a + b = b + a$. ◄

Proof Let S be the set of $b \in \mathbb{N}$ such that $a + b = b + a$ for all $a \in \mathbb{N}$. We know that $1 \in S$ by the previous lemma, so we need only show that if $b \in S$, then $\sigma(b) \in S$ as well. We check that for any $a \in \mathbb{N}$ we have

$$
\begin{aligned}
a + \sigma(b) &= a + (b+1) \\
&= (a+b) + 1 \\
&= (b+a) + 1 \\
&= b + (a+1) \\
&= b + (1+a) \\
&= (b+1) + a \\
&= \sigma(b) + a.
\end{aligned}
$$

Thus $\sigma(b) \in S$ whenever $b \in S$, and so $S = \mathbb{N}$. $\square$

We leave it to the reader's imagination the continuation of this exhilarating journey, e.g., defining multiplication recursively by

$$
m \cdot 1 = m \qquad m \cdot \sigma(n) = m \cdot n + m
$$

and then proving, for example, the commutative and associative laws of multiplication, the distributive law, etc. To fast forward slightly, however, let us next build $\mathbb{Z}$ out of $\mathbb{N}$. We first introduce a new symbol 0 and extend our operations by defining $a + 0 = 0 + a = a$ and $a \cdot 0 = 0 \cdot a = 0$, and then verifying that the various arithmetic laws listed above continue to hold. We formally introduce new symbols $-n$ for $n \in \mathbb{N}$ and extend our operations in the usual way, including defining subtraction by $a - b = a + (-b)$. We then define the integers to be the union of these three classes of numbers:

$$
\mathbb{Z} = \mathbb{N} \cup \{0\} \cup -\mathbb{N}.
$$

This union permits one more crucial component of arithmetic in $\mathbb{Z}$, the notion of *order*. In particular, we note the **Law of Trichotomy**: for each $n \in \mathbb{Z}$, exactly one of the following holds: $n \in \mathbb{N}$, $n = 0$, or $-n \in \mathbb{N}$. We can then define the symbol $>$ by writing $a > b$ if and only if $a - b \in \mathbb{N}$, and analogous definitions for the symbols $<$, $\geq$, and $\leq$. Basic properties of ordering follow, e.g., that these symbols are transitive, that $\sigma(a) > a$ for any $a \in N$, etc.

I.3   Induction and Well-Ordering

The proofs in the prior section are fine examples of proof by induction, but are phrased somewhat differently from proofs by induction one sees in an introductory proofs course and in mainstream mathematical writing. In this section, we segue slightly from the *principle* of induction to the typical *process* of proof by induction. This will tie in with another major principle of integer reasoning, the *Well-Ordering Principle*.

First, a note on semantics: broadly speaking, *deductive* reasoning is reasoning that follows a strict logical process (logic-based). *Inductive* reasoning is reasoning that argues from what came before (evidence-based). Mathematical induction is in the intersection of these: it is both logically sound and uses what came before to determine what comes next.

Proof by Mathematical Induction:

Let $P_1, P_2, P_3, \ldots, P_n, \ldots$ be a sequence of statements, one for each natural number. Suppose that:

1. P_1 is true.
2. If P_k is true for some $k \geq 1$, then P_{k+1} is also true.

Then P_n is true for all $n \in \mathbb{N}$.

Note that this is really just a slight recasting of Postulate 4. If S is the set of natural numbers n for which P_n is true, then by checking that $1 \in S$ and that $n \in S$ implies $(n + 1) \in S$, we know that every natural number is in S. In practice, here are the steps to organize a proof by mathematical induction:

1. **Prove the base case**: Prove that P_1 is true.
2. **Induction hypothesis**: Assume that for some $k \geq 1$, P_k is true.
3. **Induction step**: Prove that P_{k+1} is true assuming P_k.
4. **Conclusion**: By induction, we deduce that P_n is true for all $n \in \mathbb{N}$.

▶ **Example I.3.1** Prove that for all $n \geq 1$, we have

$$\sum_{j=1}^{n} j^3 = \frac{n^2(n + 1)^2}{4}.$$

In practice, it is wise to verify the first few examples to better understand the claim, so let's check that the first few P_n (defined as the statement to be proved for a given specific n) are indeed true statements:

$$1^3 = \frac{1^2(1 + 1)^2}{4}, \qquad 1^3 + 2^3 = \frac{2^2(2 + 1)^2}{4}, \qquad 1^3 + 2^3 + 3^3 = \frac{3^2(3 + 1)^2}{4},$$

and so on. Let's do an example with a full walkthrough of the proof template.

Proof We proceed by induction. First, for the **base case**, we verify that P_1 is true (done above). Next, for the **induction hypothesis**, assume that for some $k \geq 1$, P_k is true:

$$\sum_{j=1}^{k} j^3 = \frac{k^2(k + 1)^2}{4}.$$

Now the **induction step,** showing that the statement P_{k+1}:

$$\sum_{j=1}^{k+1} j^3 = \frac{(k+1)^2((k+1)+1)^2}{4} = \frac{(k+1)^2(k+2)^2}{4}$$

is also true. Typically, an induction step showing P_{k+1} will make critical use of P_k. We compute

$$\sum_{j=1}^{k+1} j^3 = (k+1)^3 + \sum_{j=1}^{k} j^3$$

$$= (k+1)^3 + \frac{k^2(k+1)^2}{4} \qquad \text{(by the Induction Hypothesis)}$$

$$= \frac{(k^2+4k+4)(k+1)^2}{4} \qquad \text{(wee bit o' algebra)}$$

$$= \frac{(k+2)^2(k+1)^2}{4}.$$

Thus P_{k+1} is true, and so we conclude that P_n is true for all n by the Principle of Mathematical Induction. □

That fairly classic example of an induction proof highlights the core approach, though several comments merit addressing. First, the distinction between k and n is purely psychological, designed for the author/reader of a proof by induction to not feel like they are assuming the thing they are supposed to prove. Induction proofs in the wild may simply show that $P_n \Rightarrow P_{n+1}$, without reference to a separate index. Second, several variations of the Principle of Mathematical Induction exist and lead to different proof techniques that may be of value in different contexts. Consider the following claims:

- If $1 \in S$ and $n \in S$ implies $(n+1) \in S$, then $S = N$.
- If $k \in S$, and $n \in S$ implies $(n+1) \in S$, then $\{n \in N : n \geq k\} \subseteq S$.
- If for all $n \geq 1$, $\{k \in N : k < n\} \subseteq S$ implies $n \in S$, then $S = N$.

The first of these is the regular principle of mathematical induction, and the second is a sort of "delayed induction:" if S doesn't have $1 \in S$ but is still closed under the successor function, then by virtue of containing some number k, S must also contain all natural numbers greater than k. Interestingly, we can also go the other way: we can start our induction at 0, or -1, or -42, or any integer that makes sense in the context of the problem. It merely shifts the starting point without changing the inherent structure.

▶ **Example I.3.2** Prove that for all $n \in \mathbb{N}$ such that $n > 4$, we have $2^n > n^2$.

Proof Let P_n be the statement $2^n > n^2$. We note that although P_1 is true, P_2 through P_4 are all false, but we can begin with our base case at $n = 5$, as $2^5 > 5^2$. Next, for the induction hypothesis, assume that $2^k > k^2$ for some $k > 4$. Then we need to show P_{k+1}, that $2^{k+1} > (k+1)^2$. And indeed, using the induction hypothesis that $2^k > k^2$, we have

$$2^{k+1} = 2^k + 2^k > k^2 + k^2 > k^2 + 2k + 1 = (k+1)^2,$$

where we still have to verify the middle step that $k^2 > 2k + 1$. For this, note that $k^2 \leq 2k+1$ implies $k^2 - 2k + 1 \leq 2$, so $(k-1)^2 \leq 2$. Solving gives $k \leq 1 + \sqrt{2} < 2$, contradicting that $k > 4$. This shows that P_5 is true and that P_k implies P_{k+1} for each $k \geq 5$, proving that P_n is true for all $n \geq 5$. $\qquad\qquad\square$

▶ **Example I.3.3** If m is any positive integer and n is a non-negative integer, then $m^n \geq 1$.

Proof If $n = 0$, then $m^n = 1$, so the theorem holds for $n = 0$. Now assume the theorem holds for some $n \geq 0$ and that $m \in \mathbb{N}$. Then $m^{n+1} = m^n \cdot m \geq 1 \cdot 1 = 1$, and the theorem holds for $n + 1$. $\qquad\qquad\square$

The last bullet in the list before the examples is known as **Strong Induction**, with a similar (but simpler) flowchart for an analogous proof technique. To prove P_n true for all $n \geq 1$, we assume that P_k is true for all $k < n$, and prove that P_n follows from this assumption (as opposed to proving that P_n follows from P_{n-1}). If the Principle of Induction feels like cheating, then that of Strong Induction will surely feel even worse. But in fact, the two are logically equivalent, and our remaining goal for the Appendix is to show both principles equivalent to yet one more fundamental property of the integers:

Definition I.3.4

The **Well-Ordering Principle** is the statement that for any non-empty $S \subseteq \mathbb{N}$, there is a least element of S. ◀

Despite being very different in appearance from either of the two principles of induction, we have the following:

Theorem I.3.5
The following are equivalent.

1. The Principle of Mathematical Induction holds.
2. The Principle of Strong Induction holds.
3. The Well-Ordering Principle holds.

Proof (1 $\Rightarrow$ 2) Assume first the original Principle of Mathematical Induction (PMI) holds: if $S \subseteq \mathbb{N}$ such that $1 \in S$ and $n \in S$ implies $(n + 1) \in S$, then $S = \mathbb{N}$. Now let S be a subset of $\mathbb{N}$ satisfying the hypotheses for Strong Induction: $1 \in S$ and if $1, 2, \ldots, n \in S$, then $n + 1 \in S$. We wish to use PMI to show that $S = \mathbb{N}$.

By assumption $1 \in S$. Suppose $n \in S$. If one of $1, 2, \ldots n$ is missing from S, then, since $\{1, 2, \ldots, n\}$ is a finite set, there is a least element $k \in \{1, 2, \ldots n\}$ such that $k \notin S$. But then $1, 2, \ldots, k - 1 \in S$, so by our Strong Induction hypothesis, $k \in S$, giving us a contradiction. Thus $\{1, 2, \ldots, n\} \subseteq S$, and by the hypothesis of strong induction $n + 1 \in S$. Therefore, by PMI, since $1 \in S$ and we have shown $n \in S$ implies $n + 1 \in S$, $S = \mathbb{N}$.

(2 $\Rightarrow$ 3) Let S be a non-empty subset of $\mathbb{N}$, and suppose that S does not have a least element. Let $T = \mathbb{N} - S$, the complement of S in $\mathbb{N}$. Note that $1 \in T$ since 1 is the least element of $\mathbb{N}$ and would therefore also be the least element of S if it were in S.

Suppose that for some $k \in \mathbb{N}$, if $j \leq k$, then $j \in T$. (This is certainly true for $k = 1$.) If $k + 1 \in S$, then $k + 1$ is the least element of S, which allegedly doesn't exist! Therefore, $k + 1 \in T$, so by Strong Induction, $T = \mathbb{N}$, contradicting the assumption that S is non-empty.

(3 $\Rightarrow$ 1) Assume that $\mathbb{N}$ is well-ordered, and suppose that $S \subseteq \mathbb{N}$ such that $1 \in S$ and if $k \in S$, then $k + 1 \in S$. If $S \neq \mathbb{N}$, then there is an element $k \in T = \mathbb{N} - S$. Thus T is a non-empty subset of $\mathbb{N}$ and therefore has a least element a. Since $1 \in S$, $a \neq 1$, so $a = b + 1$ for some $b \in \mathbb{N}$; and since $b < a$, $b \notin T$. Thus $b \in S$, so $b + 1 \in S$, a contradiction. Therefore $T = \mathbb{N} - S$ must be empty, so $\mathbb{N} \subseteq S$, and the Principle of Mathematical Induction holds. $\qquad\square$

A consequence of this result is that we could have taken any of induction, strong induction, or the Well-Ordering Principle as the 4th of Peano's Postulates and ended up with an equivalent system, and then proven the other two as theorems. In practice, we don't fuss too much about which one should be *the* foundational postulate.

References

1. Dedekind, R.: Was sind und Was sollen die zahlen? F. Vieweg, Braunschweig (1888)
2. Peano, G.: Arithmetices principia, nova methodo exposita. Corso, Torino (1889)

Index

© Springer Nature Switzerland AG 2022
C. McLeman et al., *Explorations in Number Theory*, Undergraduate Texts in Mathematics,
https://doi.org/10.1007/978-3-030-98931-6

Index of Notation

A

$\mathbb{N}$–set of natural numbers, 4

$\mathbb{Z}$–set of integers, 4

$\mathbb{Q}$–set of rational numbers, 4

$\mathbb{R}$–set of real numbers, 4

$\mathbb{C}$–set of complex numbers, 4

$\mathbb{Q}[i]$–set of Gaussian rationals, 6

$|z|$–the magnitude $\sqrt{a^2 + b^2}$ of a complex number $z = a + bi$, 10

$N(z)$–the norm $a^2 + b^2$ of a complex number $z = a + bi$, 10

$\overline{z}$–conjugate a complex number z, 10

$R[x]$–ring of polynomials in x with coefficients in R, 11

$\mathbb{R}[i, j, k]$–ring of quaternions, 14, 310

$A \triangle B$–symmetric difference of sets A and B, 14

$2\mathbb{Z}$–ring of even integers, 39

$a \mid b$–a divides b, 42

$a \nmid b$–a does not divide b, 42

$\gcd(a, b)$–greatest common divisor of a and b, 45, 151, 187, 314

$\text{lcm}(a, b)$–least common multiple of a and b, 45

$v_p(n)$–p-adic valuation of n, 63, 161, 249

$\pi(n)$–number of prime numbers between 1 and n, 72

$[a]$ or $[a]_{\mathcal{R}}$–equivalence class of a under equivalence relation $\mathcal{R}$, 83

$a \equiv b \pmod{n}$ –a is congruent to b modulo n, 85

$a \bmod b$–the reduction of a modulo b, 48

$a \equiv b \pmod{H}$ –a is congruent to b modulo H, 108

$\mathbb{Z}/(n)$–ring of integers modulo n, 85

$\overline{f}$–reduction of polynomial f modulo n, 90, 130

$\varphi(n)$–Euler totient function, giving the number of units mod n, 99, 104, 321

$|G|$–number of elements of (order of) a group G, 100

$R^\times$–group of units of a ring R, 101

$|g|$–order of an element g in a group, 101

aH–left coset of H containing a, 110

$[G : H]$–index of a subgroup H of a group G, 113

$x \equiv (a, b) \bmod(m, n)$–$x$ is congruent to a mod m and to b mod n, 117

$\sigma(n)$–sum of the divisors of n, 133

$\mu(n)$–Möbius μ function of n, 133

$\mathbb{Z}[i]/(\beta)$–ring of Gaussian integers modulo β, 146

$\overline{\mathbb{Q}}$–field of algebraic numbers, 178

$\overline{\mathbb{Z}}$–ring of algebraic integers, 178

$\mathbb{Q}[\sqrt{d}]$–the field of numbers of the form $a + b\sqrt{d}$, 180

$\overline{\alpha}$–the conjugate $a - b\sqrt{d}$ of $a + b\sqrt{d}$, 181

$N(\alpha)$–the norm $N(a + b\sqrt{d}) = a^2 - db^2$ of an element of $\mathbb{Q}[\sqrt{d}]$, 181

$\mathcal{N}$–a Euclidean norm on a ring, 186

UFD–Unique Factorization Domain, 192

ζ_n–the primitive n-th root of unity $e^{2\pi i/n}$ in $\mathbb{C}$, 203, 224, 301

$\left(\frac{a}{p}\right)$–Legendre symbol of a over p, 212

© Springer Nature Switzerland AG 2022
C. McLeman et al., *Explorations in Number Theory*, Undergraduate Texts in Mathematics,
https://doi.org/10.1007/978-3-030-98931-6

Printed in the United States
by Baker & Taylor Publisher Services